K. MICHALIK

Quantitative System Performance

Computer System Analysis Using
Queueing Network Models

Quantitative System Performance

Computer System Analysis Using Queueing Network Models

Edward D. Lazowska, John Zahorjan,
G. Scott Graham, and Kenneth C. Sevcik

Prentice-Hall, Inc., Englewood Cliffs, New Jersey 07632

Library of Congress Cataloging in Publication Data
Main entry under title:

Quantitative system performance.

   Bibliography: p.
   Includes index.
   1. Electronic digital computers—Evaluation.
2. Digital computer simulation.   3. Queuing theory.
I. Lazowska, Edward D.   (date)
QA76.9.E94Q36   1984   001.64′028′7   83-13791
ISBN 0-13-746975-6

© 1984 by Prentice-Hall, Inc., Englewood Cliffs, New Jersey 07632

All rights reserved. No part of this book may be reproduced in any form or by any means without permission in writing from the publisher.

Printed in the United States of America

10 9 8 7 6 5 4 3 2

ISBN   0-13-746975-6

Editorial/production supervision: Nancy Milnamow
Cover design: Edsal Enterprises
Jacket design: Lundgren Graphics, Ltd.
Manufacturing buyer: Gordon Osbourne

Prentice-Hall International, Inc., *London*
Prentice-Hall of Australia Pty. Limited, *Sydney*
Editora Prentice-Hall do Brasil, *Rio de Janeiro*
Prentice-Hall of Canada Inc., *Toronto*
Prentice-Hall of India Private Limited, *New Delhi*
Prentice-Hall of Japan, Inc., *Tokyo*
Prentice-Hall of Southeast Asia Pte. Ltd., *Singapore*
Whitehall Books Limited, *Wellington, New Zealand*

Contents

Preface   xi

I. Preliminaries   1

1. An Overview of Queueing Network Modelling   2

    1.1. *Introduction* ... 2
    1.2. *What Is a Queueing Network Model?* ... 4
    1.3. *Defining, Parameterizing, and Evaluating Queueing Network Models* ... 9
    1.4. *Why Are Queueing Network Models Appropriate Tools?* ... 14
    1.5. *Related Techniques* ... 14
    1.6. *Summary* ... 17
    1.7. *References* ... 17

2. Conducting a Modelling Study   20

    2.1. *Introduction* ... 20
    2.2. *The Modelling Cycle* ... 22
    2.3. *Understanding the Objectives of a Study* ... 27
    2.4. *Workload Characterization* ... 30
    2.5. *Sensitivity Analysis* ... 33
    2.6. *Sources of Insight* ... 35
    2.7. *Summary* ... 37
    2.8. *References* ... 39

3. Fundamental Laws   40

    3.1. *Introduction* ... 40
    3.2. *Basic Quantities* ... 40
    3.3. *Little's Law* ... 42
    3.4. *The Forced Flow Law* ... 47
    3.5. *The Flow Balance Assumption* ... 51
    3.6. *Summary* ... 52
    3.7. *References* ... 53
    3.8. *Exercises* ... 54

Contents

	4.	Queueing Network Model Inputs and Outputs	57
		4.1. *Introduction* ... 57	
		4.2. *Model Inputs* ... 57	
		4.3. *Model Outputs* ... 60	
		4.4. *Multiple Class Models* ... 62	
		4.5. *Discussion* ... 64	
		4.6. *Summary* ... 67	
		4.7. *Exercises* ... 68	
II.	**General Analytic Techniques**		69
	5.	Bounds on Performance	70
		5.1. *Introduction* ... 70	
		5.2. *Asymptotic Bounds* ... 72	
		5.3. *Using Asymptotic Bounds* ... 77	
		5.4. *Balanced System Bounds* ... 86	
		5.5. *Summary* ... 92	
		5.6. *References* ... 94	
		5.7. *Exercises* ... 95	
	6.	Models with One Job Class	98
		6.1. *Introduction* ... 98	
		6.2. *Workload Representation* ... 99	
		6.3. *Case Studies* ... 102	
		6.4. *Solution Techniques* ... 108	
		6.5. *Theoretical Foundations* ... 119	
		6.6. *Summary* ... 121	
		6.7. *References* ... 123	
		6.8. *Exercises* ... 124	
	7.	Models with Multiple Job Classes	127
		7.1. *Introduction* ... 127	
		7.2. *Workload Representation* ... 128	
		7.3. *Case Studies* ... 129	
		7.4. *Solution Techniques* ... 134	
		7.5. *Theoretical Foundations* ... 147	
		7.6. *Summary* ... 149	
		7.7. *References* ... 150	
		7.8. *Exercises* ... 150	

Contents

8. Flow Equivalence and Hierarchical Modelling — 152
   - 8.1. *Introduction* ... 152
   - 8.2. *Creating Flow Equivalent Service Centers* ... 155
   - 8.3. *Obtaining the Parameters* ... 158
   - 8.4. *Solving the High-Level Models* ... 159
   - 8.5. *An Application of Hierarchical Modelling* ... 160
   - 8.6. *Summary* ... 173
   - 8.7. *References* ... 174
   - 8.8. *Exercises* ... 175

III. Representing Specific Subsystems — 177

9. Memory — 179
   - 9.1. *Introduction* ... 179
   - 9.2. *Systems with Known Average Multiprogramming Level* ... 181
   - 9.3. *Memory Constraints* ... 184
   - 9.4. *Swapping* ... 196
   - 9.5. *Paging* ... 201
   - 9.6. *Case Studies* ... 206
   - 9.7. *Summary* ... 217
   - 9.8. *References* ... 218
   - 9.9. *Exercises* ... 219

10. Disk I/O — 222
    - 10.1. *Introduction* ... 222
    - 10.2. *Channel Contention in Non-RPS I/O Subsystems* ... 225
    - 10.3. *Channel Contention in RPS I/O Subsystems* ... 230
    - 10.4. *Additional Path Elements* ... 233
    - 10.5. *Multipathing* ... 237
    - 10.6. *Other Architectural Characteristics* ... 242
    - 10.7. *Practical Considerations* ... 245
    - 10.8. *Summary* ... 247
    - 10.9. *References* ... 248
    - 10.10. *Exercises* ... 250

11. Processors — 253
    - 11.1. *Introduction* ... 253
    - 11.2. *Tightly-Coupled Multiprocessors* ... 254

11.3. *Priority Scheduling Disciplines* ... 256
11.4. *Variations on Priority Scheduling* ... 261
11.5. *FCFS Scheduling with Class-Dependent Average Service Times* ... 262
11.6. *FCFS Scheduling with High Variability in Service Times* ... 263
11.7. *Summary* ... 266
11.8. *References* ... 268
11.9. *Exercises* ... 270

IV. Parameterization 273

12. Existing Systems 274

12.1. *Introduction* ... 274
12.2. *Types and Sources of Information* ... 275
12.3. *Customer Description* ... 279
12.4. *Center Description* ... 283
12.5. *Service Demands* ... 288
12.6. *Validating the Model* ... 291
12.7. *Summary* ... 293
12.8. *References* ... 293
12.9. *Exercises* ... 295

13. Evolving Systems 296

13.1. *Introduction* ... 296
13.2. *Changes to the Workload* ... 297
13.3. *Changes to the Hardware* ... 300
13.4. *Changes to the Operating Policies and System Software* ... 303
13.5. *Secondary Effects of Changes* ... 306
13.6. *Case Studies* ... 309
13.7. *Summary* ... 315
13.8. *References* ... 316
13.9. *Exercises* ... 318

14. Proposed Systems 320

14.1. *Introduction* ... 320
14.2. *Background* ... 321
14.3. *A General Framework* ... 323
14.4. *Tools and Techniques* ... 327
14.5. *Summary* ... 332
14.6. *References* ... 332

Contents ix

V. Perspective 335

15. Extended Applications 336

15.1. *Introduction* ... 336
15.2. *Computer Communication Networks* ... 336
15.3. *Local Area Networks* ... 339
15.4. *Software Resources* ... 342
15.5. *Database Concurrency Control* ... 343
15.6. *Operating System Algorithms* ... 347
15.7. *Summary* ... 349
15.8. *References* ... 351

16. Using Queueing Network Modelling Software 354

16.1. *Introduction* ... 354
16.2. *Components of Queueing Network Modelling Software* ... 354
16.3. *An Example* ... 360
16.4. *Summary* ... 369
16.5. *Epilogue* ... 370
16.6. *References* ... 372

VI. Appendices 375

17. Constructing a Model from RMF Data 376

17.1. *Introduction* ... 376
17.2. *Overview of MVS* ... 377
17.3. *Overview of RMF Reports* ... 378
17.4. *Customer Description* ... 383
17.5. *Center Description* ... 385
17.6. *Service Demands* ... 386
17.7. *Performance Measures* ... 388
17.8. *Summary* ... 390
17.9. *References* ... 392
17.10. *Exercises* ... 393

18. An Implementation of Single Class, Exact MVA 395

18.1. *Introduction* ... 395
18.2. *The Program* ... 395

19. An Implementation of Multiple Class, Exact MVA 398

19.1. *Introduction* ... 398
19.2. *The Program* ... 398

20. Load Dependent Service Centers   403

    20.1. *Introduction* ... 403
    20.2. *Single Class Models* ... 405
    20.3. *Multiple Class Models* ... 405
    20.4. *Program Implementation* ... 407

Index   409

Preface

This book is written for computer system performance analysts. Its goal is to teach them to apply queueing network models in their work, as tools to assist in answering the questions of cost and performance that arise throughout the life of a computer system.

Our approach to the subject arises from our collective experience in contributing to the theory of queueing network modelling, in embodying this theory in performance analysis tools, in applying these tools in the field, and in teaching computer system analysis using queueing network models in academic and industrial settings. Some important beliefs underlying our approach are:

- Although queueing network models are not a panacea, they are the appropriate tool in a wide variety of computer system design and analysis applications.
- The single most important attribute of a computer system analyst is a thorough understanding of computer systems. We assume this of our readers.
- On the one hand, mathematical sophistication is not required to analyze computer systems intelligently and successfully using queueing network models. This is the case because the algorithms for evaluating queueing network models are well developed.
- On the other hand, the purchase of a queueing network modelling software package does not assure success in computer system analysis. This is the case because defining and parameterizing a queueing network model of a particular computer system is a blend of art and science, requiring training and experience.

Queueing network modelling is a methodology for the analysis of computer systems. A methodology is a way of thinking, not a substitute for thinking.

We have divided the book into six parts. In Part I we provide four types of background material: a general discussion of queueing network modelling, an overview of the way in which a modelling study is conducted, an introduction to the interesting performance quantities in computer systems and to certain relationships that must hold among them, and a discussion of the inputs and outputs of queueing network models.

In Part II we present the techniques that are used to evaluate queueing network models — to obtain outputs such as utilizations, residence times, queue lengths, and throughputs from inputs such as workload intensities and service demands.

In Part III we explore the need for detailed models of specific subsystems, and the construction of such models for memory, disk I/O, and processor subsystems.

In Part IV we study the parameterization of queueing network models of existing systems, evolving systems, and proposed systems.

In Part V we survey some non-traditional applications, such as the analysis of computer communication networks and database concurrency control mechanisms. We also examine the structure and use of queueing network modelling software packages.

In Part VI, the appendices, we provide a case study in obtaining queueing network parameter values from system measurement data, and programs implementing the queueing network evaluation techniques described in Part II.

Case studies appear throughout the book. They are included to illustrate various aspects of computer system analysis using queueing network models. They should *not* be misconstrued as making general statements about the relative performance of various systems; the results have significance only for the specific configurations and workloads under consideration.

We have summarized a number of important modelling techniques in the form of ''Algorithms''. Our intention is to provide enough information that the reader can understand fully the essential aspects of each technique. We omit details of significance to the implementation of a technique when we feel that these details might obscure the more fundamental concepts.

It is our experience that practicing computer system analysts are relatively skilled in techniques such as workload characterization, system measurement, interpretation of performance data, and system tuning, and are at least acquainted with basic statistical methods and with simulation. Each of these subjects is well represented in the existing literature, and is given short shrift in the present book. Much interesting and important research work concerning queueing network modelling also is given short shrift; we discuss the one approach to each problem that we feel is best suited for application. For readers who desire to pursue a topic in greater detail than we have provided, each chapter concludes with a brief discussion of the relevant literature.

We owe a significant debt to Jeffrey P. Buzen and Peter J. Denning, who have been instrumental in the development of a pragmatic

Preface

philosophy of computer system analysis using queueing network models. Their influence is evident especially in our use of the *operational* framework for queueing network modelling, which conveys much greater intuition than the more traditional *stochastic* framework.

Jeffrey A. Brumfield, Jeffrey P. Buzen, Domenico Ferrari, Lenny Freilich, and Roger D. Stoesz have assisted us by reviewing our manuscript, as have several anonymous reviewers. Our work in computer system analysis using queueing network models has been supported in part by the National Science Foundation and by the Natural Sciences and Engineering Research Council of Canada. We thank our colleagues at the University of Washington and at the University of Toronto for their encouragement, and our families and friends for their forbearance.

Edward D. Lazowska, John Zahorjan,
G. Scott Graham, and Kenneth C. Sevcik

Seattle and Toronto

Part I

Preliminaries

This first part of the book provides four different sorts of background material as a prelude to our study of quantitative system performance.

In Chapter 1 we survey queueing network modelling, discussing some example applications and comparing it to more traditional approaches to computer system analysis with which the reader may be familiar.

In Chapter 2 we use case studies to explore various aspects of conducting a modelling study. Our objective is to provide some perspective on the "pieces" of the process that will be studied in the remainder of the book.

In Chapter 3 we provide a technical foundation for our work by defining a number of quantities of interest, introducing the notation that we will use in referring to these quantities, and deriving various relationships among these quantities.

In Chapter 4 we describe the inputs and the outputs of queueing network models.

Chapter 1

An Overview of Queueing Network Modelling

1.1. Introduction

Today's computer systems are more complex, more rapidly evolving, and more essential to the conduct of business than those of even a few years ago. The result is an increasing need for tools and techniques that assist in understanding the behavior of these systems. Such an understanding is necessary to provide intelligent answers to the questions of cost and performance that arise throughout the life of a system:

- *during design and implementation*
    - An aerospace company is designing and building a computer-aided design system to allow several hundred aircraft designers simultaneous access to a distributed database through graphics workstations. Early in the design phase, fundamental decisions must be made on issues such as the database accessing mechanism and the process synchronization and communication mechanism. The relative merits of various mechanisms must be evaluated prior to implementation.
    - A computer manufacturer is considering various architectures and protocols for connecting terminals to mainframes using a packet-oriented broadcast communications network. Should terminals be clustered? Should packets contain multiple characters? Should characters from multiple terminals destined for the same mainframe be multiplexed in a single packet?
- *during sizing and acquisition*
    - The manufacturer of a turn-key medical information system needs an efficient way to size systems in preparing bids. Given estimates of the arrival rates of transactions of various types, this vendor must project the response times that the system will provide when running on various hardware configurations.

1.1. Introduction

- A university has received twenty bids in response to a request for proposals to provide interactive computing for undergraduate instruction. Since the selection criterion is the "cost per port" among those systems meeting certain mandatory requirements, comparing the capacity of these twenty systems is essential to the procurement. Only one month is available in which to reach a decision.

• *during evolution of the configuration and workload*
- A stock exchange intends to begin trading a new class of options. When this occurs, the exchange's total volume of options transactions is expected to increase by a factor of seven. Adequate resources, both computer and personnel, must be in place when the change is implemented.
- An energy utility must assess the longevity of its current configuration, given forecasts of workload growth. It is desirable to know what the system bottleneck will be, and the relative cost-effectiveness of various alternatives for alleviating it. In particular, since this is a virtual memory system, tradeoffs among memory size, CPU power, and paging device speed must be evaluated.

These questions are of great significance to the organizations involved, with potentially serious repercussions from incorrect answers. Unfortunately, these questions are also complex; correct answers are not easily obtained.

In considering questions such as these, one must begin with a thorough grasp of the system, the application, and the objectives of the study. With this as a basis, several approaches are available.

One is the use of *intuition and trend extrapolation*. To be sure, there are few substitutes for the degree of experience and insight that yields reliable intuition. Unfortunately, those who possess these qualities in sufficient quantity are rare.

Another is the *experimental evaluation of alternatives*. Experimentation is always valuable, often required, and sometimes the approach of choice. It also is expensive — often prohibitively so. A further drawback is that an experiment is likely to yield accurate knowledge of system behavior under one set of assumptions, but not any insight that would allow generalization.

These two approaches are in some sense at opposite extremes of a spectrum. Intuition is rapid and flexible, but its accuracy is suspect because it relies on experience and insight that are difficult to acquire and verify. Experimentation yields excellent accuracy, but is laborious and inflexible. Between these extremes lies a third approach, the general subject of this book: *modelling*.

A model is an abstraction of a system: an attempt to distill, from the mass of details that is the system itself, exactly those aspects that are essential to the system's behavior. Once a model has been *defined* through this abstraction process, it can be *parameterized* to reflect any of the alternatives under study, and then *evaluated* to determine its behavior under this alternative. Using a model to investigate system behavior is less laborious and more flexible than experimentation, because the model is an abstraction that avoids unnecessary detail. It is more reliable than intuition, because it is more methodical: each particular approach to modelling provides a framework for the definition, parameterization, and evaluation of models. Of equal importance, using a model enhances both intuition and experimentation. Intuition is enhanced because a model makes it possible to "pursue hunches" — to investigate the behavior of a system under a wide range of alternatives. (In fact, although our objective in this book is to devise *quantitative* models, which accurately reflect the performance measures of a system, an equally effective guide to intuition can be provided by less detailed *qualitative* models, which accurately reflect the general behavior of a system but not necessarily specific values of its performance measures.) Experimentation is enhanced because the framework provided by each particular approach to modelling gives guidance as to which experiments are necessary in order to define and parameterize the model.

Modelling, then, provides a framework for gathering, organizing, evaluating, and understanding information about a computer system.

1.2. What Is a Queueing Network Model?

Queueing network modelling, the specific subject of this book, is a particular approach to computer system modelling in which the computer system is represented as a *network of queues* which is evaluated *analytically*. A network of queues is a collection of *service centers*, which represent system resources, and *customers*, which represent users or transactions. Analytic evaluation involves using *software* to solve efficiently a set of equations induced by the network of queues and its parameters. (These definitions, and the informal overview that follows, take certain liberties that will be noted in Section 1.5.)

1.2.1. Single Service Centers

Figure 1.1 illustrates a single service center. Customers arrive at the service center, wait in the *queue* if necessary, receive service from the *server*, and depart. In fact, this service center and its arriving customers constitute a (somewhat degenerate) queueing network model.

1.2. What Is a Queueing Network Model?   5

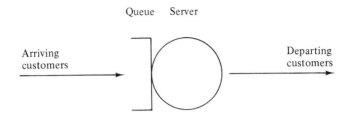

Figure 1.1 — A Single Service Center

This model has two parameters. First, we must specify the *workload intensity*, which in this case is the rate at which customers arrive (e.g., one customer every two seconds, or 0.5 customers/second). Second, we must specify the *service demand*, which is the average service requirement of a customer (e.g., 1.25 seconds). For specific parameter values, it is possible to evaluate this model by solving some simple equations, yielding *performance measures* such as *utilization* (the proportion of time the server is busy), *residence time* (the average time spent at the service center by a customer, both queueing and receiving service), *queue length* (the average number of customers at the service center, both waiting and receiving service), and *throughput* (the rate at which customers pass through the service center). For our example parameter values (under certain assumptions that will be stated later) these performance measures are:

    utilization: .625
    residence time: 3.33 seconds
    queue length: 1.67 customers
    throughput: 0.5 customers/second

Figures 1.2a and 1.2b graph each of these performance measures as the workload intensity varies from 0.0 to 0.8 arrivals/second. This is the interesting range of values for this parameter. On the low end, it makes no sense for the arrival rate to be less than zero. On the high end, given that the average service requirement of a customer is 1.25 seconds, the greatest possible rate at which the service center can handle customers is one every 1.25 seconds, or 0.8 customers/second; if the arrival rate is greater than this, then the service center will be saturated.

The principal thing to observe about Figure 1.2 is that the evaluation of the model yields performance measures that are qualitatively consistent with intuition and experience. Consider residence time. When the workload intensity is low, we expect that an arriving customer seldom will encounter competition for the service center, so will enter service immediately and will have a residence time roughly equal to its service requirement. As the workload intensity rises, congestion increases, and residence time along with it. Initially, this increase is gradual. As the

6   **Preliminaries: An Overview of Queueing Network Modelling**

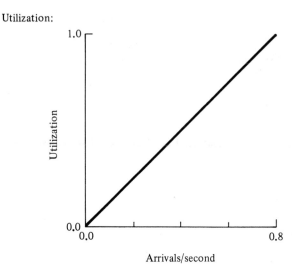

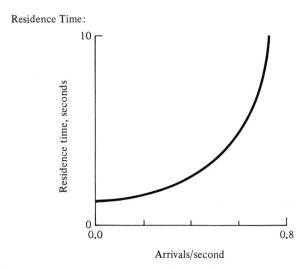

Figure 1.2a − Performance Measures for the Single Service Center

1.2. What Is a Queueing Network Model?

Queue Length:

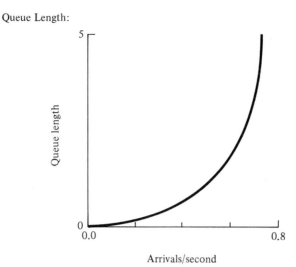

Throughput:

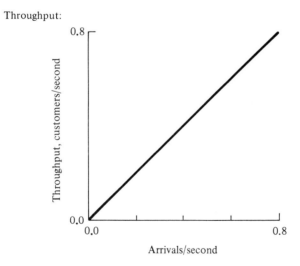

Figure 1.2b — Performance Measures for the Single Service Center

load grows, however, residence time increases at a faster and faster rate, until, as the service center approaches saturation, small increases in arrival rate result in dramatic increases in residence time.

1.2.2. Multiple Service Centers

It is hard to imagine characterizing a contemporary computer system by two parameters, as would be required in order to use the model of Figure 1.1. (In fact, however, this was done with success several times in the simpler days of the 1960's.) Figure 1.3 shows a more realistic model in which each system resource (in this case a CPU and three disks) is represented by a separate service center.

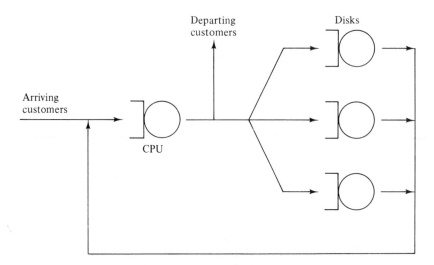

Figure 1.3 − A Network of Queues

The parameters of this model are analogous to those of the previous one. We must specify the workload intensity, which once again is the rate at which customers arrive. We also must specify the service demand, but this time we provide a separate service demand for each service center. If we view customers in the model as corresponding to transactions in the system, then the workload intensity corresponds to the rate at which users submit transactions to the system, and the service demand at each service center corresponds to the total service requirement per transaction at the corresponding resource in the system. (As indicated by the lines in the figure, we can think of customers as arriving, circulating among the service centers, and then departing. The pattern of circulation among the centers is not important, however; only the total service demand at each center matters.) For example, we might specify that transactions arrive at a rate of one every five seconds, and that each such transaction requires an average of 3 seconds of service at the CPU and 1, 2, and 4 seconds of service, respectively, at the three disks. As in the case of the single service center, for specific parameter values it is possible to evaluate this model by solving some simple equations. For our example parameter values (under certain assumptions that will be stated later) performance measures include:

>CPU utilization: .60
>average system response time perceived by users: 32.1 seconds
>average number of concurrently active transactions: 6.4
>system throughput: 0.2 transactions/second

(We consistently will use *residence time* to mean the time spent at a service center by a customer, and *response time* to correspond to the intuitive notion of perceived system response time. Most performance measures obtained from queueing network models are average values (e.g., average response time) rather than distributional information (e.g., the 90th percentile of response times). Thus the word "average" should be understood even if it is omitted.)

1.3. Defining, Parameterizing, and Evaluating Queueing Network Models

1.3.1. Definition

Defining a queueing network model of a particular system is made relatively straightforward by the close correspondence between the attributes of queueing network models and the attributes of computer systems. For example, service centers in queueing network models naturally correspond to hardware resources and their software queues in computer systems, and customers in queueing network models naturally correspond to users or transactions in computer systems.

Queueing network models have a richer set of attributes than we have illustrated thus far, extending the correspondence with computer systems. As an example of this richness, specifying the rate at which customers arrive (an approach that is well suited to representing certain transaction processing workloads) is only one of three available means to describe workload intensity. A second approach is to state the number of customers in the model. (This alternative is well suited to representing batch workloads.) A third approach is to specify the number of customers and the average time that each customer "spends thinking" (i.e., uses a terminal) between interactions. (This alternative is well suited to representing interactive workloads.) In Figure 1.4 we have modified the model of

Figure 1.3 so that the workload intensity is described using this last approach. Figure 1.5 graphs system response time and CPU utilization for this model with the original service demands (3 seconds of service at the CPU and 1, 2, and 4 seconds of service, respectively, at the three disks) when the workload consists of from 1 to 50 interactive users, each with an average think time of 30 seconds. Once again we note that the behavior of the model is qualitatively consistent with intuition and experience.

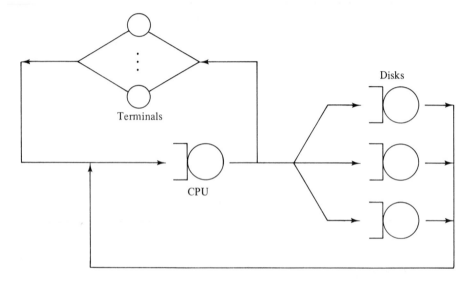

Figure 1.4 — A Model with a Terminal-Driven Workload

As another example of this richness, most computer systems have several identifiable workload components, and although the queueing network models that we have considered thus far have had a single *customer class* (all customers exhibit essentially the same behavior), it is possible to distinguish between a system's workload components in a queueing network model by making use of multiple customer classes, each of which has its own workload intensity (specified in any of the ways we have described) and service demands. For example, it is possible to model directly a computer system in which there are four workload components: transaction processing, background batch, interactive database inquiry, and interactive program development. In defining the model, we would specify four customer classes and the relevant service centers. In parameterizing the model, we would provide workload intensities for each class (for example, an arrival rate of 10 requests/minute for transaction processing, a multiprogramming level of 2 for background batch, 25 interactive database users each of whom thinks for an average of two minutes between interactions, and 10 interactive program development

1.3. Defining, Parameterizing, and Evaluating Queueing Network Models

CPU Utilization:

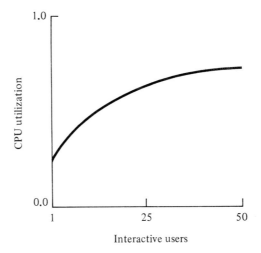

System Response Time:

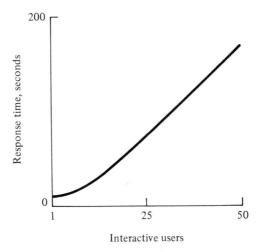

Figure 1.5 — Performance Measures for the Terminal-Driven Model

12   Preliminaries: An Overview of Queueing Network Modelling

users each of whom thinks for an average of 15 seconds between interactions). We also would provide service demands for each class at each service center. In evaluating the model, we would obtain performance measures in the aggregate (e.g., total CPU utilization), and also on a per-class basis (e.g., CPU utilization due to background batch jobs, response time for interactive database queries).

1.3.2. Parameterization

The parameterization of queueing network models, like their definition, is relatively straightforward. Imagine calculating the CPU service demand for a customer in a queueing network model of an existing system. We would observe the system in operation and would measure two quantities: the number of seconds that the CPU was busy, and the number of user requests that were processed (these requests might be transactions, or jobs, or interactions). We then would divide the busy time by the number of request completions to determine the average number of seconds of CPU service attributable to each request, the required parameter.

A major strength of queueing network models is the relative ease with which parameters can be modified to obtain answers to "what if" questions. Returning to the example in Section 1.2.2:

- What if we balance the I/O activity among the disks? (We set the service demand at each disk to $\frac{1+2+4}{3} = 2.33$ seconds and re-evaluate the model. Response time drops from 32.1 seconds to 20.6 seconds.)

- What if the workload subsequently increases by 20%? (We set the arrival rate to $0.2 \times 1.2 = 0.24$ requests/second and re-evaluate the model. Response time increases from 20.6 seconds to 26.6 seconds.)

- What if we then upgrade to a CPU 30% faster? (We set the service demand at the CPU to $3/1.3 = 2.31$ seconds and re-evaluate the model. Response time drops from 26.6 to 21.0 seconds.)

Considerable insight can be required to conduct such a *modification analysis*, because the performance measures obtained from evaluating the model can be only as accurate as the workload intensities and service demands that are provided as inputs, and it is not always easy to anticipate every effect on these parameters of a change to the configuration or workload. Consider the first "what if" question listed above. If we assume that the system's disks are physically identical then the *primary effect* of balancing disk activity can be reflected in the parameter values of the model by setting the service demand at each disk to the average value. However, there may be *secondary effects* of the change. For example, the total amount of disk arm movement may decrease. The result in

1.3. Defining, Parameterizing, and Evaluating Queueing Network Models

the system would be that the total disk service requirement of each user would decrease somewhat. If this secondary effect is anticipated, then it is easy to reflect it in the parameter values of the model, and the model, when evaluated, will yield accurate values for performance measures. If not, then the model will yield somewhat pessimistic results. Fortunately, the primary effects of modifications, which dominate performance, tend to be relatively easy to anticipate.

Models with multiple customer classes are more common than models with single customer classes because they facilitate answering many "what if" questions. (How much will interactive response time improve if the volume of background batch is decreased by 50%?) Single class models, though, have the advantage that they are especially easy to parameterize, requiring few assumptions on the part of the analyst. Using contemporary computer system measurement tools, it is notoriously difficult to determine correctly resource consumption by workload component, especially in the areas of system overhead and I/O subsystem activity. Since single class models can be parameterized with greater ease and accuracy, they are quicker and more reliable than multiple class models for answering those questions to which they are suited.

1.3.3. Evaluation

We distinguish two approaches to evaluating queueing network models. The first involves calculating bounds on performance measures, rather than specific values. For example, we might determine upper and lower bounds on response time for a particular set of parameter values (workload intensity and service demands). The virtue of this approach is that the calculations are simple enough to be carried out by hand, and the resulting bounds can contribute significantly to understanding the system under study.

The second approach involves calculating the values of the performance measures. While the algorithms for doing this are sufficiently complicated that the use of computer programs is necessary, it is important to emphasize that these algorithms are extremely efficient. Specifically, the running time of the most efficient general algorithm grows as the product of the number of service centers with the number of customer classes, and is largely independent of the number of customers in each class. A queueing network model with 100 service centers and 10 customer classes can be evaluated in only seconds of CPU time.

The algorithms for evaluating queueing network models constitute the lowest level of a queueing network modelling software package. Higher levels typically include transformation routines to map the characteristics of specific subsystems onto the general algorithms at the lowest level, a user interface to translate the "jargon" of a particular computer system into the language of queueing network models, and high-level front ends that assist in obtaining model parameter values from system measurement data.

1.4. Why Are Queueing Network Models Appropriate Tools?

Models in general, and queueing network models in particular, have become important tools in the design and analysis of computer systems. This is due to the fact that, for many applications, queueing network models achieve a favorable balance between accuracy and efficiency.

In terms of accuracy, a large body of experience indicates that queueing network models can be expected to be accurate to within 5 to 10% for utilizations and throughputs and to within 10 to 30% for response times. This level of accuracy is consistent with the requirements of a wide variety of design and analysis applications. Of equal importance, it is consistent with the accuracy achievable in other components of the computer system analysis process, such as workload characterization.

In terms of efficiency, we have indicated in the previous section that queueing network models can be defined, parameterized, and evaluated at relatively low cost. Definition is eased by the close correspondence between the attributes of queueing network models and the attributes of computer systems. Parameterization is eased by the relatively small number of relatively high-level parameters. Evaluation is eased by the recent development of algorithms whose running time grows as the product of the number of service centers with the number of customer classes.

Queueing network models achieve relatively high accuracy at relatively low cost. The incremental cost of achieving greater accuracy is high — significantly higher than the incremental benefit, for a wide variety of applications.

1.5. Related Techniques

Our informal description of queueing network modelling has taken several liberties that should be acknowledged to avoid confusion. These liberties can be summarized as follows:

1.5. Related Techniques   15

- We have not described networks of queues in their full generality, but rather a subset that can be evaluated efficiently.
- We have incorrectly implied that the only analytic technique for evaluating networks of queues is the use of software to solve a set of equations induced by the network of queues and its parameters.
- We have neglected the fact that simulation can be used to evaluate networks of queues.
- We have not explored the relationship of queueing network models to queueing theory.

The following subsections explore these issues.

1.5.1. Queueing Network Models and General Networks of Queues

This book is concerned with a subset of general networks of queues. This subset consists of the *separable* queueing networks (a name used for historical and mathematical reasons), extended where necessary for the accurate representation of particular computer system characteristics.

We restrict our attention to the members of this subset because of the efficiency with which they can be evaluated. This efficiency is mandatory in analyzing contemporary computer systems, which may have hundreds of resources and dozens of workload components, each consisting of many users or jobs.

Restriction to this subset implies certain assumptions about the computer system under study. We will discuss these assumptions in later chapters. On the one hand, these assumptions seldom are satisfied strictly. On the other hand, the inaccuracies resulting from violations of these assumptions typically are, at worst, comparable to those arising from other sources (e.g., inadequate measurement data).

General networks of queues, which obviate many of these assumptions, can be evaluated analytically, but the algorithms require time and space that grow prohibitively quickly with the size of the network. They are useful in certain specialized circumstances, but not for the direct analysis of realistic computer systems.

1.5.2. Queueing Network Models and Simulation

The principal strength of simulation is its flexibility. There are few restrictions on the behavior that can be simulated, so a computer system can be represented at an arbitrary level of detail. At the abstract end of this spectrum is the use of simulation to evaluate networks of queues. At the concrete extreme, running a benchmark experiment is in some sense using the system as a detailed simulation model of itself.

The principal weakness of simulation modelling is its relative expense. Simulation models generally are expensive to define, because this involves writing and debugging a complex computer program. (In the specific domain of computer system modelling, however, this process has been automated by packages that generate the simulation program from a model description.) They can be expensive to parameterize, because a highly detailed model implies a large number of parameters. (We will see that obtaining even the small number of parameters required by a queueing network model is a non-trivial undertaking.) Finally, they are expensive to evaluate, because running a simulation requires substantial computational resources, especially if narrow confidence intervals are desired.

A tenet of this book, for which there is much supporting evidence, is that queueing network models provide an appropriate level of accuracy for a wide variety of computer system design and analysis applications. For this reason, our primary interest in simulation is as a means to evaluate certain submodels in a study that is primarily analytic. This technique, known as *hybrid modelling*, is motivated by a desire to use analysis where possible, since the cost of evaluating a simple network of queues using simulation exceeds by orders of magnitude the cost of evaluating the same model using analysis.

1.5.3. Queueing Network Models and Queueing Theory

Queueing network modelling can be viewed as a small subset of the techniques of queueing theory, selected and specialized for modelling computer systems.

Much of queueing theory is oriented towards modelling a complex system using a single service center with complex characteristics. Sophisticated mathematical techniques are employed to analyze these models. Relatively detailed performance measures are obtained: distributions as opposed to averages, for example.

Rather than single service centers with complex characteristics, queueing network modelling employs networks of service centers with simple characteristics. Benefits arise from the fact that the application domain is restricted to computer systems. An appropriate subset of networks of queues can be selected, and evaluation algorithms can be designed to obtain meaningful performance measures with an appropriate balance between accuracy and efficiency. These algorithms can be packaged with interfaces based on the terminology of computer systems rather than the language of queueing theory, with the result that only a minimal understanding of the theory underlying these algorithms is required to apply them successfully.

1.6. Summary

This chapter has surveyed the questions of cost and performance that arise throughout the life of a computer system, the nature of queueing network models, and the role that queueing network models can play in answering these questions. We have argued that queueing network models, because they achieve a favorable balance between accuracy and cost, are the appropriate tool in a wide variety of computer system design and analysis applications.

1.7. References

This book is concerned exclusively with computer system analysis using queueing network models. Because of this relatively narrow focus, it is complemented by a number of existing books. These can be divided into three groups, distinguished by scope.

Books in the first group, such as Ferrari's [1978], discuss computer system performance evaluation in the large.

Books in the second group consider computer system modelling. Examples include books by Gelenbe and Mitrani [1980], Kobayashi [1978], Lavenberg [1983], and Sauer and Chandy [1981].

Books in the third group treat a particular aspect of computer system performance evaluation at a level of detail comparable to that of the present book: computer system measurement [Ferrari et al. 1983], the low-level analysis of system components using simple queueing formulae [Beizer 1978], the analysis of computer systems and computer communication networks using queueing theory [Kleinrock 1976], and the mathematical and statistical aspects of computer system analysis [Allen 1978; Trivedi 1982].

Queueing network modelling is a rapidly advancing discipline. With the present book as background, it should be possible to assimilate future developments in the field. Many of these will be found in the following sources:

EDP Performance Review, a digest of current information on tools for performance evaluation and capacity planning, published by Applied Computer Research.

Computer Performance, a journal published by Butterworths.

The *Journal of Capacity Management*, published by the Institute for Software Engineering.

18   Preliminaries: An Overview of Queueing Network Modelling

The *Proceedings of the CMG International Conference.* The conference is sponsored annually by the Computer Measurement Group, which also publishes the proceedings.

The *Proceedings of the CPEUG Meeting.* The meeting is sponsored annually by the Computer Performance Evaluation Users Group, which also publishes the proceedings.

The *Proceedings of the ACM SIGMETRICS Conference on Measurement and Modeling of Computer Systems.* The conference is sponsored annually by the ACM Special Interest Group on Measurement and Evaluation. The proceedings generally appear as a special issue of *Performance Evaluation Review*, the SIGMETRICS quarterly publication.

The *ACM Transactions on Computer Systems*, a journal published by the Association for Computing Machinery.

The *IEEE Transactions on Computers* and the *IEEE Transactions on Software Engineering*, two journals published by the Institute of Electrical and Electronics Engineers.

The *Proceedings of the International Symposium on Computer Performance Modelling, Measurement and Evaluation.* The symposium is sponsored at eighteen month intervals by IFIP Working Group 7.3 on Computer System Modelling.

Performance Evaluation, a journal published by North-Holland.

[Allen 1978]
 Arnold O. Allen. *Probability, Statistics, and Queueing Theory with Computer Science Applications.* Academic Press, 1978.

[Beizer 1978]
 Boris Beizer. *Micro-Analysis of Computer System Performance.* Van Nostrand Reinhold, 1978.

[Ferrari 1978]
 Domenico Ferrari. *Computer Systems Performance Evaluation.* Prentice-Hall, 1978.

[Ferrari et al. 1983]
 Domenico Ferrari, Giuseppe Serrazi, and Alessandro Zeigner. *Measurement and Tuning of Computer Systems.* Prentice-Hall, 1983.

[Gelenbe & Mitrani 1980]
 Erol Gelenbe and Israel Mitrani. *Analysis and Synthesis of Computer Systems.* Academic Press, 1980.

[Kleinrock 1976]
 Leonard Kleinrock. *Queueing Systems − Volume II: Computer Applications.* John Wiley & Sons, 1976.

1.7. References

[Kobayashi 1978]
Hisashi Kobayashi. *Modeling and Analysis — An Introduction to System Performance Evaluation Methodology.* Addison-Wesley, 1978.

[Lavenberg 1983]
Stephen S. Lavenberg (ed.). *Computer Performance Modeling Handbook.* Academic Press, 1983.

[Sauer & Chandy 1981]
Charles H. Sauer and K. Mani Chandy. *Computer Systems Performance Modeling.* Prentice-Hall, 1981.

[Trivedi 1982]
Kishor S. Trivedi. *Probability and Statistics with Reliability, Queuing, and Computer Science Applications.* Prentice-Hall, 1982.

Chapter 2

Conducting a Modelling Study

2.1. Introduction

In this chapter we take a broad look at how, when confronted with a specific computer system analysis problem, to apply the general "methodology" of queueing network modelling. This skill must be developed through experience — it cannot be absorbed passively. Recognizing this, we present a set of case studies selected to illustrate significant aspects of the methodology, sharing with the reader the experience of others.

The success of queueing network modelling is based on the fact that the low-level details of a system are largely irrelevant to its high-level performance characteristics. Queueing network models appear abstract when compared with other approaches to computer system analysis. Queueing network modelling is inherently a *top-down* process. The underlying philosophy is to begin by identifying the principal components of the system and the ways in which they interact, then supply any details that prove to be necessary. This philosophy means that a large number of assumptions will be introduced and assessed in the process of conducting a modelling study. Three principal considerations motivate these assumptions:

- *simplicity*

  There is a strong incentive to identify and eliminate irrelevant details. In fact, we will adopt a rather liberal definition of "irrelevant" in this context by generally including any system characteristic that will not have a *primary* (as opposed to *secondary*) *effect* on the results of the study. Examples include:

  — Although a system may have a large number of identifiable workload components, we may be interested in the performance of only one of them. In this case, we may choose to employ a model with only two classes, one representing the workload component of interest and the other representing the aggregate effect of all other workload components.

2.1. Introduction

- The primary effect of a CPU upgrade will be a decrease in CPU service demands. A change in the average paging and swapping activity per job may also result, but if so, this is a secondary effect.

- *adequacy of measurements*

  The measurement tools available on contemporary computer systems often fail to provide directly the quantities required to parameterize queueing network models. Queueing network models require a small number of carefully selected inputs. Measurement tools, largely for historical reasons, provide a large volume of data, most of which is of limited use for our purposes. Considerable interpretation may be required on the part of the analyst. Examples include:
  - Typically, a significant proportion of CPU activity is not attributed to specific workload components. Since the CPU tends to be a heavily utilized resource, correct attribution of its usage is important to the accuracy of a multiple class model.
  - Surprisingly, even determining the multiprogramming level of a batch workload sometimes is difficult, because some system tasks ("quiescent" or "operator" jobs) may be counted by the measurement tool.

- *ease of evaluation*

  As noted in Chapter 1, we must restrict ourselves to a subset of general networks of queues that can be evaluated efficiently. To stay within this subset, we must make compromises in the representation of certain computer system characteristics. Examples include:
  - Extremely high variability in the service requirement at a particular resource can cause performance to degrade. Direct representation of this characteristic makes queueing network models costly to evaluate, though, and examples where it is a major determinant of performance are rare. It generally is omitted from models.
  - Memory admission policies typically are complex, and the memory requirements of programs differ. The evaluation of a model is considerably eased, though, if we are willing to assume that the memory admission policy is either first-come-first-served or class-based priority, and that programs have similar memory requirements, at least within each class.

Skill in introducing and assessing assumptions is the key to conducting a successful modelling study. In general, it is important to be explicit concerning the assumptions that are made, the motivations for their introduction, and the arguments for their plausibility. This allows the analyst's reasoning to be examined, and facilitates evaluating the sensitivity of the results to the assumptions.

The material in this chapter has a spectrum of interpretations ranging from fairly shallow to fairly subtle. The reader with little experience will find a collection of brief case study descriptions indicating the applicability of queueing network models. The reader with moderate experience will learn something of the ways in which queueing network modelling studies are conducted. The reader with considerable experience will discover insights concerning various aspects of conducting a modelling study that can be used to great advantage. Because of this spectrum of interpretations, we will ask you to review this chapter during Part V of the book.

2.2. The Modelling Cycle

The most common application of queueing network modelling involves projecting the effect on performance of changes to the configuration or workload of an existing system. There are three phases to such a study. In the *validation* phase, a *baseline model* of the existing system is constructed and its sufficiency established. In the *projection* phase, this model is used to forecast the effect on performance of the anticipated modifications. In the *verification* phase, the actual performance of the modified system is compared to the model's projections. Taken together, these three phases are referred to as the *modelling cycle*, illustrated in Figure 2.1.

The validation phase begins with the definition of the model, which includes selection of those system resources and workload components that will be represented, identification of any system characteristics that may require special attention (e.g., priority scheduling, paging), choice of model structure (e.g., separable, hybrid), and procedures for obtaining the necessary parameters from the available measurement data.

Next, the system is measured to obtain *workload measures*, from which model inputs will be calculated, and *performance measures*, which will be compared to model outputs. In some cases these are the same; for instance, device utilizations are workload measures (they are used to calculate service demands) and also performance measures (they are used to assess the accuracy of the model). On the other hand, the multiprogramming level of a batch workload is strictly a workload measure, and system response time is strictly a performance measure.

The workload measures then are used to parameterize the model, a step that may require various transformations. The model is evaluated, yielding outputs. These are compared to the system's performance measures. Discrepancies indicate flaws in the process, such as system characteristics that were ignored or represented inappropriately, or model inputs whose values were established incorrectly. Unfortunately, the absence of

2.2. The Modelling Cycle

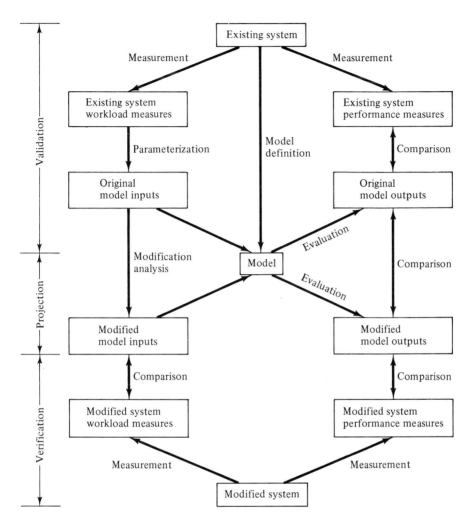

Figure 2.1 — The Modelling Cycle

such discrepancies does not guarantee that the model will project properly the effect of system or workload modifications. Confidence in a model's predictive abilities may come from two sources. The first is repetitive validation over a number of measurement intervals, perhaps involving selected modifications. For example, if the objective of a modelling study is to assess the benefits of additional memory, it may be possible to repeat the validation phase while various amounts of existing memory are disabled. The second is completion of the verification phase, discussed below.

Preliminaries: Conducting a Modelling Study

In the projection phase, model inputs are modified to reflect the anticipated changes to the system or workload. This is a complex process, to which we will devote considerable attention later in the book (Chapter 13). The model then is evaluated. The difference between the modified model outputs and the original model outputs is the projected effect of the modification.

Finally, in the verification phase, the modified system is measured and two comparisons are made. First, its performance measures are compared to the model outputs. Second, its workload measures are compared to the model inputs. Discrepancies between the projections of the model and the performance of the system can arise from two sources: the omission or mis-representation of (retrospectively) significant system characteristics, and the evolution of the system in a way that differs from that which was anticipated. Understanding and evaluating these sources of discrepancy is crucial to gaining confidence in queueing network modelling as a computer system analysis technique. The accuracy of a model's performance projections can be no greater than the accuracy of the workload projections furnished as input.

To illustrate the modelling cycle we describe two case studies undertaken at a computing complex consisting of a number of IBM 370/168s, 370/168-APs (dual processors), and 3033s running the MVS operating system along with applications such as TSO (interactive processing), IMS (database management), JES (spooling), and TCAM/VTAM (terminal management). The objective of each study was to determine the impact of a significant workload modification.

In the first study, the question under consideration was: "Can the workloads presently running on two separate 370/168 uniprocessors be combined on a single 3033?" (A 3033 is considered to have 1.6 to 1.8 times the processing power of a 168.) On each of the original systems, the principal application was an IMS workload. In addition, one of the systems had a background batch workload, and each had various system tasks.

In the validation phase, each of the original systems was measured and modelled. IMS response time was the performance measure of greatest interest, since response time degradation was the anticipated effect of the modification.

In the projection phase, a single model was defined in which each of the original workloads (IMS-1, IMS-2, and batch) was individually represented, with CPU service demand adjusted to account for the speed differences of the CPUs. It was assumed that the I/O subsystem of the 3033 would be the combination of the I/O subsystems of the 168s, so I/O subsystem parameters were not changed in any way.

2.2. The Modelling Cycle

performance measure	workload component	model output	measurement data
CPU utilization	IMS-1	43%	40%
	IMS-2	31%	32%
	batch	3%	3%
	total	77%	75%
response time	IMS-1	0.84 secs.	1.3 secs.
	IMS-2	0.79 secs.	0.89 secs.
throughput	batch	2 jobs/hr.	1.7 jobs/hr.

Table 2.1 − The Modelling Cycle: Case Study 1

In the verification phase, the workloads were combined on the 3033. Performance measures were compared to the model outputs. Table 2.1 displays the results, which are typical of those that can be expected in a study such as this: the projections of the model are sufficiently accurate to be of great utility in planning, and the discrepancy in utilizations is less than the discrepancy in response times.

The second study involved the five loosely-coupled systems described below:

system	CPU type	workload
1	3033	JES for all systems
2	370/168-AP	interactive graphics, batch
3	370/168-AP	batch
4	3033	TSO, IMS, batch
5	3033	batch

The question under consideration was "Can the workload of System 5 be distributed among the four other systems without significant adverse effects on performance, allowing System 5 to be released for cost reduction?"

In the validation phase, Systems 2 through 5 were measured and modelled. (System 1 was excluded from the study.)

In the projection phase, the batch multiprogramming level in the models of Systems 2, 3, and 4 was increased to correspond to the addition of 27% of the workload of System 5. (Management hoped to place 19% of System 5's workload on System 1 and 27% on each of Systems 2, 3, and 4.) This simple approach was possible because of the similarity of the batch workloads on the various systems.

In the verification phase, System 5's workload was distributed among the remaining systems. For each system individually, performance measures were compared to the model outputs. In each case, the anticipated effect of the modification was an increase in the resource consumption of the batch workload (its multiprogramming level had increased), and a degradation in the performance of the other workload components. Tables 2.2, 2.3, and 2.4 display the results for Systems 2, 3, and 4 respectively. Once again, the results are typical of those that can be expected. When used in studies involving system modification, queueing network models may project relative performance with greater accuracy than absolute performance. Consider the response time of the interactive graphics workload in Table 2.2. The original model yielded 4.8 seconds, where 5.2 seconds was measured. The modified model yielded 5.0 seconds. It makes sense to interpret this as a projected response time degradation of 4% ($\frac{5.0-4.8}{4.8}$). In fact, the measured response time degradation was 7.5%.

perf. measure	workload component	original system	original model	modified model	modified system
CPU util.	graphics	76%	74%	74%	72%
	batch	11%	10%	13%	13%
	total	87%	84%	87%	85%
resp. time	graphics	5.2 secs.	4.8 secs.	5.0 secs.	5.6 secs.
t'put.	batch	28/hr.	27/hr.	35/hr.	30/hr.

Table 2.2 − The Modelling Cycle: Case Study 2, System 2

perf. measure	workload component	original system	original model	modified model	modified system
CPU util.	batch	63%	64%	76%	73%
t'put.	batch	101/hr.	104/hr.	130/hr.	120/hr.

Table 2.3 − The Modelling Cycle: Case Study 2, System 3

perf. measure	workload component	original system	original model	modified model	modified system
CPU util.	TSO	65%	67%	65%	63%
	IMS	3%	2%	2%	2%
	batch	15%	15%	21%	20%
	total	83%	84%	88%	85%
resp. time	TSO	4.3 secs.	4.4 secs.	5.0 secs.	5.9 secs.

Table 2.4 − The Modelling Cycle: Case Study 2, System 4

Although we have presented the modelling cycle in an orderly fashion, conducting a modelling study is by no means a strictly sequential process. There are strong dependencies among the various components of the validation and projection phases. Compatibility must be achieved between the definition of the model, the measurements used to parameterize the model, and the techniques used to evaluate the model. Achieving this compatibility, and reconciling it with the objectives of a particular modelling study, is inherently iterative in nature.

2.3. Understanding the Objectives of a Study

It is obvious that the validation phase of a modelling study requires a thorough understanding of the computer system under consideration. Perhaps it is less obvious that a thorough understanding of the objectives of the study is of equal importance. In fact, though, this latter understanding is a key component of the top-down philosophy of queueing network modelling. Many system characteristics that would need to be represented in a fully general model may be irrelevant in a particular study. Identifying these characteristics leads to a simpler model and a simpler modelling study.

A typical example of this phenomenon involved a computer manufacturer about to announce a new CPU in a minicomputer architectural family. During the design of this CPU, extensive low-level performance studies had been carried out, yielding measures such as the average execution rate for various instruction mixes. Prospective customers, however, would be interested in higher-level characterizations such as "In a specific configuration, how does it compare to existing CPUs in the architectural family in terms of the number of users it can support?"

The manufacturer had a set of fifteen benchmarks that had been used in the past for this sort of characterization. Each of the benchmarks had four workload components: editing, file creation, file modification, and a compile-link-execute sequence. The benchmarks differed in the number of "users" in each workload component. These "users" were generated by means of *remote terminal emulation (RTE)*, a technique in which the system of interest is coupled to a second system which simulates interactive users and gathers performance data.

Unfortunately, it was impossible to configure the prototype of the new CPU with the I/O subsystem of interest for the purpose of conducting RTE experiments. Instead, the following strategy was devised:

- Configure an existing, faster CPU in the architectural family with the I/O subsystem of interest.
- Conduct RTE experiments on this configuration for each of the fifteen benchmarks.
- Use a queueing network model to project the performance of each of these benchmarks when the new, slower CPU is substituted. Establish the CPU service demand in the model by taking into account the ratio of the instruction execution rates of the two CPUs.

Given this strategy, the obvious approach would be to define a rather general model of the system. The inputs to this model would include the workload intensities and service demands of each of the four workload components. The model would be capable of reflecting the different characteristics of the fifteen benchmarks by suitable adjustments to the inputs. After this model had been validated, the CPU service demand for each workload component would be scaled appropriately, and the model then would be used to project the performance of the benchmarks on the new system, again by suitable adjustments to the model inputs.

This approach has a significant hidden complexity. The system under consideration includes a sophisticated memory management policy that employs both paging and swapping. The amount of service demanded by each user at the paging and swapping devices is not intrinsic; rather, it depends upon the particular mix of workload components in each benchmark. Thus, the different characteristics of the fifteen benchmarks cannot be reflected in the model simply by adjusting the workload intensities. Instead, a general queueing network model of the system would need to include, as part of its definition, a procedure for estimating variations in the paging and swapping service demands as functions of the mix of workload components.

Devising such a procedure certainly is feasible, but it adds considerably to the complexity of the modelling study, and it provides a level of generality that is not required. Bearing in mind that the objective of this study was restricted to estimating the relative performance of each of the fifteen benchmarks on the two configurations, we can achieve a significant simplification by assuming that the paging and swapping activity of each user, while sensitive to changes in the mix of workload components, are insensitive to changes in CPU speed. This assumption allows the paging and swapping service demands of each workload component to be measured for each of the benchmarks during the RTE experiments, and provided as inputs to the queueing network model, rather than being estimated using a procedure supplied as part of the model definition.

The two approaches to this computer system analysis problem are contrasted in Figure 2.2. The assumption on which the simplified approach

2.3. Understanding the Objectives of a Study

relies is not valid universally, but any inaccuracies that result are strictly secondary, and in fact are probably smaller in magnitude than those that inevitably would arise in attempting to estimate variations in paging and swapping service demands as functions of the mix of workload components. (We will return to this study in Section 2.5, adding further details.)

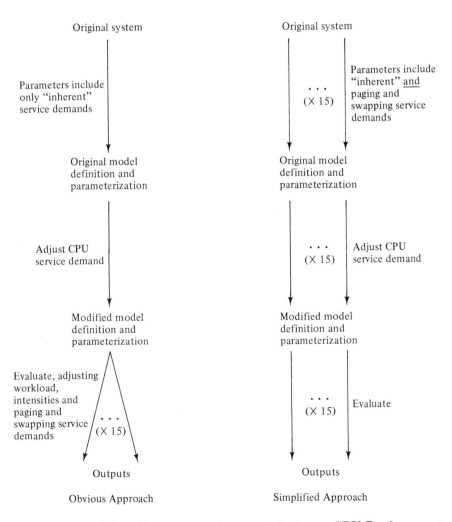

Figure 2.2 — Two Approaches to Modelling a CPU Replacement

2.4. Workload Characterization

In discussing the validation phase of the modelling cycle, we identified *measurement* as the process of obtaining workload measures for the computer system of interest, and *parameterization* as the process of transforming those workload measures into the inputs of a queueing network model. These activities, while not necessarily straightforward, are often considerably less difficult than *workload characterization*: the process of selecting the workload or workloads on which to base the performance study.

Difficult questions arise even in considering an existing computing environment: What constitutes a "typical" workload? How should a measurement interval be selected? Should data from several measurement intervals be averaged? These uncertainties are compounded in considering an environment that cannot be measured directly (e.g., in contemplating the movement of an existing workload to a new system, or the introduction of a new workload to an existing system).

Every approach to computer system analysis — intuition and trend extrapolation, experimental evaluation of alternatives, or modelling — requires workload characterization. Strangely, the imprecision inherent in workload characterization argues for the use of queueing network models. In principle, greater accuracy might be obtained (at significantly greater cost) through experimentation or through simulation modelling. In practice, however, the dominant source of error is apt to lie with the workload characterization, even when queueing network models are employed.

The following case study serves three purposes. The first is to illustrate the use of queueing network modelling in a situation where benchmarking is the traditional approach. The second is to demonstrate *hierarchical workload characterization* as a way to achieve flexibility. By this, we mean progressing in an orderly fashion from a high-level characterization (identification of workload components) through an intermediate level (machine-independent characterizations of each of the components) to a low level (service demands). The third is to show that useful insights can be obtained despite serious imprecision in the workload characterization.

In 1979, a university began a program to acquire medium-scale interactive computer systems for instructional use. In response to a *request for proposals (RFP)*, roughly twenty bids were received, most involving multiple systems. The relative performance of candidate systems was to be a major factor in the acquisition decision. Two approaches to evaluating this relative performance were considered. The first was to construct a multi-user benchmark characteristic of the anticipated workload, then use a remote terminal emulator to run that benchmark on each

2.4. Workload Characterization

candidate system. The second was to perform limited measurements on the candidate systems, then use queueing network modelling to compare performance. The latter approach was appropriate because of the limited time and manpower available for the study, the large number of candidate systems, and the high degree of uncertainty that existed concerning the anticipated workload.

The first step in the study was to characterize the anticipated workload in high-level terms: What were the identifiable workload components? What was the relative volume of each component? What were the significant characteristics of a typical transaction belonging to each component?

Instructional computing previously had been handled in batch mode. The migration of this function to interactive facilities, and its subsequent expansion, was to be a multi-year process involving multiple acquisitions. It was assumed that the initial interactive workload would be similar in composition to the existing instructional batch workload, with the addition of an editing component.

Measurements indicated that the existing workload had only two significant components. Nearly 80% of all transactions were Fortran compilations. Nearly 20% of all transactions were the execution of pre-compiled library routines to process student-created datasets. A simple characterization of the compilations was the average number of lines of source code: roughly 100. A simple characterization of the executions was their average service demand on the existing system: 4.55 seconds of CPU service and 5.35 seconds of disk service. (The average size of the student-created datasets processed by these transactions was 100 lines.)

It was assumed that an editing session would precede each compilation or execution, so that the overall mix of workload components would be 40% compilations, 10% executions, and 50% editing sessions. Since most editing would be performed by inexperienced typists using line-oriented editors to make a small number of changes to a file, it was assumed that the dominant resource demands would occur in accessing and saving files. The average size of the file being edited, 100 lines, thus was a simple characterization of the editing sessions.

The second step in the study was to translate this high-level workload characterization into parameters for models of each of the candidate systems. Determining workload intensities was not an issue. Each of the three workload components was treated as a transaction workload with an arrival rate equal to the established proportion of the total arrival rate. Model outputs were tabulated for a range of total arrival rates. Determining service demands for each workload component on each system (i.e., the average service required at each device by a transaction belonging to each workload component) involved running three extremely simple experiments on each system. For compilations, a 100-line program was compiled on an otherwise idle system and CPU and disk busy times were measured. This experiment captured the effects of hardware speed, compiler efficiency, and overhead in initiating and terminating compilations. For executions, the CPU and disk service requirements that had been measured on the existing batch system were scaled. The scaling factor for CPU service was obtained by running a single computational benchmark on the existing system and on each candidate system. The scaling factor for disk service was obtained using a single Fortran I/O benchmark. For editing sessions, the default editor available on each candidate system was used on an otherwise idle system to access a 100-line file, modify a single line, and save the file. CPU and disk busy times were measured.

Table 2.5 shows the results of these experiments for three candidate systems: a VAX-11/780, a Prime 750, and a Prime 550. Note the dramatically different efficiencies observed for the two Fortran compilers available on the Primes. Note also the relative inefficiency of the interface between the editor and the file system on the VAX.

system	workload component	service demand, secs.	
		CPU	disk
Digital VAX-11/780	compilation	2.0	1.0
	execution	11.9	10.7
	editing session	0.5	0.8
Prime 750	compilation		
	compiler A	0.8	0.2
	compiler B	7.0	1.0
	execution	13.7	7.1
	editing session	0.15	0.05
Prime 550	compilation		
	compiler A	1.3	0.75
	compiler B	11.3	3.75
	execution	27.9	21.4
	editing session	0.3	0.1

Table 2.5 — Service Demands for Three Systems

Based on these values, queueing network models of the candidate systems were parameterized and evaluated. (Representing multiple disks involved distributing the calculated disk service demand among several service centers. Parameterization was simplified by the fact that it was not necessary to consider overhead due to memory contention, which typically grows with workload intensity. It was a stipulation of the RFP that systems be overconfigured with respect to memory.) Figures 2.3 and 2.4 show typical results of the study: average response time versus total transaction arrival rate for compilations and executions, respectively, for the VAX-11/780, the Prime 750 with compiler A, and the Prime 750 with compiler B. Note that the performance of the Prime depends critically on the choice of compiler, and that this choice affects all users, not just those doing compilations. (A reminder: these results have significance only for the specific configurations and workloads under consideration.)

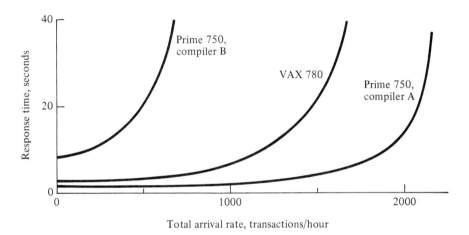

Figure 2.3 − Compilation Response Time Versus Total Arrival Rate

Variations can be investigated with ease. The effect of a disk load imbalance can be explored by shifting the proportion of the service demand allocated to each service center. The sensitivity of the results to the workload characterization can be studied; e.g., the relative arrival rates of the three workloads could be altered.

2.5. Sensitivity Analysis

Every computer system analyst encounters situations in which questionable assumptions must be introduced. *Sensitivity analysis* can be used to determine the extent to which such assumptions cast doubt on the conclusions of the study. A sensitivity analysis can take many forms. Two of the most common are:

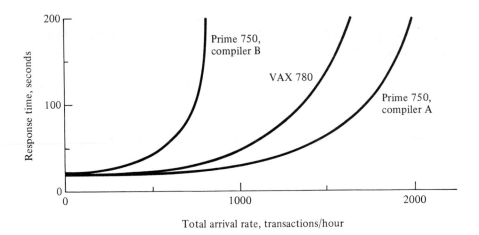

Figure 2.4 — Execution Response Time Versus Total Arrival Rate

- The analyst may test the robustness of the results to the assumption in question. Doing so involves evaluating the model a number of times for variations in the assumption, and comparing the results.
- The analyst may obtain bounds on the expected performance, by evaluating the model for extreme values of the assumption.

Inadequate measurement data frequently is the culprit that prompts a sensitivity analysis. To illustrate the role of sensitivity analysis in coping with this situation we return to the CPU replacement case study introduced in Section 2.3. As illustrated in Figure 2.2, the approach adopted entailed fifteen separate experiments, one per benchmark. Each experiment consisted of three phases: the existing system was measured while executing one of the benchmarks, a queueing network model was constructed and validated, and this model was used to project benchmark performance with the new CPU, by manipulating the CPU service demand parameter of each workload component.

Difficulty was encountered during the validation phase because a significant proportion of the system's I/O activity was not attributed to specific workload components by the available measurement tools. For example, it was possible to determine the total number of swaps during a measurement interval, and also the average disk service demand per swap, but it was not possible to determine which user or workload component was the "victim" of the swap. Had the study been based on a single class model, this would not have been a problem. However, the objective was to assess the impact of the CPU replacement on each of the four workload components individually, so a multiple class model was required.

Various methods of allocating this measured I/O activity among the four workload components yielded different values for some of the input parameters of the model. Not surprisingly, different response time projections from the model resulted. As an example, for one of the benchmarks the measured response time for file modification transactions was 10 seconds, while for three different but equally reasonable methods of allocating measured I/O activity among the four workload components, the model projected response times of 6, 7, and 11 seconds. (Similarly spurious results were obtained from this model for the response times of the three other workload components.)

Consider the set of inputs for which the model projected a response time of 6 seconds. When the CPU service demand parameter was adjusted to reflect the substitution of the slower CPU, this model projected that response time would be 7.2 seconds. It makes no sense to claim that the response time for file modification transactions on the new system will be 7.2 seconds, because the measured response time on the existing, faster system was 10 seconds. Nor does it make sense to claim that response time will increase by 20% ($\frac{7.2 - 6.0}{6.0}$), because there is no reason to believe that the projected effect of the CPU substitution is insensitive to the method used to allocate measured I/O activity among the workload components. We can hypothesize such an insensitivity, though, and then test this hypothesis. Table 2.6 displays projected response times for the system with the existing CPU and the new CPU, for the three approaches to I/O activity allocation. Although the absolute response time values differ for the three approaches, the projected percentage changes do not. Thus, we can conclude that the effect of the CPU substitution will be in increase of roughly 20% in the response time of file modification transactions, from 10 seconds (the measured value) to 12 seconds. (Similar results were obtained for the other three workload components.)

2.6. Sources of Insight

A major virtue of queueing network modelling is that the modelling cycle yields many insights about the computer system under study. These insights occur during workload characterization, model definition, system measurement, model parameterization, and modification analysis. It is important to bear in mind that the model outputs obtained during the projection phase of the modelling cycle are only one of many sources of insight. Consider the following case study.

method of allocating I/O activity	workload component	response time, seconds		
		model of original CPU	model of new CPU	projected change
A	editing file creation file mod. compile-link-execute	... ... 6 ...	... ... 7.2 ...	... ... + 20% ...
B	editing file creation file mod. compile-link-execute	... ... 7 ...	... ... 8.3 ...	... ... + 18% ...
C	editing file creation file mod. compile-link-execute	... ... 11 ...	... ... 13.1 ...	... ... + 19% ...

Table 2.6 — Response Times for Three Assumptions

An insurance company decentralized its claims processing by establishing identical minicomputer systems at twenty geographically distributed sites. As the workload grew, these systems ceased to provide adequate response, and a two-step capacity expansion program was begun: an immediate upgrade at every site to one of two software-compatible systems available from the original vendor, followed by a three year process of "unconstrained" system acquisition and software conversion. Queueing network modelling was used to evaluate the alternatives for each step. In this section, we consider the choice of a "transition system" for each site.

Working together, the vendor (IBM) and the insurance company had estimated that performance would "improve by a factor of 1.5 to 2.0" if the existing system (a 3790 in each case) were replaced with the less expensive of the two transition systems (an 8130), and "improve by a factor of 2.0 to 3.5" if it were replaced by the more expensive of the transition systems (an 8140). (Note the considerable ambiguity in these statements.) The charter of the modelling study was to determine at which of the twenty sites the more expensive system would be required in order to achieve acceptable performance during the three-year transition period.

The information provided in support of the study included measurements of the existing 3790 system taken at several sites under "live" workload, measurements of the 3790 and the more expensive transition system (the 8140) during benchmark trials in which varying numbers of clerks entered transactions from scripts, and information from the vendor comparing the CPU and disk speeds of the three systems. The "live" workload tests revealed that although there were three distinct workload components, one of these, which had been identified in advance as being of primary interest, was responsible for roughly 75% of the transactions and 90% of the resource consumption. A single class model was therefore deemed appropriate. The benchmark tests confirmed the vendor's estimates of relative hardware speeds, although they were too limited (in terms of the range of workload intensities considered) to yield any insight about overall performance. From consideration of all of the available information it was possible to calculate the service demands shown below:

	service demands, seconds	
system	CPU	disk
3790 (existing)	4.6	4.0
8130	5.1	1.9
8140	3.1	1.9

As indicated, the two transition systems were equipped with identical disks that were roughly twice as fast as the disks on the existing system. The transition systems differed in their CPUs: the 8130 CPU was, in fact, slightly slower than that of the existing 3790, while the 8140 CPU was roughly 50% faster.

Now we make a key observation. On the existing system, the workload is CPU-bound. Furthermore, since response times are unacceptable, we can assume that the workload intensity is sufficiently high that the CPU is approaching saturation. The faster disks of the 8130 are of little value under these circumstances, while its slower CPU is a significant liability. Without further examination, we can conclude that replacing the 3790 with the 8130 will cause a degradation in response time.

On the basis of this analysis, the insurance company performed benchmark tests on the 8130. These tests confirmed the analysis, with the result that all sites were upgraded to 8140s. (This study will be considered further in Chapter 5.)

2.7. Summary

The most challenging aspect of computer system analysis using queueing network models is not the technical details of defining, parameterizing, and evaluating the models. Rather, it is the process of tailoring the general "methodology" of queueing network modelling to a specific computer system analysis context. Unfortunately, while the former is easily taught, the latter is best learned through experience. In this chapter we have attempted to share with the reader the experience of others, by presenting a set of case studies selected to illustrate significant aspects of the methodology. Among the points that we have emphasized are:

- Queueing network modelling inherently is a top-down process in which the low-level details of a system are presumed to be irrelevant to its high-level performance characteristics.
- Because queueing network models are abstract, many assumptions are made in conducting a modelling study. These assumptions are motivated by simplicity, adequacy of measurements, and ease of evaluation. It is important to be explicit concerning the assumptions that are made, the motivations for their introduction, and the arguments for their plausibility.
- Conducting a modelling study is an iterative process because of dependencies that exist among the definition of the model, the measurements used to parameterize the model, the techniques used to evaluate the model, and the objectives of a particular modelling study.
- Confidence in a model's predictive abilities can be acquired through repetitive validation over a number of measurement intervals, perhaps involving selected minor modifications.
- This confidence can be reinforced through the verification process: measuring a modified system, then comparing its performance measures to the model outputs and its workload measures to the model inputs.
- When used in studies involving system modification, queueing network models may project relative performance with greater accuracy than absolute performance.
- A clear understanding of the objectives of a modelling study can contribute to simplicity in the model and in the modelling effort.
- Concentrating on representing the primary effects of a system or workload modification also can contribute to simplicity.
- Workload characterization is a challenging, inherently imprecise process. Useful insights can be obtained despite this imprecision. Characterizing a workload hierarchically helps to achieve flexibility.
- Sensitivity analysis can be used to determine the extent to which questionable assumptions cast doubt on the conclusions of a study. Two common forms of sensitivity analysis are testing the robustness of model outputs to variations of assumptions, and obtaining bounds on model outputs for extreme values of assumptions.

- Valuable insights are gained throughout the modelling cycle, not merely during the projection phase.

2.8. References

The identification of simplicity, adequacy of measurements, and ease of evaluation as factors motivating the introduction of assumptions is due to Kienzle and Sevcik [1979], who also suggested the division of the modelling cycle into validation, projection, and verification phases.

The MVS case studies described in Section 2.2 were conducted by Lo [1980]. The CPU performance comparison described in Sections 2.3 and 2.5 was conducted by Myhre [1979]. The system acquisition case study described in Section 2.4 was conducted by Lazowska [1980]. (Figures 2.3 and 2.4 are taken from this paper.) The insurance claims processing case study described in Section 2.6 was conducted by Sevcik, Graham, and Zahorjan [1980].

[Kienzle & Sevcik 1979]
 M.G. Kienzle and K.C. Sevcik. A Systematic Approach to the Performance Modelling of Computer Systems. *Proc. IFIP W.G.7.3 International Symposium on Computer Performance Modelling, Measurement and Evaluation* (1979), 3-27.

[Lazowska 1980]
 Edward D. Lazowska. The Use of Analytic Modelling in System Selection. *Proc. CMG XI International Conference* (1980), 63-69.

[Lo 1980]
 T.L. Lo. Computer Capacity Planning Using Queueing Network Models. *Proc. IFIP W.G.7.3 International Symposium on Computer Performance Modelling, Measurement and Evaluation* (1980), 145-152. Copyright © 1980 by the Association for Computing Machinery.

[Myhre 1979]
 Scott A. Myhre. A Queueing Network Solution Package Based on Mean Value Analysis. M.Sc. Thesis, Department of Computer Science, University of Washington, February 1979.

[Sevcik et al. 1980]
 K.C. Sevcik, G.S. Graham, and J. Zahorjan. Configuration and Capacity Planning in a Distributed Processing System. *Proc. 16th CPEUG Meeting* (1980), 165-171.

Chapter 3

Fundamental Laws

3.1. Introduction

This chapter provides the technical foundation for much of the remainder of the book. It has three objectives. The first is to define a number of quantities of interest and to introduce the notation that we will use in referring to these quantities. The second is to derive various algebraic relationships among these quantities, some of which, because of their importance, will be identified as *fundamental laws*. The third is to explore thoroughly the most important of these fundamental laws, *Little's law* (named for J.D.C. Little), which states that the average number of requests in a system must equal the product of the throughput of that system and the average time spent in that system by a request.

Because of the volume of notation introduced, this chapter may appear formidable. It is not. The material is summarized in three small tables in Section 3.6, which we suggest you copy for convenient reference.

3.2. Basic Quantities

If we were to observe the abstract system shown in Figure 3.1 we might imagine measuring the following quantities:

T, the length of *time* we observed the system

A, the number of request *arrivals* we observed

C, the number of request *completions* we observed

From these measurements we can define the following additional quantities:

3.2. Basic Quantities

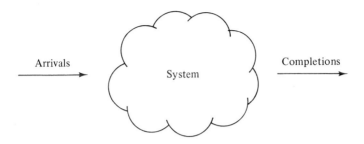

Figure 3.1 − An Abstract System

λ, the *arrival rate*: $\lambda \equiv \dfrac{A}{T}$

If we observe 8 arrivals during an observation interval of 4 minutes, then the arrival rate is 8/4 = 2 requests/minute.

X, the *throughput*: $X \equiv \dfrac{C}{T}$

If we observe 8 completions during an observation interval of 4 minutes, then the throughput is 8/4 = 2 requests/minute.

If the system consists of a single resource, we also can measure:

B, the length of time that the resource was observed to be *busy*

Two more defined quantities now are meaningful:

U, the *utilization*: $U \equiv \dfrac{B}{T}$

If the resource is busy for 2 minutes during a 4 minute observation interval, then the utilization of the resource is 2/4, or 50%.

S, the average *service requirement* per request: $S \equiv \dfrac{B}{C}$

If we observe 8 completions during an observation interval and the resource is busy for 2 minutes during that interval, then on the average each request requires 2/8 minutes of service.

We now can derive the first of our fundamental laws. Algebraically, $\dfrac{B}{T} = \dfrac{C}{T}\dfrac{B}{C}$. From the three preceding definitions, $\dfrac{B}{T} \equiv U$, $\dfrac{C}{T} \equiv X$, and $\dfrac{B}{C} \equiv S$. Hence:

> **The Utilization Law:** $U = XS$

That is, the utilization of a resource is equal to the product of the throughput of that resource and the average service requirement at that resource. As an example, consider a disk that is serving 40 requests/second, each of which requires .0225 seconds of disk service. The utilization law tells us that the utilization of this disk must be $40 \times .0225 = 90\%$.

3.3. Little's Law

The utilization law in fact is a special case of Little's law, which we now will derive in a more general setting. Figure 3.2 is a graph of the total number of arrivals and completions occurring at a system over time. Each step in the higher step function signifies the occurrence of an arrival at that instant; each step in the lower signifies a completion. At any instant, the vertical distance between the arrival and completion functions represents the number of requests present in the system. Over any time interval, the area between the arrival and completion functions represents the accumulated time in system during that interval, measured in request-seconds (or request-minutes, etc.). For example, if there are three requests in the system during a two second period, then six request-seconds are accumulated. This area is shaded in Figure 3.2 for an observation interval of length $T = 4$ minutes. We temporarily denote accumulated time in system by W. We define:

N, the average *number of requests* in the system: $N \equiv \dfrac{W}{T}$

> If a total of 2 request-minutes of residence time are accumulated during a 4 minute observation interval, then the average number of requests in the system is $2/4 = 0.5$.

R, the average system *residence time* per request: $R \equiv \dfrac{W}{C}$

> If a total of 2 request-minutes of residence time are accumulated during an observation interval in which 8 requests complete, then the average contribution of each completing request (informally, the average system residence time per request) is $2/8 = 0.25$ minutes.

Algebraically, $\dfrac{W}{T} = \dfrac{C}{T}\dfrac{W}{C}$. But $\dfrac{W}{T} \equiv N$, $\dfrac{C}{T} \equiv X$, and $\dfrac{W}{C} \equiv R$.

3.3. Little's Law

Hence:

> **Little's Law:** $N = XR$

That is, the average number of requests in a system is equal to the product of the throughput of that system and the average time spent in that system by a request.

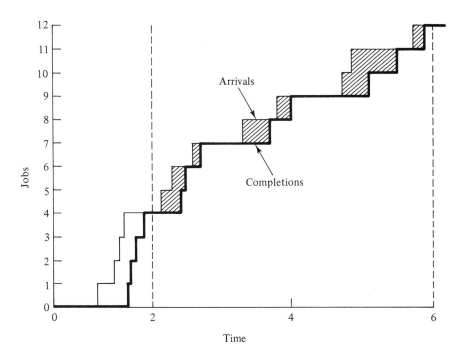

Figure 3.2 − **System Arrivals and Completions**

A subtle but important point in our derivation of Little's law is that the quantity R does not necessarily correspond to our intuitive notion of average residence time or response time − the expected time from arrival to departure. This discrepancy is due to end effects: it is hard to know how to account for requests that are present just prior to the start or just after the end of an observation interval. For the time being, suffice it to say that if the number of requests passing through the system during the observation interval is significantly greater than the number present at the beginning or end, then R corresponds closely to our intuition, and if the observation interval begins and ends at instants when system is empty, then this correspondence is exact. (End effects arise elsewhere; for example, considerations similar to those affecting R also affect our earlier definition of S, the average service requirement per request.)

Little's law is important for three reasons. First, because it is so widely applicable (it requires only very weak assumptions), it will be valuable to us in checking the consistency of measurement data. Second, in studying computer systems we frequently will find that we know two of the quantities related by Little's law (say, the average number of requests in a system and the throughput of that system) and desire to know the third (the average system residence time, in this case). Third, Little's law is central to the algorithms for evaluating queueing network models, which we will introduce in Part II.

Given a computer system, Little's law can be applied at many different levels: to a single resource, to a subsystem, or to the system as a whole. The key to success is consistency: the definitions of population, throughput, and residence time must be compatible with one another. In Figure 3.3 we illustrate this by applying Little's law to a hypothetical timesharing system at four different levels, as indicated by the four boxes in the figure.

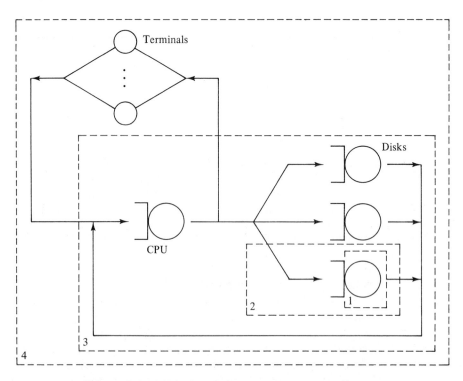

Figure 3.3 — Little's Law Applied at Four Levels

3.3. Little's Law

Box 1 is perhaps the most subtle; it illustrates the application of Little's law to a single resource, *not including* its queue. In this example, population corresponds to the utilization of the resource (there are either zero or one requests present at any instant in time; the resource is utilized whenever there is one request present; thus resource utilization is equal to the proportion of time there is one request present, which is also equal to the average number of requests present), throughput corresponds to the rate at which the resource is satisfying requests, and residence time corresponds to the average service requirement per request at the resource (remember, queueing delay is not included in this application of Little's law; once a request acquires the resource, it remains at that resource for its service time). This application of Little's law constitutes an alternative derivation of the utilization law. To repeat the example used previously, suppose that the resource is a disk, that the disk is serving 40 requests/second ($X = 40$), and that the average request requires .0225 seconds of disk service ($S = .0225$). Then Little's law ($U = XS$) tells us that the utilization of the disk must be $40 \times .0225 = 90\%$.

Box 2 illustrates the application of Little's law to the same resource, this time including its queue. Now, population corresponds to the total number of requests either in queue or in service, throughput remains the rate at which the resource is satisfying requests, and residence time corresponds to the average time that a request spends at the resource per visit, both queueing time and service time. Suppose that the average number of requests present is 4 ($N = 4$) and that the disk is serving 40 requests/second ($X = 40$). Then Little's law ($N = XR$) tells us that the average time spent at the disk by a request must be $4/40 = 0.1$ seconds. Note that we can now compute the average queueing time of a request (a total of 0.1 seconds are spent both queueing and receiving service, of which .0225 seconds are devoted to receiving service, so the average queueing time must be .0775 seconds) and also the average number of requests in the queue (an average total of 4 requests are either queueing or receiving service, and on the average there are 0.9 requests receiving service, so the average number awaiting service in the queue must be 3.1).

Box 3 illustrates the application of Little's law to the *central subsystem* − the system without its terminals. Our definition of "request" changes at this level: we are no longer interested in visits to a particular resource, but rather in system-level interactions. Population corresponds to the number of customers in the central subsystem, i.e., those users not thinking. Throughput corresponds to the rate at which interactions flow between the terminals and the central subsystem. Residence time corresponds to our conventional notion of response time: the period of time from when a user submits a request until that user's response is returned. Suppose that system throughput is 1/2 interaction per second ($X = 0.5$) and that, on the average, there are 7.5 "ready" users ($N = 7.5$). Then Little's law ($N = XR$) tells us that average response time must be $7.5/0.5 = 15$ seconds.

Finally, box 4 illustrates the application of Little's law to the entire system, including its terminals. Here, population corresponds to the total number of interactive users, throughput corresponds to the rate at which interactions flow between the terminals and the system, and residence time corresponds to the sum of system response time and user think time. Suppose that there are 10 users, average think time is 5 seconds, and average response time is 15 seconds. Then Little's law tells us that the system throughput must be $\frac{10}{15+5} = 0.5$ interactions/second. If we denote think time by Z then we can write this incarnation of Little's law as $N = X(R+Z)$. As with the utilization law, this application is so ubiquitous that we give it its own name and notation, expressing R in terms of the other quantities:

$$\text{The Response Time Law:} \quad R = \frac{N}{X} - Z$$

As an example application of the response time law, suppose that a system has 64 interactive users, that the average think time is 30 seconds, and that system throughput is 2 interactions/second. Then the response time law tells us that response time must be $\frac{64}{2} - 30 = 2$ seconds.

In earlier chapters we have noted that throughputs and utilizations typically are projected with greater accuracy than residence times. We now are in a position to understand why this must be. Suppose we were to construct a queueing network model of the system in the previous example. The number of users (64) and the average think time (30 seconds) would be parameters of the model, along with the service demands at the various resources in the system. Throughput and response time would be outputs of the model. Suppose that the model projected a throughput of 1.9 interactions/second, an error of just 5%. Since the response time law must be satisfied by the queueing network model, a compensating error in projected response time must result:

$$R = \frac{64}{1.9} - 30$$

Thus the model must project a response time of 3.7 seconds, an error of 85%.

3.4. The Forced Flow Law

In discussing Little's law, we allowed our field of view to range from an individual resource to an entire system. At different levels of detail, different definitions of "request" are appropriate. For example, when considering a disk, it is natural to define a request to be a disk access, and to measure throughput and residence time on this basis. When considering an entire system, on the other hand, it is natural to define a request to be a user-level interaction, and to measure throughput and residence time on this basis.

The relationship between these two views of a system is expressed by the *forced flow law*, which states that the flows (throughputs) in all parts of a system must be proportional to one another. Suppose that during an observation interval we count not only system completions, but also the number of completions at each resource. We define the *visit count* of a resource to be the ratio of the number of completions at that resource to the number of system completions, or, more intuitively, to be the average number of visits that a system-level request makes to that resource. If we let a variable with the subscript k refer to the k-th resource (a variable with no subscript continues to refer to the system as a whole), then we can write this definition as:

V_k, the *visit count* of resource k: $V_k \equiv \dfrac{C_k}{C}$

> If during an observation interval we measure 10 system completions and 150 completions at a specific disk, then on the average each system-level request requires $150/10 = 15$ disk operations.

If we rewrite this definition as $C_k = V_k C$ and recall that the completion count divided by the length of the observation interval is defined to be the throughput, then the throughput of resource k is given by:

$$\boxed{\text{The Forced Flow Law:} \quad X_k = V_k X}$$

An informal statement of the forced flow law is that the various components of a system must do comparable amounts of work (measured in "transaction's worth") in a given time interval. As an example, suppose we are told that each job in a batch processing system requires an average of 6 accesses to a specific disk, and that the disk is servicing 12 requests from batch jobs per second. Then we know that the system throughput of batch jobs must be $12/6 = 2$ jobs/second. If, in addition, we are told that another disk is servicing 18 batch job requests per second, then we know that each batch job requires on average $18/2 = 9$ accesses to this second disk.

Little's law becomes especially powerful when combined with the forced flow law. As an example, suppose that we are asked to determine average system response time for an interactive system with the following known characteristics:

25 terminals ($N = 25$)
18 seconds average think time ($Z = 18$)
20 visits to a specific disk per interaction ($V_{disk} = 20$)
30% utilization of that disk ($U_{disk} = .30$)
25 millisecond average service requirement per visit
to that disk ($S_{disk} = .025$ secs.)

We would like to apply the response time law: $R = \dfrac{N}{X} - Z$. We know the number of terminals and the average think time, but are missing the throughput. We do, however, know the visit count at one specific disk (that is, the average number of visits made to that disk by an interactive request), so if we knew the throughput at that disk we would be able to apply the forced flow law to obtain system-level throughput. To obtain disk throughput we can use the utilization law, since we know both utilization and service requirement at this device. We calculate the following quantities:

$$\text{disk throughput: } X_{disk} = \frac{U_{disk}}{S_{disk}} = \frac{.30}{.025} = 12 \text{ requests/sec.}$$

$$\text{system throughput: } X = \frac{X_{disk}}{V_{disk}} = \frac{12}{20} = 0.6 \text{ interactions/sec.}$$

$$\text{response time: } R = \frac{N}{X} - Z = \frac{25}{0.6} - 18 = 23.7 \text{ secs.}$$

Note that we can describe an interaction's disk service requirement in either of two ways: by saying that an interaction makes a certain number of visits to the disk and requires a certain amount of service on each visit, or by specifying the total amount of disk service required by an interaction. These two points of view are equivalent, and whichever is more convenient should be chosen. We define:

D_k, the *service demand* at resource k: $D_k \equiv V_k S_k$

If a job makes an average of 20 visits to a disk and requires an average of 25 milliseconds of service per visit, then that job requires a total of $20 \times 25 = 500$ milliseconds of disk service, so its service demand is 500 milliseconds at that disk.

From now on we will use S_k to refer to the service requirement per visit at resource k, and D_k to refer to the total service requirement at that resource. We define D, with no subscript, to be the sum of the D_k: the total service demanded by a job at all resources.

3.4. The Forced Flow Law

Again, consistency is crucial to success. Consider using the utilization law to calculate the utilization of a resource. We can express throughput in terms of visits to that resource (X_k), in which case service requirement must be expressed as service requirement per visit (S_k). Using the forced flow law, we can also express throughput in terms of system-level interactions (X), in which case service requirement must be expressed on a per-interaction basis (D_k). In other words, $U_k = X_k S_k = XD_k$.

In Chapter 1 we observed that service demands are one of the parameters required by queueing network models. If we observe a system for an interval of length T, we can easily obtain the utilizations of the various resources, U_k, and the system-level completion count, C. The service demands at the various resources then can be calculated as $D_k = \frac{B_k}{C} = \frac{U_k T}{C}$. It is fortunate that queueing network models can be parameterized in terms of the D_k rather than the corresponding V_k and S_k, since the former typically are much more easily obtained from measurement data than the latter.

As a final illustration of the versatility of Little's law in conjunction with the forced flow law, consider Figure 3.4, which represents a timesharing system with a memory constraint: swapping may occur between interactions, so a request may be forced to queue for a memory partition prior to competing for the resources of the central subsystem. As indicated by the boxes, we once again are going to apply Little's law at several different levels. The following actual measurement data was obtained by observing the timesharing workload on a system with several distinct workloads:

average number of timesharing users: 23 ($N = 23$)
average response time perceived by a user: 30 seconds ($R = 30$)
timesharing throughput: 0.45 interactions/second ($X = .45$)
average number of timesharing requests occupying memory: 1.9 ($N_{in\ mem} = 1.9$)
average CPU service requirement per interaction: 0.63 seconds ($D_{CPU} = .63$)

Now, consider the following questions:

- What was the average think time of a timesharing user? Applying the response time law at the level of box 4 in the figure, $R = \frac{N}{X} - Z$, so $Z = \frac{23}{.45} - 30$, or 21 seconds.
- On the average, how many users were attempting to obtain service, i.e., how many users were not "thinking" at their terminals? Applying Little's law at the level of box 3, $N_{want\ mem} = XR = .45 \times 30$, or 13.5 users. Of the 23 users on this system, an average of 13.5 were

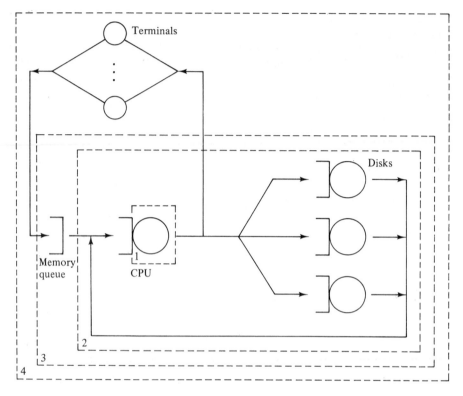

Figure 3.4 − Little's Law Applied to a Memory Constrained System

attempting to obtain service at any one time. We know from measurement data that only 1.9 were occupying memory on the average, so the remaining 11.6 must have been queued awaiting access to memory.

- On the average, how much time elapses between the acquisition of memory and the completion of an interaction? Applying Little's law at the level of box 2, $N_{in\ mem} = XR_{in\ mem}$, so $R_{in\ mem} = 1.9/0.45$, or 4.2 seconds. In other words, of the 30 second response time perceived by a user, nearly 26 seconds are spent queued awaiting access to memory.

- What is the contribution to CPU utilization of the timesharing workload? Applying the utilization law to the CPU (box 1), $U_{CPU} = XD_{CPU} = .45 \times .63$, or 28% of the capacity of the CPU. Notice that in this application of the utilization law, throughput was defined in terms of system-level interactions and service requirement was defined on a per-interaction basis.

3.5. The Flow Balance Assumption

Frequently it will be convenient to assume that systems satisfy the *flow balance* property, namely, that the number of arrivals equals the number of completions, and thus the arrival rate equals the throughput:

> **The Flow Balance Assumption:** $A = C$, therefore $\lambda = X$

The flow balance assumption can be tested over any measurement interval, and it can be strictly satisfied by careful choice of measurement interval.

When used in conjunction with the flow balance assumption, Little's law and the forced flow law allow us to calculate device utilizations for systems whose workload intensities are described in terms of an arrival rate. In Figure 3.5 we show a queueing network model similar to that used to represent the VAX-11/780 in the case study described in Section 2.4. There are three devices (a CPU and two disks) and three transaction classes with the following characteristics:

transaction class	arrival rate trans./hr.	service demand, seconds/transaction		
		CPU	disk 1	disk 2
compilation	480	2.0	0.75	0.25
execution	120	11.9	5.0	5.7
editing session	600	0.5	0.2	0.6

To calculate the utilization of a device in this system we apply the utilization law separately to each transaction class, then sum the results. As an example, consider the CPU. If compilation transactions are arriving to the system at a rate of 480/hour and each one brings 2.0 seconds of work to the CPU, then CPU utilization due to compilation transactions must equal $\frac{480}{3600} \times 2.0 = 27\%$. Similar arguments for execution and editing transactions yield CPU utilizations of 40% and 8%, respectively. Thus total CPU utilization must be 75%.

How is it possible to analyze the classes independently without accounting for their mutual interference? Assuming that the system is able to handle the offered load (i.e., assuming that the calculated utilization of no device is greater than 100%), the flow balance assumption is reasonable. Thus the throughput of the system will be the same as the arrival rate to the system. The forced flow law guarantees that the various devices in the system will do comparable amounts of work (measured

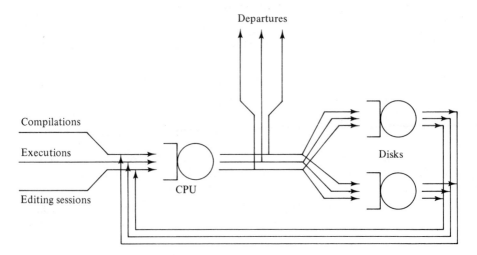

Figure 3.5 — Calculating Utilizations Using Flow Balance

in "transaction's worth") in a given time. Interference between transactions does not affect this. Rather, it causes an increase in the average number of transactions resident in the system, which causes a corresponding increase in response time (by Little's law). In Part II we will learn how to quantify the extent of this interference.

3.6. Summary

In this chapter we have defined a number of quantities of interest, introduced the notation that we will use in referring to these quantities, and derived various algebraic relationships among these quantities. These developments are reviewed in the following tables, which we suggest you copy for convenient reference.

Table 3.1 summarizes the notation that we have established. The table includes a subscript on those quantities that require one, either explicit or implicit. In some cases, the quantity must refer to a specific resource. In other cases, the quantity may refer either to a specific resource or to a specific subsystem. Table 3.2 summarizes the fundamental laws. Table 3.3 summarizes the additional algebraic relationships among the various quantities that we have defined. We also have introduced and used the flow balance assumption: $A = C$, therefore $\lambda = X$.

T	length of an observation interval
A_k	number of arrivals observed
C_k	number of completions observed
λ_k	arrival rate
X_k	throughput
B_k	busy time
U_k	utilization
S_k	service requirement per visit
N	customer population
R_k	residence time
Z	think time of a terminal user
V_k	number of visits
D_k	service demand

Table 3.1 — Notation

The Utilization Law:	$U_k = X_k S_k = X D_k$
Little's Law:	$N = XR$
The Response Time Law:	$R = \dfrac{N}{X} - Z$
The Forced Flow Law:	$X_k = V_k X$

Table 3.2 — Fundamental Laws

3.7. References

Buzen and Denning's *operational analysis* has heavily influenced our philosophy in general, and this chapter in particular. Much of the notation and the identification of laws and assumptions is taken from their work. Of special note are [Buzen 1976] (from which we have even borrowed the title of this chapter) and [Denning & Buzen 1978].

Little's law is named for J.D.C. Little, who first proved it in 1961 [Little 1961].

$$\lambda_k \equiv \frac{A_k}{T}$$

$$X_k \equiv \frac{C_k}{T}$$

$$U_k \equiv \frac{B_k}{T}$$

$$S_k \equiv \frac{B_k}{C_k} = \frac{U_k T}{C_k}$$

$$V_k \equiv \frac{C_k}{C}$$

$$D_k \equiv V_k S_k = \frac{B_k}{C} = \frac{U_k T}{C}$$

Table 3.3 — Additional Relationships

[Buzen 1976]
 Jeffrey P. Buzen. Fundamental Operational Laws of Computer System Performance. *Acta Informatica* 7,2 (1976), 167-182.

[Denning & Buzen 1978]
 Peter J. Denning and Jeffrey P. Buzen. The Operational Analysis of Queueing Network Models. *Computing Surveys* 10,3 (September 1978), 225-261.

[Little 1961]
 J.D.C. Little. A Proof of the Queueing Formula $L = \lambda W$. *Operations Research* 9 (1961), 383-387.

3.8. Exercises

1. Consider the specific computer system with which you are most familiar. How would you calculate the basic service demand D_k at the CPU? At each disk device? How would you calculate the average number of jobs in memory?

2. Software monitor data for an interactive system shows a CPU utilization of 75%, a 3 second CPU service demand, a response time of 15 seconds, and 10 active users. What is the average think time of these users?

3.8. Exercises

3. An interactive system with 80 active terminals shows an average think time of 12 seconds. On average, each interaction causes 15 paging disk accesses. If the service time per paging disk access is 30 ms. and this disk is 60% busy, what is the average system response time?
4. Suppose an interactive system is supporting 100 users with 15 second think times and a system throughput of 5 interactions/second.
   a. What is the response time of the system?
   b. Suppose that the service demands of the workload evolve over time so that system throughput drops to 50% of its former value (i.e., to 2.5 interactions/second). Assuming that there still are 100 users with 15 second think times, what would their response time be?
   c. How do you account for the fact that response time in (b) is more than twice as large as that in (a)?
5. Consider a system modelled as shown in Figure 3.6. A user request submitted to the system must queue for memory, and may begin processing (in the central subsystem) only when it has obtained a memory partition.

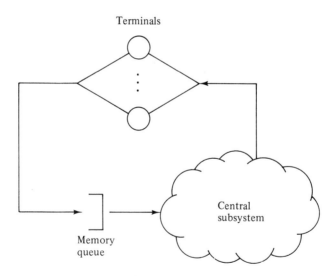

Figure 3.6 — A Memory Constrained System

   a. If there are 100 active users with 20 second think times, and system response time (the sum of memory queueing and central subsystem residence times) is 10 seconds, how many customers are competing for memory on average?

b. If memory queueing time is 8 seconds, what is the average number of customers loaded in memory?

6. In a 30 minute observation interval, a particular disk was found to be busy for 12 minutes. If it is known that jobs require 320 accesses to that disk on average, and that the average service time per access is 25 milliseconds, what is the system throughput (in jobs/second)?

7. Consider a very simple model of a computer system in which only the CPU is represented. Use Little's law to argue that the minimum average response time for this system is obtained by scheduling the CPU so that it always serves the job with the shortest expected remaining service time (i.e., the job that is expected to finish soonest if placed in service).

8. Consider the following measurement data for an interactive system with a memory constraint:

length of measurement interval:	1 hour
average number of users:	80
average response time:	1 second
average number of memory-resident requests:	6
number of request completions:	36,000
utilizations of:	
CPU	75%
Disk 1	50%
Disk 2	50%
Disk 3	25%

a. What was throughput (in requests / second)?
b. What was the average "think time"?
c. On the average, how many users were attempting to obtain service (i.e., not "thinking")?
d. On the average, how much time does a user spend waiting for memory (i.e., not "thinking" but not memory-resident)?
e. What is the average service demand at Disk 1?

Chapter 4

Queueing Network Model Inputs and Outputs

4.1. Introduction

We are prepared now to state precisely the inputs and outputs of queueing network models. We noted in Chapter 1 that, in order to achieve an appropriate balance between accuracy and cost, we are restricting our attention to the subset of general networks of queues that consists of the *separable* queueing networks, extended where necessary for the accurate representation of particular computer system characteristics. Sections 4.2 - 4.4 describe the inputs and outputs of separable queueing networks. For notational simplicity we first present this material in the context of models with a single customer class (Sections 4.2 and 4.3), and then generalize to multiple class models (Section 4.4). In Section 4.5, we discuss certain computer system characteristics that cannot be represented directly using the inputs available for separable models, and certain performance measures that cannot be obtained directly from the available outputs. These motivate the extensions of separable networks that will be explored later in the book.

4.2. Model Inputs

The basic entities in queueing network models are *service centers*, which represent system resources, and *customers*, which represent users or jobs or transactions. Table 4.1 lists the inputs of single class queueing network models, which describe the relationships among customers and service centers. In the subsections that follow, these parameters are discussed in some detail.

4.2.1. Customer Description

The workload intensity may be described in any of three ways, named to suggest the computer system workloads they are best suited to representing:

customer description	The *workload intensity*, one of: λ, the *arrival rate* (for *transaction* workloads), or N, the *population* (for *batch* workloads), or N and Z, the *think time* (for *terminal* workloads)
center description	K, the number of *service centers* For each service center k: its *type*, either *queueing* or *delay*
service demands	For each service center k: $D_k \equiv V_k S_k$, the *service demand*

Table 4.1 — Single Class Model Inputs

- A *transaction* workload has its intensity specified by a parameter λ, indicating the rate at which requests (customers) arrive. A transaction workload has a population that varies over time. Customers that have completed service leave the model.

- A *batch* workload has its intensity specified by a parameter N, indicating the average number of active jobs (customers). (N need not be an integer.) A batch workload has a fixed population. Customers that have completed service can be thought of as leaving the model and being replaced instantaneously from a backlog of waiting jobs.

- A *terminal* workload has its intensity specified by two parameters: N, indicating the number of active terminals (customers), and Z, indicating the average length of time that customers use terminals ("think") between interactions. (Again, N need not be an integer.)

A terminal workload is similar to a batch workload in that its total population is fixed. In fact, a terminal workload with a think time of zero is in every way equivalent to a batch workload. On the other hand, a terminal workload is similar to a transaction workload in that the population of the *central subsystem* (the system excluding the terminals) varies, provided that the terminal workload has a non-zero think time. Note that N is an upper bound on the central subsystem population of a terminal workload, whereas no upper bound exists for a transaction workload.

We sometimes refer to models with transaction workloads as *open* models, since there is an infinite stream of arriving customers. Models with batch or terminal workloads are referred to as *closed* models, since customers "re-circulate". This distinction is made because the algorithms used to evaluate open models differ from those used for closed models. It highlights the similarity between batch and terminal workloads.

4.2. Model Inputs

4.2.2. Center Description

Service centers may be of two types: *queueing* and *delay*. These are represented as shown in Figure 4.1.

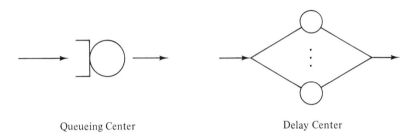

Queueing Center   Delay Center

Figure 4.1 — Queueing and Delay Service Centers

Customers at a queueing center compete for the use of the server. Thus the time spent by a customer at a queueing center has two components: time spent waiting, and time spent receiving service. Queueing centers are used to represent any system resource at which users compete for service, e.g., the CPU and I/O devices. As shown in the figure, a queueing center is drawn as a queue plus a server.

Because customers in a single class model are indistinguishable, it is not necessary to specify the scheduling discipline at a queueing center. The same performance measures will result from any scheduling discipline in which exactly one customer is in service whenever there are customers at the center.

Customers at a delay center each (logically) are allocated their own server, so there is no competition for service. Thus the residence time of a customer at a delay center is exactly that customer's service demand there. The most common use of a delay center is to represent the think time of terminal workloads. However, delay centers are useful in any situation in which it is necessary to impose some known average delay. For instance, a delay center could be used to represent the delay incurred by sending large amounts of data over a dedicated low speed transmission line. As shown in the figure, an icon suggestive of concurrent activity is used to represent a delay center.

4.2.3. Service Demands

The service demand of a customer at center k, D_k, is the total amount of time the customer requires in service at that center. Thus the set of service demands (one for each center) characterizes the behavior of the customer in terms of processing requirements. In a single class model, customers are indistinguishable with respect to their service demands, which can be thought of as representing the "average customer" in the actual system.

D_k can be calculated directly as B_k/C (the measured busy time of device k divided by the measured number of system completions), or may be thought of as the product of V_k, the number of visits that a customer makes to center k, and S_k, the service requirement per visit. It is possible to parameterize queueing network models at this more detailed level. However, a surprising characteristic of separable queueing networks is that their solutions depend only on the product of V_k and S_k at each center, and not on the individual values. Thus a model in which customers make 100 visits to the CPU, each for 10 milliseconds of service, is equivalent to one in which customers make a single visit for one second of service. For simplicity (to reduce the number of parameters and to facilitate obtaining their values) we generally will choose to parameterize our models in terms of D_k. Note that we define D to be the total service demand of a customer at all centers: $D \equiv \sum_{k=1}^{K} D_k$.

4.3. Model Outputs

Table 4.2 lists the outputs obtained by evaluating a single class queueing network model. Comments appear in the subsections that follow.

system measures	R average system response time X system throughput Q average number in system
center measures	U_k utilization of center k R_k average residence time at center k X_k throughput of center k Q_k average queue length at center k

Table 4.2 — Single Class Model Outputs

The values of these outputs depend upon the values of all of the model inputs. It will be especially useful to be able to specify that an output value corresponds to a particular workload intensity value. To do so, we follow the output with the parenthesized workload intensity: $X(N)$ is the throughput for a batch or terminal class with population N, $Q_k(\lambda)$ is the average queue length at center k for a transaction class with arrival rate λ, etc.

4.3. Model Outputs

4.3.1. Utilization

The utilization of a center may be interpreted as the proportion of time the device is busy, or, equivalently, as the average number of customers in service there. (The latter interpretation is the only one that makes sense for a delay center.)

4.3.2. Residence Time

Just as D_k is the total service demand of a customer at center k (in contrast to S_k, the service requirement per visit), R_k is the total residence time of a customer at center k (as opposed to the time spent there on a single visit). If the model is parameterized in terms of V_k and S_k, then the time spent per visit at center k can be calculated as R_k/V_k.

System response time, R, corresponds to our intuitive notion of response time; for example, the interval between submitting a request and receiving a response on an interactive system. Obviously, system response time is the sum of the residence times at the various centers:
$$R = \sum_{k=1}^{K} R_k.$$

4.3.3. Throughput

If a model is parameterized in terms of D_k then we can obtain system throughput, X, but do not have sufficient information to calculate device throughputs, X_k. (This is a small price to pay for the convenience that results from the less detailed parameterization.) If a model is parameterized in terms of V_k and S_k, then device throughputs can be calculated using the forced flow law, as $X_k = V_k X$.

4.3.4. Queue Length

The average queue length at center k, Q_k, includes all customers at that center, whether waiting or receiving service. The number of customers waiting can be calculated as $Q_k - U_k$, since U_k can be interpreted as the average number of customers receiving service at center k.

Q denotes the average number in system. For a batch class, $Q = N$. For a transaction class, $Q = XR$ (by Little's law). For a terminal class, $Q = N - XZ$ ($Q = XR$, and $R = N/X - Z$.) In general, the average population of any subsystem can be obtained either by multiplying the throughput of the subsystem by the residence time there, or by summing the queue lengths at the centers belonging to the subsystem.

4.3.5. Other Outputs

Various other outputs can be computed at some additional cost. As one example, we occasionally will wish to know the *queue length distribution* at a center: the proportion of time that the queue length has each possible value. We denote the proportion of time that the queue length at center k has the value i by $P[Q_k = i]$.

4.4. Multiple Class Models

4.4.1. Inputs

Multiple class models consist of C *customer classes*, each of which has its own workload intensity (λ_c, N_c, or N_c and Z_c) and its own service demand at each center ($D_{c,k}$). Within each class, the customers are indistinguishable. (Note that we have re-used the symbol C, which denoted customer completions in Chapter 3. Confusion will not arise.)

Multiple class models consisting entirely of open (transaction) classes are referred to as open models. Models consisting entirely of closed (batch or terminal) classes are referred to as closed. Models consisting of both types of classes are referred to as *mixed*.

The overall workload intensity of a multiple class model is described by a vector with an entry for each class: $\vec{\lambda} \equiv (\lambda_1, \lambda_2, \ldots, \lambda_C)$ if the model is open, $\vec{N} \equiv (N_1, N_2, \ldots, N_C)$ if it is closed (in point of fact, Z_c also must be included for terminal classes), and $\vec{I} \equiv (N_1 \text{ or } \lambda_1, N_2 \text{ or } \lambda_2, \ldots, N_C \text{ or } \lambda_C)$ if it is mixed.

As was the case for single class models, we do not specify the scheduling discipline at a queueing center. Roughly, the assumption made is that the scheduling discipline is *class independent*, i.e., it does not make use of information about the class to which a customer belongs. The same performance measures will result from any scheduling discipline that satisfies this assumption, along with the earlier assumption that exactly one customer is in service whenever there are customers at the center.

Table 4.3 summarizes the inputs of multiple class models. By analogy to the single class case, we define D_c to be the total service demand of a class c customer at all centers: $D_c \equiv \sum_{k=1}^{K} D_{c,k}$.

4.4.2. Outputs

All performance measures can be obtained on a per-class basis (e.g., $U_{c,k}$ and X_c) as well as on an aggregate basis (e.g., U_k and X). For utilization, queue length, and throughput, the aggregate performance measure

4.4. Multiple Class Models

customer description	C, the number of *customer classes* For each class c: its *workload intensity*, one of: λ_c, the *arrival rate* (for *transaction* workloads), or N_c, the *population* (for *batch* workloads), or N_c and Z_c, the *think time* (for *terminal* workloads)
center description	K, the number of *service centers* For each service center k: its *type*, either *queueing* or *delay*
service demands	For each class c and center k: $D_{c,k} \equiv V_{c,k} S_{c,k}$, the *service demand*

Table 4.3 — Multiple Class Model Inputs

equals the sum of the per-class performance measures (e.g., $U_k = \sum_{c=1}^{C} U_{c,k}$). For residence time and system response time, however, the per-class measures must be weighted by relative throughput, as follows:

$$R = \sum_{c=1}^{C} \frac{R_c X_c}{X} \qquad R_k = \sum_{c=1}^{C} \frac{R_{c,k} X_c}{X}$$

This makes intuitive sense, and can be demonstrated formally using Little's law (see Exercise 2).

Table 4.4 summarizes the outputs of multiple class models. The following reminders, similar to comments made in the context of single class models, should be noted in studying the table:

- The basic outputs are average values (e.g., average response time) rather than distributional information (e.g., the 90th percentile of response time). Thus the word "average" should be understood even if it is omitted.
- X_k and $X_{c,k}$ are meaningful only if the model is parameterized in terms of $V_{c,k}$ and $S_{c,k}$, rather than $D_{c,k}$.
- To specify that an output value corresponds to a particular workload intensity value, we follow the output symbol with the parenthesized workload intensity.

system measures	aggregate	R average system response time X system throughput Q average number in system
	per class	R_c average class c system response time X_c class c system throughput Q_c average class c number in system
center measures	aggregate	U_k utilization of center k R_k average residence time at center k X_k throughput at center k Q_k average queue length at center k
	per class	$U_{c,k}$ class c utilization of center k $R_{c,k}$ average class c residence time at center k $X_{c,k}$ class c throughput at center k $Q_{c,k}$ average class c queue length at center k

Table 4.4 − **Multiple Class Model Outputs**

4.5. Discussion

The specific inputs and outputs available for separable queueing network models, as just described, are dictated by a set of mathematical assumptions imposed to ensure efficiency of evaluation. The purpose of the present section is to consider the practical impact of these assumptions on the accuracy of our models. Specifically:

- What important computer system characteristics cannot be represented directly using separable models?
- Given these apparent inadequacies, how can we explain the success of separable models in computer system analysis?
- How does the analyst approach the inevitable situations in which separable models truly are inadequate?

Naturally, complete answers to these questions must await the remainder of the book. The present section contains a foreshadowing of these answers, to provide insight and guide intuition. For simplicity, our discussion will be set largely in the single class context.

4.5. Discussion

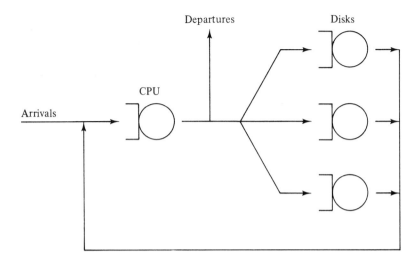

Figure 4.2 − **The Canonical Computer System Model**

Figure 4.2 illustrates the canonical separable queueing network model of a centralized system, which appears throughout the book. This model has the inputs and outputs discussed earlier in this chapter. Service centers are used to represent the CPU and the active I/O storage devices, e.g., disks. On the one hand, this model bears a close structural resemblance to a computer system. On the other hand, there are certain computer system characteristics that cannot be represented directly using the available inputs, and certain performance measures that cannot be obtained directly from the available outputs. These include:

- *simultaneous resource possession* − We have no direct way to express the fact that a customer may require simultaneous service at multiple resources. As an example, in order to transfer data to or from disk it may be necessary to concurrently use the disk, a controller, and a channel.

- *memory constraints* − Using a transaction workload, we are assuming implicitly that an arbitrarily large number of customers can be memory resident simultaneously. Using a batch workload, we are assuming implicitly that the multiprogramming level is constant. Using a terminal workload, we are assuming implicitly that all terminal users can be resident in memory simultaneously. In practice, it often occurs that the number of simultaneously active jobs varies over time but is limited by some memory constraint.

- *blocking* — In systems such as store-and-forward communications networks, the state of one resource can affect the processing of customers at another.
- *adaptive behavior* — A timesharing system may dynamically allocate scratch files to lightly loaded disks. A communications network may make dynamic routing decisions based on the populations at various nodes.
- *process creation and synchronization* — Since the number of customers in a class must either remain constant (closed classes) or be unbounded (open classes), it is not possible to represent explicitly a process executing a *fork* (spawning a sub-process) when it reaches a particular point in its computation. Similarly, since customers are independent of one another it is not possible to model directly synchronization points in the computation of two or more processes.
- *high service time variability* — In practice, extremely high service burst length variability can degrade the performance of a system.
- *priority scheduling* — Since a priority scheduler makes use of class dependent information, it will yield different performance measures than the class independent scheduling disciplines assumed in multiple class queueing network models.
- *response time distributions* — The list of useful model outputs obtainable directly at reasonable cost does not include the distribution of response times.

How is it, then, that separable queueing network models are successful at representing the behavior of complex contemporary computer systems, and at projecting the impact of modifications to their hardware, software, and workload?

First, consider the process of defining and parameterizing a model of an existing system. Much of the relevant complexity of the system that we appear to be ignoring is, in fact, captured *implicitly* in the measurement data used to parameterize the model. As an example, consider the effect of I/O path contention. Our canonical model represents only disks, not intermediate path elements such as channels and controllers. However, in parameterizing the model we will set the service demand at each disk center k, D_k, equal to the measured total disk busy time per job, which we will calculate as $U_k\ T/C$ (C here is the measured number of completions). In measuring the disk, we will find it busy not only during seek, latency, and data transfer, but also during those periods when it is attempting to obtain a free path. In other words, the effect of I/O path contention is incorporated *indirectly*, through the disk service demand parameters. A model parameterized in this way can be expected to do a good job of representing the behavior of the system during the measurement interval.

Next, consider using such a model to project the effect of modifications. In many cases, indirect representations of system characteristics based on measurement data can be assumed to be insensitive to the proposed modification. For example, the primary effect of a CPU upgrade can be represented in a model by adjusting CPU service demands. Any effect of this modification on disk service demands — either "intrinsic" demands (seek, latency, and data transfer times) or the component due to path contention — is strictly secondary in nature. It is in these cases that separable models prove adequate on their own.

Sometimes, of course, the objective of a study is to answer detailed questions about modifications that can be expected to affect the indirect representations of system characteristics. For example, if I/O path contention were known to be a significant problem, an analyst might want to use a queueing network model to project the performance improvement that would result from path modifications. In cases such as this, separable models can be augmented with procedures that calculate revised estimates for those portions of various service demands that are indirect representations of relevant system characteristics. These are the "extensions" alluded to in Chapter 1. This approach achieves the necessary accuracy, while preserving the ability to evaluate the model efficiently. Such techniques exist for each of the system characteristics mentioned earlier in this section.

4.6. Summary

We have enumerated and discussed the inputs and the outputs of separable queueing network models. These were summarized for the single class case in Tables 4.1 and 4.2, respectively, and for the multiple class case in Tables 4.3 and 4.4, respectively.

We have noted that the availability of inputs and outputs is dictated by assumptions imposed to ensure the efficient evaluation of the model. We have considered the practical impact of these assumptions on the accuracy of the models.

In many cases, separable models are adequate by themselves, because complex system characteristics are captured implicitly in the measurement data used to parameterize them. Part II of the book is devoted to evaluation algorithms for models of this sort.

In other cases, separable models must be augmented with procedures that calculate revised estimates for those portions of various service demands that are indirect representations of relevant system characteristics. Part III of the book is devoted to such procedures.

4.7. Exercises

1. Consider the system with which you are most familiar:
   a. How would you obtain parameter values for a single class model from the available measurement data?
   b. How would you obtain parameter values for a multiple class model from the available measurement data?
   c. What aspects of your system important to its performance seem to be omitted from the simple single or multiple class models that you might define?

2. Show that $R = \sum_{c=1}^{C} \frac{R_c X_c}{X}$, that is, that the average response time in a system with multiple job classes is a throughput-weighted average of the individual average response times.

3. In creating a model of a computer system, there are two extreme positions we can take as to the workload representation:
   a. assume all jobs are identical, in which case a single class model is appropriate, or
   b. assume each job is significantly different from every other job, and represent the workload with a class per job.

   What are the advantages and disadvantages of each approach?

Part II

General Analytic Techniques

Part II of the book discusses algorithms for evaluating separable queueing network models. To *evaluate* a queueing network model is to obtain outputs such as utilizations, residence times, queue lengths, and throughputs, from inputs such as workload intensities and service demands.

In Chapter 5 we show how to obtain bounds on performance, using extremely straightforward reasoning and simple computations that can be performed by hand.

In Chapters 6 and 7 we present more sophisticated algorithms that yield specific performance measures, rather than bounds. Chapter 6 is devoted to models with one job class. Chapter 7 extends this discussion to models with multiple job classes.

In Chapter 8 we introduce *flow equivalent service centers*, which can be used to represent the behavior of entire subsystems. Such *hierarchical modelling* is one important way to extend separable queueing network models to represent system characteristics that violate the assumptions required for separability — characteristics such as those listed at the end of Chapter 4. Chapter 8 thus forms a bridge to Part III of the book.

Chapter 5

Bounds on Performance

5.1. Introduction

We begin this part of the book with a chapter devoted to the simplest useful approach to computer system analysis using queueing network models: *bounding analysis*. With very little computation it is possible to determine upper and lower bounds on system throughput and response time as functions of the system workload intensity (number or arrival rate of customers). We describe techniques to compute two classes of performance bounds: *asymptotic bounds* and *balanced system bounds*. Asymptotic bounds hold for a wider class of systems than do balanced system bounds. They also are simpler to compute. The offsetting advantage of balanced system bounds is that they are tighter, and thus provide more precise information than asymptotic bounds.

There are several characteristics of bounding techniques that make them interesting and useful:

- The development of these techniques provides valuable insight into the primary factors affecting the performance of computer systems. In particular, the critical influence of the system bottleneck is highlighted and quantified.
- The bounds can be computed quickly, even by hand. Bounding analysis therefore is suitable as a first cut modelling technique that can be used to eliminate inadequate alternatives at an early stage of a study.
- In many cases, a number of alternatives can be treated together, with a single bounding analysis providing useful information about them all.

In contrast to the bounding techniques discussed here, the more sophisticated analysis techniques presented in subsequent chapters require considerably more computation — to the point that it is infeasible to perform the analysis by hand.

Bounding techniques are most useful in *system sizing* studies. Such studies involve rather long-range planning, and consequently often are

5.1. Introduction

based on preliminary estimates of system characteristics. With such imprecision in knowledge of the system, quick bounding studies may be more appropriate than more detailed analyses leading to specific estimates of performance measures. System sizing studies typically involve consideration of a large number of candidate configurations. Often a single resource (such as the CPU) is the dominant concern, because the remainder of the system can be configured to match the power of this resource. Bounding analysis permits considering *as one alternative* a group of candidate configurations that have the same critical resource but differ with respect to the pattern of demands at the other service centers.

Bounding techniques also can be used to estimate the potential performance gain of alternative upgrades to existing systems. In Section 5.3 we indicate how graphs of the bounds can provide insight about the extent of service demand reduction required at the bottleneck center if it is to be possible to meet stated performance goals. (Service demand at a center can be reduced either by shifting some work away from the center or by substituting a faster device at the center.)

Our discussion of bounding analysis is restricted to the single class case. Multiple class generalizations exist, but they are not used widely. One reason for this is that bounding techniques are most useful for capacity studies of the bottleneck center, for which single class models suffice. Additionally, a major attraction of bounding techniques in practice is their simplicity, which would be lost if multiple classes were included in the models.

The models we consider in the remainder of this chapter can be described by the following parameters:

- K, the number of service centers;
- D_{max}, the largest service demand at any single center;
- D, the sum of the service demands at the centers;
- the *type* of the customer class (*batch*, *terminal*, or *transaction*);
- Z, the average think time (if the class is of terminal type).

For models with transaction type workloads, the throughput bounds indicate the maximum customer arrival rate that can be processed by the system, while the response time bounds reflect the largest and smallest possible response times that these customers could experience as a function of the system arrival rate. For models with batch or terminal type workloads, the bounds indicate the maximum and minimum possible system throughputs and response times as functions of the number of customers in the system. We refer to throughput upper and response time lower bounds as *optimistic* bounds (since they indicate the best possible performance), and we refer to throughput lower and response time upper bounds as *pessimistic* bounds (since they indicate the worst possible performance). While we treat only bounds on system throughput and response time in the following sections, the fundamental laws of Chapter 3 can be used to transform these into bounds on other performance measures, such as service center throughputs and utilizations.

5.2. Asymptotic Bounds

Asymptotic bounding analysis provides optimistic and pessimistic bounds on system throughput and response time in single class queueing networks. As their name suggests, they are derived by considering the (asymptotically) extreme conditions of light and heavy loads. The validity of the bounds depends on only a single assumption: that the service demand of a customer at a center does not depend on how many other customers currently are in the system, or at which service centers they are located.

The type of information provided by asymptotic bounds depends on whether the system workload is open (transaction type) or closed (batch or terminal type). We begin with the simpler case, that of transaction type workloads.

5.2.1. Transaction Workloads

For transaction workloads, the throughput bound indicates the maximum possible arrival rate of customers that the system can process successfully. If the arrival rate exceeds this bound, a backlog of unprocessed customers grows continually as jobs arrive. Thus, in the long run, an arriving job has to wait an indefinitely long time (since there may be any number of jobs already in queue when it arrives). In this case we say that the system is *saturated*. The throughput bound thus is the arrival rate that separates feasible processing from saturation.

The key to determining the throughput bound is the utilization law: $U_k = X_k S_k$ for each center k. If we denote the arrival rate to the system as λ, then $X_k = \lambda V_k$, and the utilization law can be rewritten as $U_k = \lambda D_k$, where D_k is the service demand at center k. To derive the throughput bound, we simply note that as long as all centers have unused capacity (i.e., have utilizations less than one), an increased arrival rate can be accommodated. However, when any of the centers becomes saturated (i.e., has utilization one), the entire system becomes saturated, since no increase in the arrival rate of customers can be handled successfully. Thus, the throughput bound is the smallest arrival rate λ_{sat} at which any center saturates. Clearly, the center that saturates at the lowest arrival rate is the *bottleneck* center − the center with the largest service demand. Let *max* be the index of the bottleneck center. Then:

5.2. Asymptotic Bounds

so:
$$U_{max}(\lambda) = \lambda D_{max} \leq 1$$

$$\lambda_{sat} = \frac{1}{D_{max}}$$

Thus, for arrival rates greater than or equal to $1/D_{max}$ the system is saturated, while the system is capable of processing arrival rates less than $1/D_{max}$.

Asymptotic response time bounds indicate the largest and smallest possible response times experienced by customers when the system arrival rate is λ. Because the system is unstable if $\lambda > \lambda_{sat}$ we limit our investigation to the case where the arrival rate is less than the throughput bound. There are two extreme situations:

- In the best possible case, no customer ever interferes with any other, so that no queueing delays are experienced. In that case the system response time of each customer is simply the sum of its service demands, which we denote by D.

- In the worst possible case, n customers arrive together every n/λ time units (the system arrival rate is $\frac{n}{n/\lambda} = \lambda$). Customers at the end of the batch are forced to queue for customers at the front of the batch, and thus experience large response times. As the batch size n increases, more and more customers are waiting an increasingly long time. Thus, for any postulated pessimistic bound on response times for system arrival rate λ, it is possible to pick a batch size n sufficiently large that the bound is exceeded. We conclude that there is no pessimistic bound on response times, regardless of how small the arrival rate λ might be.

These results are somewhat unsatisfying. Fortunately, the throughput and response time bounds provide more information in the case of closed (batch and terminal) workload types.

5.2.2. Batch and Terminal Workloads

Figures 5.1a and 5.1b show the general form of the asymptotic bounds on throughput and response time for batch and terminal workloads, respectively. The bounds indicate that the precise values of the actual throughputs and response times must lie in the shaded portions of the figures. The general shapes and positions of these values are indicated by the curves in the figures.

Batch Throughput:

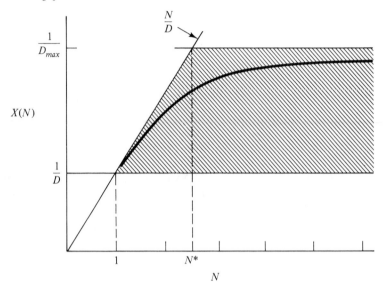

Batch Response Time:

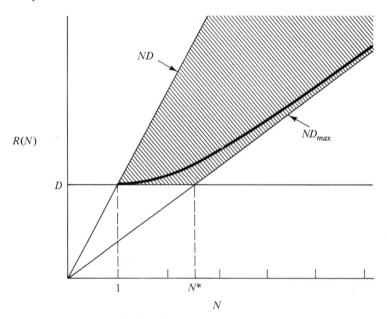

Figure 5.1a — Asymptotic Bounds on Performance

5.2. Asymptotic Bounds

Terminal Throughput:

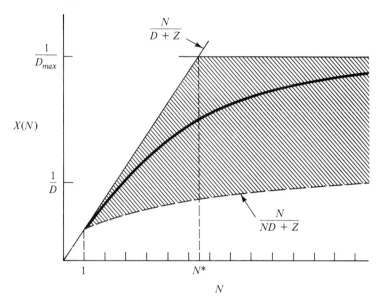

Terminal Response Time:

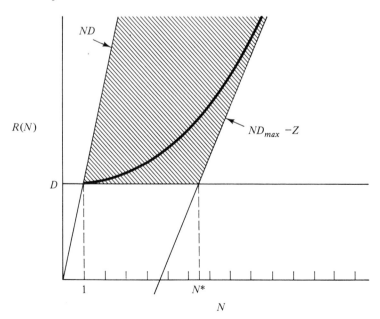

Figure 5.1b − Asymptotic Bounds on Performance

To derive the bounds shown in the figures, we first consider the bounds on throughput, and then use Little's law to transform them into corresponding bounds on response time. Our analysis is stated in terms of terminal workloads. By taking the think time, Z, to be zero, we obtain results for batch workloads.

We begin with the heavy load (many customer) situation. As the number of customers in the system (N) becomes large, the utilizations of all centers grow, but clearly no utilization can exceed one. From the utilization law we have for each center k that:

$$U_k(N) = X(N) D_k \leqslant 1$$

Each center limits the maximum possible throughput that the system can achieve. Since the bottleneck center (*max*) is the first to saturate, it restricts system throughput most severely. We conclude that:

$$X(N) \leqslant \frac{1}{D_{max}}$$

Intuitively this is clear, because if each customer requires on average D_{max} time units of service at the bottleneck center, then in the long run customers certainly cannot be completed any faster than one every D_{max} time units.

Next consider the light load (few customers) situation. At the extreme, a single customer alone in the system attains a throughput of $1/(D+Z)$, since each interaction consists of a period of service (of average length $D = \sum_{k=1}^{K} D_k$) and a think time (of average length Z). As more customers are added to the system there are two bounding situations:

- The smallest possible throughput occurs when each additional customer is forced to queue behind all other customers already in the system. In this case, with N customers in the system, $(N-1)D$ time units are spent queued behind other customers, D time units are spent in service, and Z time units are spent thinking, so that the throughput *of each customer* is $1/(ND + Z)$. Thus, system throughput is $N/(ND + Z)$.

- The largest possible throughput occurs when each additional customer is not delayed at all by any other customers in the system. In this case no time is spent queueing, D time units are spent in service, and Z time units are spent thinking. Thus, the throughput *of each customer* is $1/(D+Z)$, and system throughput is $N/(D+Z)$.

The above observations can be summarized as the asymptotic bounds on system throughput:

$$\frac{N}{ND + Z} \leqslant X(N) \leqslant \min\left(\frac{1}{D_{max}}, \frac{N}{D+Z}\right)$$

5.3. Using Asymptotic Bounds

Note that the optimistic bound consists of two components, the first of which applies under heavy load and the second of which applies under light load. As illustrated by Figure 5.1, there is a particular population size N^* such that for all N less than N^* the light load optimistic bound applies, while for all N larger than N^* the heavy load bound applies. This crossover point occurs where the values of the two bounds are equal:

$$N^* = \frac{D+Z}{D_{max}}$$

We can obtain bounds on response time $R(N)$ by transforming our throughput bounds using Little's law. We begin by rewriting the previous equation:

$$\frac{N}{ND+Z} \leqslant \frac{N}{R(N)+Z} \leqslant \min\left(\frac{1}{D_{max}}, \frac{N}{D+Z}\right)$$

Inverting each component to express the bounds on $R(N)$ yields:

$$\max\left(D_{max}, \frac{D+Z}{N}\right) \leqslant \frac{R(N)+Z}{N} \leqslant \frac{ND+Z}{N}$$

or:

$$\max(D, ND_{max} - Z) \leqslant R(N) \leqslant ND$$

5.2.3. Summary of Asymptotic Bounds

Table 5.1 summarizes the asymptotic bounds. Algorithm 5.1 indicates the steps by which the asymptotic bounds can be calculated for batch and terminal workloads. (The calculations for transaction workloads are trivial.) Note that all bounds are straight lines with the exception of the pessimistic throughput bound for terminal workloads. Consequently, once D and D_{max} are known, calculation of the asymptotic bounds expressed as functions of the number of customers in the network takes only a few arithmetic operations. The amount of computation is independent of both the number of centers in the model and the range of customer populations of interest.

5.3. Using Asymptotic Bounds

In this section we present three applications of asymptotic bounds: a case study in which asymptotic bounds proved useful, an assessment of the effect of alleviating a bottleneck, and an example of modification analysis.

	workload type	bounds
X	batch	$\dfrac{1}{D} \leq X(N) \leq \min\left(\dfrac{N}{D}, \dfrac{1}{D_{max}}\right)$
	terminal	$\dfrac{N}{ND+Z} \leq X(N) \leq \min\left(\dfrac{N}{D+Z}, \dfrac{1}{D_{max}}\right)$
	transaction	$X(\lambda) \leq 1/D_{max}$
R	batch	$\max(D, ND_{max}) \leq R(N) \leq ND$
	terminal	$\max(D, ND_{max} - Z) \leq R(N) \leq ND$
	transaction	$D \leq R(\lambda)$

Table 5.1 − Summary of Asymptotic Bounds

5.3.1. Case Study

Asymptotic bound analysis was enlightening in the case study introduced in Section 2.6. (That section may be reviewed for additional background.)

An insurance company had twenty geographically distributed sites based on IBM 3790s that were providing unacceptable response times. The company decided to enter a three year selection, acquisition, and conversion cycle, but an interim upgrade was required. IBM 8130s and 8140s both were capable of executing the existing applications software, and consequently were considered for use during the three year transition period. After discussions with the vendor, the company believed that the use of 8130s would result in performance improving by a factor of 1.5 to 2 over the 3790s, while the use of 8140s would lead to performance improving by a factor of 2 to 3.5. (No precise statement of the significance of the "performance improvement factor" was formulated.)

A modelling study was initiated to determine those sites at which the less expensive 8130 system would suffice. It was known that the 8130 and 8140 systems both included a disk that was substantially faster than that of the 3790. With respect to CPU speed, the 8130 processor was slightly slower than the 3790, while the 8140 was approximately 1.5 times

5.3. Using Asymptotic Bounds

(Steps are presented assuming a terminal workload; to treat a batch workload, set Z to zero.)

1. Calculate $D = \sum_{k=1}^{K} D_k$ and $D_{max} = \max_k D_k$.

2. Calculate the intersection point of the components of the optimistic bounds:
$$N^* = \frac{D+Z}{D_{max}}$$

3. Bounds on throughput pass through the points:

    optimistic bound:

    $(0, 0)$ and $(1, \frac{1}{D+Z})$ for $N \leqslant N^*$

    $(0, \frac{1}{D_{max}})$ and $(1, \frac{1}{D_{max}})$ for $N \geqslant N^*$

    pessimistic bound:

    This bound is not linear in N, and so must be calculated for each population of interest using the equation in Table 5.1.

4. Bounds on average response time pass through the points:

    optimistic bound:

    $(0, D)$ and $(1, D)$ for $N \leqslant N^*$

    $(0, -Z)$ and $(1, D_{max}-Z)$ for $N \geqslant N^*$

    pessimistic bound:

    $(0, 0)$ and $(1, D)$

Algorithm 5.1 — Closed Model Asymptotic Bounds

faster. Through a combination of this information, "live" measurements of existing 3790 systems, and benchmark experiments on two of the systems (3790 and 8140), the following service demands were determined:

	service demands, seconds	
system	CPU	disk
3790 (observed)	4.6	4.0
8130 (estimated)	5.1	1.9
8140 (estimated)	3.1	1.9

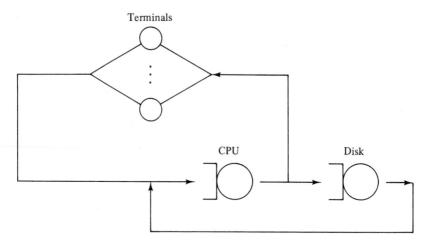

Figure 5.2 — Case Study Model

With the service demands established, a bounding model was used to assess the performance to be expected from each of the three systems. Figure 5.2 depicts the queueing network. (Although some sites had two physical disk drives, the disk controller did not permit them to be active simultaneously. For this reason, having only a single disk service center in the model is appropriate.) The parameters are:

- K, the number of service centers (2);
- D_{max}, the largest service demand (4.6 seconds for the 3790, 5.1 for the 8130, and 3.1 for the 8140);
- D, the sum of the service demands (8.6, 7.0, and 5.0, respectively);
- the type of the customer class (terminal);
- Z, the average think time (an estimate of 60 seconds was used).

Applying Algorithm 5.1 to the model of each of the three systems leads to the optimistic asymptotic bounds graphed in Figure 5.3. (The pessimistic bounds have been omitted for clarity.) These reveal that, at heavy loads, performance of the 8130 will be *inferior* to that of the 3790. This is a consequence of the fact that the 8130 has a slower CPU, which is the bottleneck device. Thus, rather than a performance gain of 1.5 to 2, a performance degradation could be expected in moving from 3790s to 8130s whenever the number of active terminals exceeded some threshold. Figure 5.3 indicates a performance gain in moving from 3790s to 8140s, although not the expected factor of two or more.

On the basis of the study, additional benchmark tests were done to re-assess the advisability of involving 8130s in the transition plan. These studies confirmed that the performance of 8130s would be worse than that of 3790s when the number of terminals was roughly fifteen or more,

5.3. Using Asymptotic Bounds

Throughput:

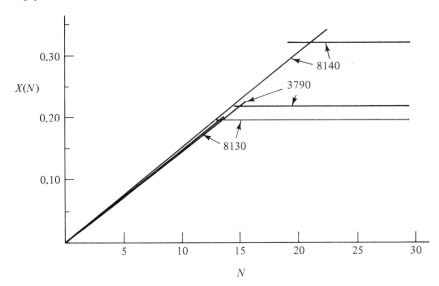

Response Time:

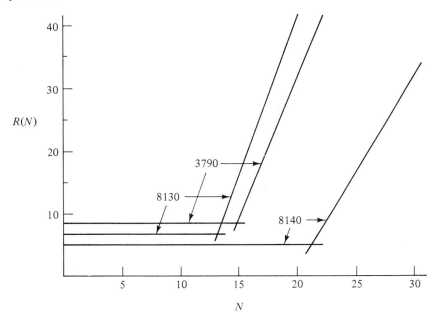

Figure 5.3 — **Asymptotic Bounds in the Case Study**

and that the performance gain of 8130s over 3790s at lighter loads would be negligible. Consequently, there was no performance reason to invest in 8130s for any sites. Eventually the company decided to install 8140s at all sites during the transition period. Without the simple modelling study, the company might have ordered 8130s without doing benchmark tests on them, with disappointing results. (A note of caution: the conclusions reached in this study would not necessarily hold in a context involving a different workload.)

5.3.2. Effect of Bottleneck Removal

So far we have been most concerned with the bottleneck center, which constrains throughput to be at most $1/D_{max}$. What happens if we alleviate that bottleneck, either by replacing the device with a faster one or by shifting some of the work to another device? In either case, D_{max} is reduced and so the throughput optimistic bound, $1/D_{max}$, increases. A limit to the extent of this improvement is imposed by the center with the second highest service demand originally. We call this center the *secondary bottleneck*, as contrasted with the *primary bottleneck*.

Consider a model with three service centers ($K=3$) and a terminal workload with average think time equal to 15 seconds ($Z=15$) and service demands of 5, 4, and 3 seconds at the centers ($D_1=5$, $D_2=4$, and $D_3=3$). Figure 5.4 shows the optimistic asymptotic bounds for this example, supplemented by lines indicating the heavy load optimistic bounds on performance corresponding to each center. Such a graph provides a visual representation of the extent of performance improvement possible by alleviating the primary bottleneck. As the load at the bottleneck center is reduced, the heavy load optimistic bound on throughput moves upwards, while the heavy load optimistic bound on average response time pivots downward (about the point (0, 0) for batch workloads and about the point (0, $-Z$) for terminal workloads). The light load asymptotes also change, but they are much less sensitive to the service demand at any single center than are the heavy load asymptotes.

An important lesson to be learned is the futility of improving any center but the bottleneck with respect to enhancing performance at heavy load. Reducing the service demand at centers other than the bottleneck improves only the light load asymptote, and the improvement usually is insignificant. Figure 5.5 compares the effects on the asymptotic bounds of independently doubling the speed (halving the service demand) at the primary and secondary bottlenecks for this example system. Observe that, at heavy load, performance gains only are evident when the demand at the primary bottleneck is reduced.

5.3. Using Asymptotic Bounds

Throughput:

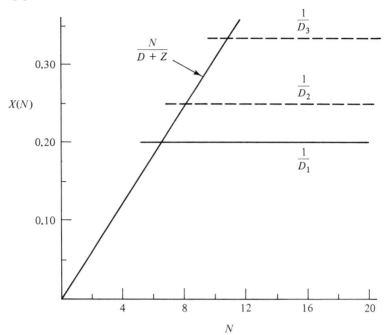

Response Time:

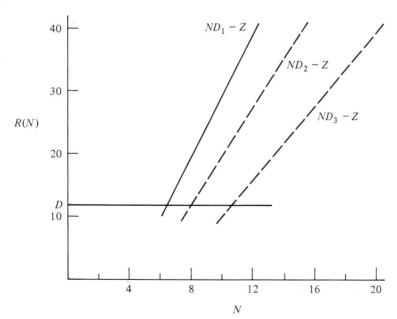

Figure 5.4 — Secondary and Tertiary Asymptotic Bounds

84   General Analytic Techniques: Bounds on Performance

Throughput:

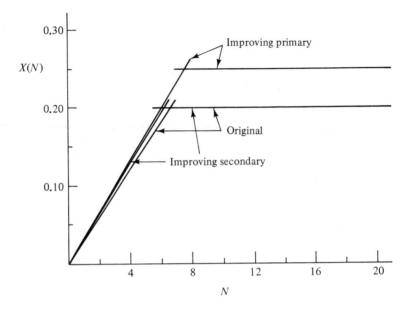

Response Time:

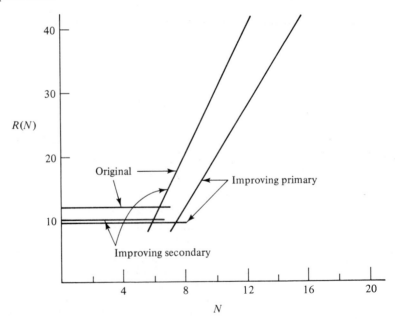

Figure 5.5 — Relative Effects of Reducing Various Service Demands

5.3. Using Asymptotic Bounds

5.3.3. Modification Analysis Example

Here we examine the use of asymptotic bounds to assess the impact of modifications to an existing system. Consider a simplified interactive system for which the following measurements have been obtained:

T = 900 seconds   length of the measurement interval

B_1 = 400 seconds   CPU busy
B_2 = 100 seconds   slow disk busy
B_3 = 600 seconds   fast disk busy

C = 200 jobs   completed jobs
C_2 = 2,000   slow disk operations
C_3 = 20,000   fast disk operations

Z = 15 seconds   think time

The service demands per job are $D_1=2.0$, $D_2=0.5$, and $D_3=3.0$. The visit counts to the disks are $V_2=10$ and $V_3=100$. The service times per visit to the disks are $S_2=.05$ and $S_3=.03$. We consider four improvements that can be made to the system. These are listed below, along with an indication of how each would be reflected in the parameters of the model:

1. Replace the CPU with one that is twice as fast. $D_1 \leftarrow 1$

2. Shift some files from the faster disk to the slower disk, balancing their demands. We consider only the primary effect, which is the change in disk speed, and ignore possible secondary effects such as the fact that the average size of blocks transferred may differ between the two disks. The new disk service demands are derived as follows. $V_2 + V_3 = 110$. Because $S_2=.05$ and $S_3=.03$, this is the same as:

$$\frac{V_2 S_2}{.05} + \frac{V_3 S_3}{.03} = 110$$

Since we wish to have $D_2 = V_2 S_2 = V_3 S_3 = D_3$:

$$D_2 \left[\frac{1}{.05} + \frac{1}{.03} \right] = 110$$

and $D_2 = D_3 = 2.06$. Dividing by the appropriate service times, we obtain the new visit counts: $V_2=41$ and $V_3=69$.

3. Add a second fast disk (center 4) to handle half the load of the busier existing disk. Once again, we consider only the primary effects of the change. $K \leftarrow 4$, $D_3 \leftarrow 1.5$, $D_4 \leftarrow 1.5$

4. The three changes made together: the faster CPU and a balanced load across two fast disks and one slow disk. Service demands become $D_1=1$, $D_2=1.27$, $D_3=1.27$, and $D_4=1.27$. These were derived in a manner similar to that employed above. We know that $V_2 + V_3 + V_4 = 110$. To ensure that $D_2 = D_3 = D_4$:

$$\frac{V_2 S_2}{.05} + \frac{V_3 S_3}{.03} + \frac{V_4 S_4}{.03} = 110$$

$$D_2 \left[\frac{1}{.05} + \frac{1}{.03} + \frac{1}{.03} \right] = 110$$

$$D_2 = D_3 = D_4 = \left[\frac{.0015}{.13} \right] 110 = 1.27$$

Figure 5.6 shows the optimistic asymptotic bounds for the original system (labelled "None"), for each modification individually (labelled "(1)", "(2)", and "(3)", respectively), and for the three in combination (labelled "(1) and (2) and (3)"). Intuitively, the first change might appear to be the most significant, yet Figure 5.6 shows that this is not true. Because the fast disk is the original bottleneck, changes 2 and 3 are considerably more influential. Note that change 2 yields almost as much improvement as change 3 although it requires no additional hardware. The combination of the three modifications yields truly significant results.

The modification analysis done in this section has involved only asymptotic bounds on performance. In Chapter 13 we will consider modification analysis once again, using more sophisticated techniques to evaluate our models.

5.4. Balanced System Bounds

With a modest amount of computation beyond that required for asymptotic bounds, tighter bounds can be obtained. These bounds are called *balanced system bounds* because they are based upon systems that are "balanced" in the sense that the service demand at every center is the same, i.e., $D_1 = D_2 = D_3 = \ldots = D_K$. Figures 5.7a and 5.7b show the general form of balanced system bounds (together with the asymptotic bounds) for batch (5.7a) and terminal (5.7b) workloads.

We first establish some special properties of balanced systems. We then show how these properties can be exploited to determine bounds on performance that complement the asymptotic bounds and lead to more precise knowledge of system behavior. The derivation of balanced system bounds is shown for batch workloads only. The reader is asked to work through the derivation for transaction workloads in Exercise 5. Bounds

5.4. Balanced System Bounds

Throughput:

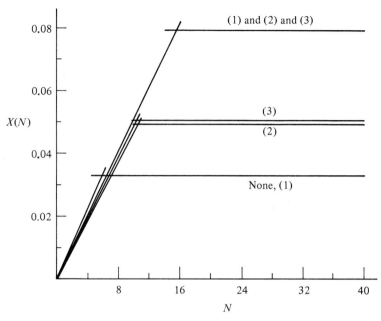

Response Time:

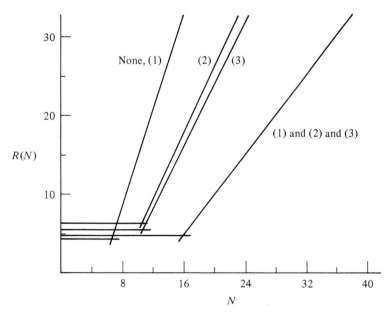

Figure 5.6 — Example of the Effects of Various Changes

General Analytic Techniques: Bounds on Performance

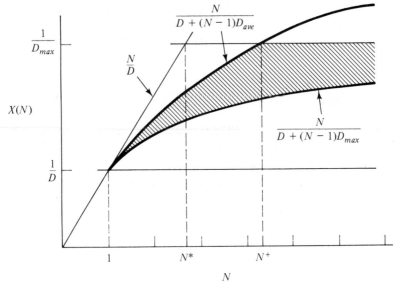

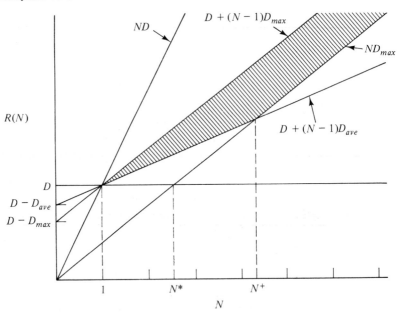

Figure 5.7a — Balanced System Bounds on Performance

5.4. Balanced System Bounds

Terminal Throughput:

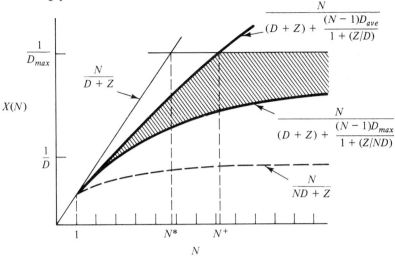

Terminal Response Time:

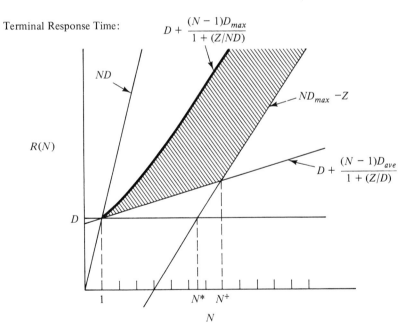

Figure 5.7b — Balanced System Bounds on Performance

for each of batch, terminal, and transaction workload types are given in Table 5.2.

The analysis of balanced systems is a special case of the techniques to be presented in Chapter 6. Formally, this analysis requires that various assumptions be made about the system being modelled. (These assumptions will be described in Chapter 6.) This is in contrast to asymptotic bounds, which require only that the service demand of a customer at a center does not depend on how many other customers are currently in the system or at which centers they are located.

For balanced systems, the techniques to be presented in Chapter 6 have a particularly simple form. The utilization of every service center is given by:

$$U_k(N) = \frac{N}{N+K-1}$$

(We do not attempt to justify this now, either intuitively or formally.) By the utilization law, system throughput is then:

$$X(N) = \frac{U_k}{D_k} = \frac{N}{N+K-1} \times \frac{1}{D_k}$$

where D_k is the service demand at every center.

Let D_{max}, D_{ave}, and D_{min} denote respectively the maximum, average, and minimum of the service demands at the centers of the model we wish to evaluate. We bound the throughput of that system by the throughputs of two related balanced systems: one with service demand D_{min} at every center, and the other with service demand D_{max} at every center:

$$\frac{N}{N+K-1} \times \frac{1}{D_{max}} \leqslant X(N) \leqslant \frac{N}{N+K-1} \times \frac{1}{D_{min}}$$

These inequalities hold because, of all systems with K centers, N customers, and maximum service demand D_{max}, the one with the lowest throughput is the balanced system with demand D_{max} at each center. Similarly, of all systems with K centers, N customers, and minimum demand D_{min}, the one with the highest throughput is the balanced system with demand D_{min} at each center. Corresponding bounds on average response times are:

$$(N+K-1) D_{min} \leqslant R(N) \leqslant (N+K-1) D_{max}$$

5.4. Balanced System Bounds

Tighter balanced system bounds can be obtained by constraining not only the maximum service demand, D_{max}, but also the total demand, D (or equivalently, the average demand, D_{ave}). Of all systems with a given total service demand $D = \sum_{k=1}^{K} D_k$, the one with the highest throughput (and the lowest average response time) is the one in which all service demands are equal (i.e., $D_k = D/K$, $k = 1, \ldots, K$). This confirms our intuition that the increase in delay resulting from an increase in load is greater than the decrease in delay resulting from an equivalent decrease in load. Therefore, optimistic bounds are given by:

$$X(N) \leq \frac{N}{N+K-1} \times \frac{1}{D_{ave}} = \frac{N}{D + (N-1) D_{ave}}$$

and:

$$D + (N-1) D_{ave} \leq R(N)$$

Note that the optimistic balanced system bound intersects the heavy load component of the optimistic asymptotic bound (at a point that we will denote by N^+). Beyond this point, the balanced system bound is defined to coincide with the asymptotic bound.

Analogously, of all systems with total demand D and maximum demand D_{max}, the one with the lowest throughput has D/D_{max} centers with demand D_{max}, and zero demand at the remaining centers. (The fact that D/D_{max} may not be an integer hampers intuition, but not the validity of the bounds.) Therefore, pessimistic bounds are:

$$\frac{N}{N + \frac{D}{D_{max}} - 1} \times \frac{1}{D_{max}} = \frac{N}{D + (N-1) D_{max}} \leq X(N)$$

and:

$$R(N) \leq D + (N-1) D_{max}$$

Table 5.2 summarizes the balanced system bounds for batch, terminal, and transaction workloads. Algorithm 5.2 indicates how these bounds can be calculated for batch and terminal workloads. (The calculations for transaction workloads are trivial.) For batch workloads, the bounds on average response time are straight lines. Also, the optimistic bound on average response time for terminal workloads is a straight line. However, balanced system bounds on throughput and the pessimistic balanced system bound on response time for terminal workloads are not linear in N, and thus must be computed separately for each value of N of interest.

	workload type	bounds
X	batch	$\dfrac{N}{D + (N-1)D_{max}} \leq X(N)$ $\leq \min(\dfrac{1}{D_{max}}, \dfrac{N}{D + (N-1)D_{ave}})$
	terminal	$\dfrac{N}{D + Z + \dfrac{(N-1)D_{max}}{1 + Z/(ND)}} \leq X(N)$ $\leq \min(\dfrac{1}{D_{max}}, \dfrac{N}{D + Z + \dfrac{(N-1)D_{ave}}{1 + Z/D}})$
	transaction	$X(\lambda) \leq 1/D_{max}$
R	batch	$\max(ND_{max}, D + (N-1)D_{ave}) \leq R(N)$ $\leq D + (N-1)D_{max}$
	terminal	$\max(ND_{max} - Z, D + \dfrac{(N-1)D_{ave}}{1 + Z/D}) \leq R(N)$ $\leq D + \dfrac{(N-1)D_{max}}{1 + Z/(ND)}$
	transaction	$\dfrac{D}{1 - \lambda D_{ave}} \leq R(\lambda) \leq \dfrac{D}{1 - \lambda D_{max}}$

Table 5.2 — Summary of Balanced System Bounds

5.5. Summary

In this chapter we have introduced techniques for obtaining bounds on the performance measures of systems. The bounds are summarized in Tables 5.1 and 5.2, and procedures for calculating them are given in Algorithms 5.1 and 5.2. Asymptotic bounds and balanced system bounds are important for a number of reasons:

5.5. Summary

1. Calculate the asymptotic bounds using Algorithm 5.1.
2. Determine the point at which the optimistic balanced system bound intersects the optimistic asymptotic bound. For a batch workload:

$$N^+ = \frac{D - D_{ave}}{D_{max} - D_{ave}}$$

   For a terminal workload:

$$N^+ = \frac{(D+Z)^2 - D\,D_{ave}}{(D+Z)D_{max} - D\,D_{ave}}$$

   The optimistic balanced system bound need be calculated only from 1 to N^+ since it is defined to coincide with the asymptotic bound beyond N^+.

3. Calculate balanced system bounds on average response time. For a batch workload, the bounds are lines through the points:

   optimistic bound :

   $(0, D-D_{ave})$ and $(1, D)$

   pessimistic bound :

   $(0, D-D_{max})$ and $(1, D)$

   For a terminal workload, the bounds are lines through the points:

   optimistic bound :

   $(0, D - \frac{D_{ave}}{1 + Z/D})$ and $(1, D)$

   pessimistic bound :

   The pessimistic bound for terminal workloads is not linear in N, so must be calculated for each population of interest using the equation in Table 5.2.

4. Calculate balanced system bounds on throughput for the range of N of interest using the equations in Table 5.2. (Again, these are not linear in N.)

Algorithm 5.2 — Closed Model Balanced System Bounds

- Because they are so simple to calculate, even by hand (they require only a few arithmetic operations once D and D_{max} are known), they are a quick way to obtain a rough feel for the behavior of a system.
- They reveal the critical influence of the bottleneck service center. Changes to the system that do not affect the bottleneck center do not alter the heavy load bounds on performance. Hence, throughput curves for *all* systems with bottleneck demand D_{max} are constrained to lie below the line $1/D_{max}$. To improve performance beyond this limit, it is necessary to reduce the demand at the bottleneck center in some way.
- Diagrams that show secondary bottlenecks as well as the primary one provide insight into the extent of improvements realizable by various modifications to the system that reduce the demand on the primary bottleneck.
- In the early phases of system design and system sizing, bounding studies offer the advantage that a group of configurations may be able to be treated as a single alternative. This is the case because of the critical influence of the bottleneck center, noted above.

Using fundamental laws, bounds on center utilizations and throughputs can be calculated from the asymptotic and balanced system bounds on system throughput. The system throughput bounds of Tables 5.1 and 5.2 are transformed into bounds on center k utilization simply by multiplying through by D_k (since the utilization law states that $U_k(N) = X(N) D_k$). Similarly, bounds on center k throughput are obtained by multiplying through by V_k (due to the forced flow law: $X_k(N) = X(N) V_k$).

In the chapters that follow, we present methods for calculating specific values of performance measures rather than bounds. These values will form smooth curves that are asymptotic to the light and heavy load optimistic asymptotic bounds and to the pessimistic balanced system bounds.

5.6. References

Muntz and Wong [1974] carried out the first study devoted specifically to the asymptotic properties of closed queueing networks. Denning and Kahn [1975] obtained related results independently. Denning and Buzen [1978] describe asymptotic analysis in discussing bottlenecks as part of operational analysis. Beizer [1978] also includes bounding analysis as part of his performance evaluation methodology. Balanced system bound analysis was developed by Zahorjan et al. [1982]. The case study of Section 5.3.1 was carried out by Sevcik et al. [1980].

[Beizer 1978]
   Boris Beizer. *Micro-Analysis of Computer System Performance.* Van Nostrand Reinhold, 1978.
[Denning & Buzen 1978]
   Peter J. Denning and Jeffrey P. Buzen. The Operational Analysis of Queueing Network Models. *Computing Surveys 10*,3 (September 1978), 225-261.
[Denning & Kahn 1975]
   Peter J. Denning and Kevin C. Kahn. Some Distribution Free Properties of Throughput and Response Time. Report CSD-TR-159, Computer Science Department, Purdue University, May 1975.
[Muntz & Wong 1974]
   R.R. Muntz and J.W. Wong. Asymptotic Properties of Closed Queueing Network Models. *Proc. 8th Princeton Conference on Information Sciences and Systems* (1974).
[Sevcik et al. 1980]
   K.C. Sevcik, G.S. Graham, and J. Zahorjan. Configuration and Capacity Planning in a Distributed Processing System. *Proc. 16th CPEUG Meeting* (1980), 165-171.
[Zahorjan et al. 1982]
   J. Zahorjan, K.C. Sevcik, D.L. Eager, and B.I. Galler. Balanced Job Bound Analysis of Queueing Networks. *CACM 25*,2 (February 1982), 134-141.

5.7. Exercises

1. In a system serving both batch jobs and terminal users, the following observations were made during a 30 minute interval:

active terminals	40
think time	20 seconds
interactive response time	5 seconds
disk service time per access	20 milliseconds
disk accesses per batch job	100
disk accesses per terminal interaction	5
disk utilization	60%

   a. What is batch throughput?

   b. Using only the information given above, calculate the maximum batch throughput possible if interactive response times of 15 seconds are to be achievable. What assumption must you make in answering this question?

2. Consider an interactive system with a CPU and two disks. The following measurement data was obtained by observing the system:

observation interval	30 minutes
active terminals	30
think time	12 seconds
completed transactions	1,600
fast disk accesses	32,000
slow disk accesses	12,000
CPU busy	1,080 seconds
fast disk busy	400 seconds
slow disk busy	600 seconds

a. Determine the visit counts (V_k), service times per visit (S_k), and service demands (D_k) at each center.

b. Give optimistic and pessimistic asymptotic bounds on throughput and response time for 5, 10, 20, and 40 active terminals.

Consider the following modifications to the system:

    1: Move all files to the fast disk.
    2: Replace the slow disk by a second fast disk.
    3: Increase the CPU speed by 50% (with the original disks).
    4: Increase the CPU speed by 50% and balance the disk load across two fast disks.

c. For the original system and for modifications 1 through 4, graph optimistic and pessimistic asymptotic bounds on throughput and response time as functions of the number of active terminals.

d. For the original system and for modification 3, specify the maximum number of terminals that can be active such that the asymptotic bounds do not preclude the possibility of an 8 second average response time.

e. If 40 terminals were active on the original system, how much would the CPU have to be speeded up so that the bounds would not rule out the possibility of achieving 10 second average response times?

f. If 80 terminals were active on the original system, what minimum modifications to the system would be required so that the bounds would not rule out the possibility of achieving 15 second average response times?

3. An installation with a CPU intensive workload is considering moving from a centralized system with a single large CPU to a decentralized system with several smaller CPUs.

5.7. Exercises

    a. Suppose that 10 processors each 1/10-th the speed of the large processor can be operated at the same cost as the large processor. Use asymptotic throughput and response time bounds to investigate the conditions under which such a change clearly would be beneficial or detrimental (considering performance issues only).

    b. Suppose that 15 processors each 1/10-th the speed of the large processor can be operated at the same cost. How does this affect your answer to (a)?

4. Consider a model with three service centers and service demands $D_1 = 5$ seconds, $D_2 = 4$ seconds, and $D_3 = 3$ seconds.

    a. Graph the optimistic and pessimistic asymptotic throughput and response time bounds for this model with a batch workload.

    b. On the same graphs, include balanced system bounds for the model.

    c. What is the relationship between the two sets of bounds in terms of the range of possible values to which they restrict performance measures? What is their relationship in terms of computational effort?

    d. Repeat your calculations for a terminal class with 15 second think times.

5. The assumptions introduced in deriving balanced system bounds for transaction workloads do not result in an improvement over the asymptotic bound for system throughput; we still have $X(\lambda) \leq 1/D_{max}$. However, they do yield an improved response time bound. The key to this improvement is the equation:

$$R_k(\lambda) = \frac{D_k}{1 - U_k(\lambda)}$$

    a. Using this equation, derive optimistic and pessimistic response time bounds based on balanced systems in which the service demands at all centers are set to D_{min} (optimistic) and D_{max} (pessimistic).

    b. Derive improved bounds by using the fact that the sum of the service demands in the original system is D. (Check your results against Table 5.2.)

    c. Compute the value of λ_{sat} for a system with three service centers with service demands of 8, 4, and 2 seconds. Sketch the two sets of response time bounds you just derived for arrival rates λ between 0 and λ_{sat}.

Chapter 6

Models with One Job Class

6.1. Introduction

In this chapter we examine *single class queueing network models*. Single class models are refinements of bounding models that provide *estimates* of performance measures, rather than simply bounds. For instance, instead of determining that the throughput of a certain system is between 1.1 and 2.0 jobs/minute (for a given population size), a single class model would provide an estimate of the actual throughput, such as 1.7 jobs/minute.

In single class models, the customers are assumed to be indistinguishable from one another. Although single class models always are simplifications, they nonetheless can be accurate representations of real systems. There are a number of situations in which a single class model might be used:

- *increased information* — The results of a bounding study might not provide sufficiently detailed information. Single class models are the next step in a progression of increasingly detailed models.
- *single workload of interest* — The computer system under consideration may be running only a single workload of significance to its performance. Therefore, it may not be necessary to represent explicitly the other workload components.
- *homogeneous workloads* — The various workload components of a computer system may have similar service demands. A reasonable modelling abstraction is to consider them all to belong to a single customer class.

Conversely, there are a number of situations in which it might be inappropriate to model a computer system workload by a single customer class. These situations typically arise either because distinct workload components exhibit markedly differing resource usage, or because the aim of the modelling study requires that inputs or outputs be specified in terms of the individual workload components rather than in terms of the aggregate workload. Typical instances of each are:

- *multiple distinct workloads* — On a system running both batch and timesharing workloads, the batch workload might be CPU bound while the timesharing workload is I/O bound. A queueing network model with a customer population consisting of a single class representing an "average" job might not provide accurate projections, since jobs in the actual system do not behave as though they were nearly indistinguishable.
- *class dependent model inputs* — In a mixed batch/timesharing system, the timesharing workload is expected to grow by 100% over the next 2 years, while the batch workload is expected to grow by only 10%. Since in a single class model there is only a single class of "average" customers, it is not possible to set the input parameters such that workload components exhibit differing growth rates. Thus, a single class model is not an appropriate representation.
- *class dependent model outputs* — In a batch environment running both production and development programs, projections about the time in system of each workload component, rather than just an estimate of "average" time in system, might be desired. Since there is only one class of customers in a single class model, outputs are given in terms of that class only, and it is difficult to interpret these measures in terms of the original classes of the system. Thus a multiple class model is required.

Systems having workloads with substantially differing characteristics, as exhibited by the examples above, may be modelled more reasonably by multiple class than by single class queueing networks. These more sophisticated models are discussed in Chapter 7.

The next two sections of this chapter deal with the practical application of single class queueing networks as models of computer systems. Section 6.2 discusses the use of the workload intensity parameter to mimic the job mix behavior of a computer system. Section 6.3 describes a number of case studies in which single class models have been employed.

This discussion of the practice of single class models is followed by a discussion of their theory. In Section 6.4 the algorithms required to evaluate the models are developed and illustrated with examples. Section 6.5 presents the theoretical underpinnings upon which the models rest.

6.2. Workload Representation

The workload representation of a single class queueing network model is given by two model inputs: the set of *service demands*, and the *workload intensity*. In using a single class model, one inherently makes the assumption that all jobs running in the system are sufficiently similar that their differences do not have a major effect on system performance. Thus, calculating the set of service demands is fairly straightforward, as only a single set is required. (In contrast, with multiple class models one first must decide how many classes to represent, and then must calculate a distinct set of service demands for each class.)

Establishing the workload intensity has two aspects: selecting an appropriate workload type (transaction, batch, or terminal), and setting the appropriate workload intensity parameter(s) for that type. Selecting an appropriate workload type typically is straightforward, since the three workload types of queueing network models correspond directly to the three predominant workload types of computer systems. One technical distinction that arises is that between *open* models (those with transaction classes) and *closed* models (those with batch or terminal classes). Since the number of customers that may be in an open model at any time is unbounded, while the number of customers that may be in a closed model is bounded by the population of the closed class, the response times of open models tend to be larger than those of corresponding closed models with the same system throughput. This occurs because in open models the potential for extremely large queue lengths exists, while in closed models, because of the finite population, it does not. This difference usually is significant only when some device in the system is near saturation.

This brings us to the question of how to set the workload intensity parameter. In queueing network models, the workload intensity is a fixed quantity (an arrival rate, a population, or a population and a think time). In contrast, in a computer system the workload intensity may vary. Despite this discrepancy, queueing network models are useful in a wide variety of situations:

- *heavy load assumption* − It may be interesting to study the behavior of a system under the maximum possible load. By hypothesis, the load is sufficiently heavy that there always are jobs waiting to enter memory. Thus, when one job completes and releases memory, it immediately is replaced by another job. The workload therefore is represented as a batch class with a constant number of customers equal to the maximum multiprogramming level of the system.

- *non-integer workload intensity* − The measurement data for a system might show that the average multiprogramming level (or active number of terminal users) is not an integer. Some algorithms for evaluating queueing network models allow non-integer customer populations. Other algorithms do not. For the latter, the model can

6.2. Workload Representation

be evaluated for the neighboring integer workload intensity values and the non-integer solution obtained by interpolation. For instance, if the measured multiprogramming level were 4.5, the solutions of the model with batch populations of 4 and 5 could be computed, and their average taken as the projection for 4.5 customers.

- *workload intensity distribution* — Measurement data might provide a distribution of observed workload intensities, e.g., proportions of time $P[N=n]$ that there were n active terminal users on the system. This distribution could be used to weight the solutions obtained for a model with each observed number of users. Table 6.1 gives an example.

n	$P[N=n]$	$U_{CPU}(n)$	$X(n)$	$R(n)$
0	.1	0	0	0
1	.2	.032	.0525	.787
2	.3	.062	.1031	1.546
3	.3	.092	.1515	2.273
4	.1	.119	.1974	2.961

$$U_{CPU} = \sum_{n=1}^{4} P[N=n] U_{CPU}(n) = .0645$$

$$R = \sum_{n=1}^{4} \left[\frac{X(n)P[N=n]}{\sum_{j=1}^{4} X(j)P[N=j]} R(n) \right] = 2.492$$

Table 6.1 — Use of Distributional Information

- *sizing studies* — Because the solutions of single class models can be obtained extremely quickly, it is feasible to evaluate a model for a large number of workload intensities. Thus, questions such as "What is the maximum transaction arrival rate that can be supported with average response time below 3 seconds?" can be answered by varying the arrival rate of a model (e.g., setting $\lambda = 1, 2, ...$) and observing the reported response times.
- *robustness studies* — Similarly, since it often is the case that workload growth cannot be forecast accurately, it generally is useful to evaluate a model for a range of workload intensities surrounding the expected one. This allows the analyst to assess the impact on projected performance of a growth in the workload that exceeds expectations.

6.3. Case Studies

Three applications of single class queueing network models are described in this section. The first is a classic study in which an extremely simple model gave surprisingly accurate performance projections. The second is an application in which the effects of modifying certain hardware and software characteristics were investigated. The third illustrates a recent use of a single class model for capacity planning.

6.3.1. A Model of an Interactive System

We first consider what may be the earliest application of queueing network modelling to computer systems. We include this study despite its age (it was performed in 1965) because of its historical interest and because it demonstrates vividly that extremely simple models can be accurate predictors of performance.

The system under study was an IBM 7094 running the Compatible Time-Sharing System (CTSS). CTSS was an experimental interactive system based on swapping. Only a single user could be "active" at a time. The entire system − CPU, disks, and memory − was "time-sliced" among users as a unit.

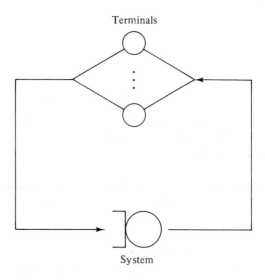

Figure 6.1 − Interactive System Model

The purpose of the study was to investigate the response time behavior of the system as a function of the number of users. To do so, the model of Figure 6.1 was constructed. It contains a terminal workload,

6.3. Case Studies

representing the user population, and a single service center representing the system (CPU and disks). This single service center representation is sufficient because, with only one user active at a time, there can be no overlap in processing at the CPU and the disks (individual users on this system did not exploit this capability). Thus, in terms of average response time, it does not matter (in the model) whether a user spends time at the CPU or the disks, but simply that the appropriate amount of time transpires.

Notice that by using a single service center to represent the system, we have solved a simple memory constraint problem. Had the model contained separate CPU and disk service centers, it would have been less accurate because it would have allowed customers to be processing at both simultaneously, while in the actual system this was not possible. This technique of collapsing a number of service centers into a single service center to represent memory constraints can be extended in quite powerful ways, as will be explained in Chapter 9.

The model was parameterized from measurements taken during system use, which provided average think time, average CPU and disk processing times, and average memory requirement. The service demand at the system service center was set equal to the sum of the measured processing times and the disk service required for swapping a job of average size. The number of customers in the model then was varied, and response time estimates for each population were obtained. Figure 6.2 compares the model projections with measured response times.

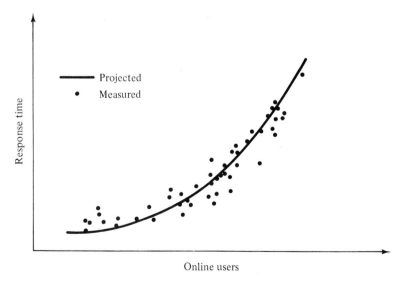

Figure 6.2 − Measured and Projected Response Times

6.3.2. A Model with Modification Analysis

In this case study a single class model was used to evaluate the benefits of several proposed changes to a hardware and software configuration. The system under consideration was an IBM System/360 Model 65J with three channels. Channels one and two were connected to 8 and 16 IBM 2314-technology disks, respectively. Channel three was connected to a drum, which was used exclusively by the operating system. Because the use of this drum was overlapped entirely with the processing of user jobs, it was omitted from the model. (Customers in the model represent user jobs, which never visited the drum.)

The model of this system is shown in Figure 6.3. It is parameterized by specifying service demands for the CPU, disks, and channels, as well as the workload type and intensity. The model differs from our "standard" model (cf. Section 4.5) because of the inclusion of service centers representing the channels. In general, a model of this sort can lead to significant error (as will be explained shortly). However, because of the characteristics of this system, good accuracy was obtained.

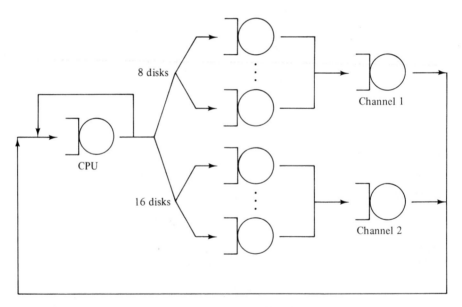

Figure 6.3 — System Model

The base CPU service requirement per job was estimated by dividing the total CPU busy time (both system and user time) over the measurement interval by the number of jobs that completed in the interval. Thus, system CPU overhead (such as that required to handle CPU scheduling and user I/O) was allocated equally among all jobs.

6.3. Case Studies

Parameterizing the I/O subsystem (the disks and channels) was more complicated. The disk technology of the system required that both the disk and the channel be held during rotational latency (the period during which the data is rotating to the read/write heads of the disk) and data transfer, while seeks could proceed at each disk independently of its channel. In the model, the channel service demands were set to the sum of the average latency and data transfer times, while the disk service demands were set to the average seek time. Thus, all components of I/O service time were represented exactly once. (If all three components of service were represented at the disks, customers in the model would experience latency and transfer service twice, and projected performance measures would be seriously in error.)

There is a danger in representing multiple component I/O subsystems in this manner. Unlike the actual system, no customer in the model ever holds both a disk and a channel simultaneously. Thus, there is the potential for artificial parallelism in the model, since a disk center that logically is being used for the latency and transfer portion of one job's service might be used at the same time to seek by another job. Accounting for this inaccuracy in general is a difficult problem. (Chapter 10 discusses I/O modelling in more detail.) However, in the case of this particular system, the effect of the potential parallelism in the model was negligible because the utilizations of the disk devices were fairly well balanced, and the total number of disks was much larger than the average multiprogramming level. Thus, the probability that a customer would require service from a disk already in use by another customer was small, and consequently so was the amount of artificial parallelism.

Measurement of the system showed that the average multiprogramming level varied significantly during the measurement interval. To account for this variability, the model was evaluated once for each observed multiprogramming level. Performance projections were obtained by weighting the distinct solutions by the percentage of time the corresponding multiprogramming levels were observed in the system.

The purpose of the modelling study was to evaluate the effects of the following proposed changes to the system:
- Replace eight of the 2314-technology disks on one channel with six IBM 3330 disks. The effect of this change was reflected in the model by altering the service demands of the affected channel and disk service centers, since 3330s seek and transfer data faster than 2314s, and also have *rotational position sensing (RPS)* capability, which allows the disk to disconnect from the channel during rotational latency, reconnecting only when the required sector is about to come under the read/write heads.

- Replace *extended core storage (ECS)* with faster memory. This would result in an effectively faster CPU, since the processing rate was limited by the memory access time. As most of the programs executed out of ECS were system routines, the effect of this change was reflected in the model by reducing the portion of the CPU service demand corresponding to supervisor state (system) processing.
- Implement an operating system improvement. This improvement was expected to reduce overhead by 8%. Thus this change was reflected in the model by decreasing the portion of the CPU service demand corresponding to operating system processing.

The model was parameterized to reflect various combinations of the proposed system improvements, and the effect on user (problem state) CPU utilization was noted. (The use of U_{CPU} as the performance metric is an odd aspect of this study, since U_{CPU} can be made to increase simply by slowing down the processor. More typical metrics are system throughput and system response time.) The operating system improvement alone was projected to yield a 5% increase in U_{CPU}. In conjunction with the ECS replacement, the gain was projected to be 25%. When the operating system improvement was combined with the disk upgrade, a similar 25% gain was projected. This pair of modifications actually was implemented; subsequent measurements showed that U_{CPU} had increased by about 20%, even though the basic CPU service demand had diminished due to an unanticipated change in the workload. Thus, the model provided a close projection of true system behavior.

This example shows that quite simple models can be used to answer performance questions of interest. It is important to notice how little of the detail of the computer system is represented in the model; only those aspects of the system that were crucial to performance and under consideration for modification were represented. For example, there is no explicit representation of memory in the model. This simplicity is a great advantage of queueing network models.

6.3.3. Capacity Planning

The purpose of this study was to evaluate the impact on response time of an anticipated 3% quarterly growth in the volume of the current workload. The system was an Amdahl 470 with 8 MB of main store, 16 channels, and 40 disks. The system was running IBM's MVS operating system and IMS database system, running a transaction processing workload. IMS was supporting five *message processing regions*: areas of main memory allocated and scheduled by IMS, each of which can accommodate one user request. If more than five requests were outstanding, the remainder queued for an available region.

6.3. Case Studies

Many different transaction types existed in the system. However, they were increasing in volume at about the same rate, so a single class model was sufficient to investigate the performance question of interest. (If various transaction types had been growing at differing rates, a multiple class model would have been required.) The model of the system is shown in Figure 6.4. It contains a single transaction workload, representing the aggregate of all the transaction types in the system, a memory queue, reflecting the fact that only five message processing regions were available, a CPU service center, and 40 disk service centers.

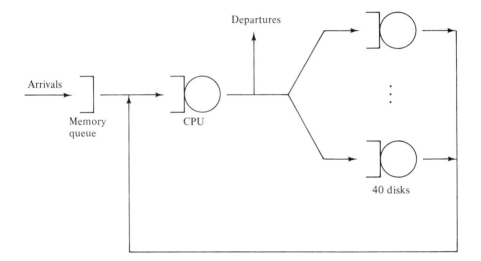

Figure 6.4 — System Model

Because this model contains a memory queue, it is not separable, and so cannot be evaluated directly by the techniques to be introduced later in this chapter. In Chapter 9 we discuss general methods for evaluating models of this type. For now, it is sufficient to observe that the solution of an open model with a saturated memory queue is roughly equivalent to the solution of a corresponding closed model in which the open class of customers has been replaced by a closed class with multiprogramming level equal to maximum possible number of simultaneously active jobs. This model is separable, so can be evaluated easily.

Parameters for the model were obtained from information gathered by software monitors:

- The arrival rate of customers was set equal to the measured transaction arrival rate.

- The service demand at the CPU was set equal to:
$$D_{CPU} = U_{CPU}\, T\, /\, C$$
where U_{CPU} was the measured *CPU* utilization, T was the length of the measurement interval, and C was the number of transactions that completed during the interval.
- The service demand at each disk k was set equal to:
$$D_k = U_k\, T\, /\, C$$

Notice that because of the way the service demands were calculated, both overhead and inherent service requirements were included. In the case of the CPU, this means that both user and system processing time were accounted for. In the case of the disks, this means that seek, rotational latency, data transfer, and any time lost because of I/O path contention were included. This approach to accounting for overhead can be quite useful when it is anticipated that the ratio of overhead to useful processing time will be relatively insensitive to the proposed modifications being investigated. The advantage of this approach is the simple way in which service demands can be computed. (For example, we do not need to determine the duration of each component of disk service time.) The disadvantage is that anticipated changes in the ratios of overhead to inherent service times cannot be modelled without more detailed information. For the modifications considered in this study, it was not felt that this was a significant drawback.

Having set the parameters, the model was evaluated to obtain response time projections. Figure 6.5 graphs projected response time against year for four different memory sizes: the existing configuration, adequate to support five message processing regions, and expanded configurations supporting six, seven, and eight message processing regions. On the basis of this study, it was concluded that, with the addition of memory, the system would be adequate for at least two years.

6.4. Solution Techniques

The solution of a queueing network model is a set of performance measures that describe the time averaged (or long term) behavior of the model. Computing these measures for general networks of queues is quite expensive and complicated. However, for separable queueing network models, solutions can be obtained simply.

The specific procedures followed to analyze separable queueing networks differ for open and closed models. We consider each in turn.

6.4. Solution Techniques

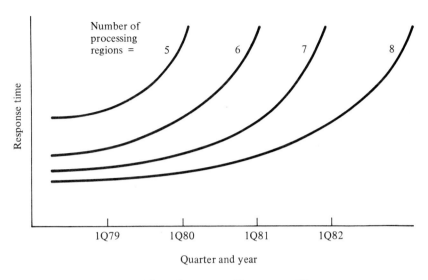

Figure 6.5 — Projected Response Times

6.4.1. Open Model Solution Technique

For open models (those with transaction workloads), one of the key output measures, system throughput, is given as an input. Because of this, the solution technique for these models is especially simple. We list here the formulae that apply for each performance measure of interest.

- *processing capacity*

    The processing capacity of an open model, λ_{sat}, is the arrival rate at which it saturates. This is given by:

    $$\lambda_{sat} = \frac{1}{\max_k D_k} = \frac{1}{D_{max}}$$

    In the derivations that follow, we assume that $\lambda < \lambda_{sat}$.

- *throughput*

    By the forced flow law, if λ customers/second enter the network, then the system output rate must also be λ customers/second. Similarly, if each customer requires on average V_k visits to device k, the throughput at device k must be λV_k visits/second. Thus:

    $$X_k(\lambda) = \lambda V_k$$

- **utilization**

 By the utilization law, device utilization is equal to throughput multiplied by service time. Thus:

 $$U_k(\lambda) = X_k(\lambda) S_k = \lambda D_k$$

 (In the case of delay centers, the utilization must be interpreted as the average number of customers present.)

- **residence time**

 The residence time at center k, $R_k(\lambda)$, is the total time spent at that center by a customer, both queueing and receiving service. For service centers of delay type, there is no queueing component, so $R_k(\lambda)$ is simply the service time multiplied by the number of visits:

 $$R_k(\lambda) = V_k S_k = D_k \quad \text{(delay centers)}$$

 For queueing centers, R_k is the sum of the total time spent in service and the total time spent waiting for other customers to complete service. The former component is $V_k S_k$. The latter component is the time spent waiting for customers already in the queue when a customer arrives. Letting $A_k(\lambda)$ designate the average number of customers in queue as seen by an arriving customer, the queueing component is $V_k \left[A_k(\lambda) S_k \right]$. (By assumption, to be discussed in Section 6.5, the expected time until completion of the job in service when a new job arrives is equal to the service time of the job.) Thus, for queueing centers the residence time is given by:

 $$R_k(\lambda) = V_k \left[S_k + S_k A_k(\lambda) \right]$$
 $$= D_k \left[1 + A_k(\lambda) \right]$$

 An implication of the assumptions made in constructing separable networks is that the queue length seen upon arrival at center k, $A_k(\lambda)$, is equal to the time averaged queue length $Q_k(\lambda)$, giving:

 $$R_k(\lambda) = D_k \left[1 + Q_k(\lambda) \right]$$

 which, using Little's law to re-express Q_k, is:

 $$R_k(\lambda) = D_k \left[1 + \lambda R_k(\lambda) \right]$$
 $$= \frac{D_k}{1 - U_k(\lambda)} \quad \text{(queueing centers)}$$

 This equation exhibits the intuitively appealing property that as $U_k(\lambda) \to 0$, $R_k(\lambda) \to D_k$, and as $U_k(\lambda) \to 1$, $R_k(\lambda) \to \infty$.

6.4. Solution Techniques

- *queue length*

  By Little's law:
  $$Q_k(\lambda) = \lambda R_k(\lambda)$$
  $$= \begin{cases} U_k & \text{(delay centers)} \\ \dfrac{U_k(\lambda)}{1 - U_k(\lambda)} & \text{(queueing centers)} \end{cases}$$

- *system response time*

  System response time is the sum of the residence times at all service centers:
  $$R(\lambda) = \sum_{k=1}^{K} R_k(\lambda)$$

- *average number in system*

  The average number in system can be calculated using Little's law, or by summing the queue lengths at all centers:
  $$Q(\lambda) = \lambda R(\lambda) = \sum_{k=1}^{K} Q_k(\lambda)$$

These formulae are summarized as Algorithm 6.1.

---

processing capacity : $\lambda_{sat} = 1 / D_{max}$

throughput : $X(\lambda) = \lambda$

utilization : $U_k(\lambda) = \lambda D_k$

residence time : $R_k(\lambda) = \begin{cases} D_k & \text{(delay centers)} \\ \dfrac{D_k}{1 - U_k(\lambda)} & \text{(queueing centers)} \end{cases}$

queue length : $Q_k(\lambda) = \lambda R_k(\lambda)$
$$= \begin{cases} U_k(\lambda) & \text{(delay centers)} \\ \dfrac{U_k(\lambda)}{1 - U_k(\lambda)} & \text{(queueing centers)} \end{cases}$$

system response time : $R(\lambda) = \sum_{k=1}^{K} R_k(\lambda)$

average number in system : $Q(\lambda) = \lambda R(\lambda) = \sum_{k=1}^{K} Q_k(\lambda)$

Algorithm 6.1 — Open Model Solution Technique

Open Model Example

Figure 6.6 shows a simple open model with three service centers, and illustrates the calculation of various performance measures. (All times are in seconds.)

6.4.2. Closed Model Solution Techniques

The technique we use to evaluate closed queueing networks (those with terminal or batch classes) is known as *mean value analysis* (*MVA*). It is based on three equations:

- *Little's law applied to the queueing network as a whole*:

$$X(N) = \frac{N}{Z + \sum_{k=1}^{K} R_k(N)} \quad (6.1)$$

where $X(N)$ is the system throughput and $R_k(N)$ the residence time at center k, when there are N customers in the network. (As usual, if the customer class is batch type, we take $Z = 0$.) Note that system throughput can be computed from input parameter data if the device residence times $R_k(N)$ are known.

- *Little's law applied to the service centers individually*:

$$Q_k(N) = X(N) R_k(N) \quad (6.2)$$

Once again, the residence times must be known before Little's law can be applied to compute queue lengths.

- *The service center residence time equations*:

$$R_k(N) = \begin{cases} D_k & \text{(delay centers)} \\ D_k\left[1 + A_k(N)\right] & \text{(queueing centers)} \end{cases} \quad (6.3)$$

where $A_k(N)$ is the average number of customers seen at center k when a new customer arrives.

Note that, as with open networks, the key to computing performance measures for closed networks is the set of $A_k(N)$. If these were known, the $R_k(N)$ could be computed, followed by $X(N)$ and the $Q_k(N)$. In the case of open networks we were able to substitute the time averaged queue lengths, $Q_k(N)$, for the arrival instant queue lengths, $A_k(N)$. In the case of closed networks, this substitution is not possible. To see that $A_k(N)$ does not equal $Q_k(N)$ in closed networks, consider the network consisting of two queueing service centers and a single customer with a service

6.4. Solution Techniques

Model Inputs:

$$V_{CPU} = 121 \qquad V_{Disk1} = 70 \qquad V_{Disk2} = 50$$
$$S_{CPU} = .005 \qquad S_{Disk1} = .030 \qquad S_{Disk2} = .027$$
$$D_{CPU} = 0.605 \qquad D_{Disk1} = 2.1 \qquad D_{Disk2} = 1.35$$
$$\lambda = 0.3 \text{ jobs/sec.}$$

Model Structure:

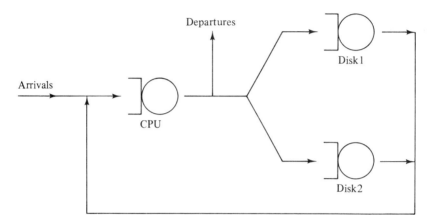

Selected Model Outputs:

$$\lambda_{sat} = \frac{1}{D_{max}} = \frac{1}{2.1} = .476 \text{ jobs/sec.}$$

$$X_{CPU}(.3) = \lambda V_{CPU} = (.3)(121) = 36.3 \text{ visits/sec.}$$

$$U_{CPU}(.3) = \lambda D_{CPU} = (.3)(.605) = .182$$

$$R_{CPU}(.3) = \frac{D_{CPU}}{1 - U_{CPU}(.3)} = \frac{.605}{.818} = .740 \text{ secs.}$$

$$Q_{CPU}(.3) = \frac{U_{CPU}(.3)}{1 - U_{CPU}(.3)} = \frac{.182}{.818} = .222 \text{ jobs}$$

$$R(.3) = R_{CPU}(.3) + R_{Disk1}(.3) + R_{Disk2}(.3)$$
$$= .740 + 5.676 + 2.269 = 8.685 \text{ secs.}$$

$$Q(.3) = \lambda R(\lambda) = (.3)(8.685) = 2.606 \text{ jobs}$$

Figure 6.6 — Open Model Example

demand of 1 second at each center. Since there is only one customer, the time averaged queue lengths at the service centers are simply their utilizations, so $Q_1(1) = Q_2(1) = 1/2$. However, the arrival instant queue lengths $A_1(1)$ and $A_2(1)$ both are zero, because with a single customer in the network no customers could possibly be in queue ahead of an arriving customer. In general, the key distinction is that the arrival instant queue lengths are computed conditioned on the fact that some customer is arriving to the center (and so cannot itself be in the queue there), while the time averaged queue lengths are computed over randomly selected moments (so all customers potentially could be in the queue).

As mentioned above, evaluating a model requires that we first compute the $A_k(N)$. There are two basic techniques, exact and approximate. We emphasize that this distinction refers to how the solution relates to the model, rather than to the computer system itself. The accuracy of the solution relative to the performance of the computer system depends primarily on the accuracy of the parameterization of the model, and not on which of the two solution techniques is chosen.

We next examine each of the two solution methods, beginning with the exact technique.

6.4.2.1. Exact Solution Technique

The exact MVA solution technique is important for two reasons:
- It is the basis from which the approximate technique is derived.
- There are no known bounds on the inaccuracy of the approximate technique. While typically it is accurate to within a few percent relative to the true solution, it cannot be guaranteed that in any particular situation the results will not be worse.

The exact solution technique involves computing the arrival instant queue lengths $A_k(N)$ exactly, then applying equations (6.1)-(6.3). The characteristic of closed, separable networks that makes them amenable to this approach is that the $A_k(N)$ have a particularly simple form:

$$A_k(N) = Q_k(N-1) \qquad (6.4)$$

In other words, the queue length seen at arrival to a queue when there are N customers in the network is equal to the time averaged queue length there with one less customer in the network. This equation has an intuitive justification. At the moment a customer arrives at a center, it is certain that this customer itself is not already in queue there. Thus, there are only $N-1$ other customers that could possibly interfere with the new arrival. The number of these that actually are in queue, on average, is simply the average number there when only those $N-1$ customers are in the network.

6.4. Solution Techniques

The exact MVA solution technique, shown as Algorithm 6.2, involves the iterative application of equations (6.1)-(6.4). These equations allow us to calculate the system throughput, device residence times, and time averaged device queue lengths when there are n customers in the network, given the time averaged device queue lengths with $n-1$ customers. The iteration begins with the observation that all queue lengths are zero with zero customers in the network. From that trivial solution, equations (6.1)-(6.4) can be used to compute the solution for one customer in the network. Since the time averaged queue lengths with one customer in the network are equal to the arrival instant queue lengths with two customers in the network, the solution obtained for a population of one can be used to compute the solution with a population of two. Successive applications of the equations compute solutions for populations 3, 4, ..., N.

---

for $k \leftarrow 1$ to K do $Q_k \leftarrow 0$
for $n \leftarrow 1$ to N do
begin

  for $k \leftarrow 1$ to K do $R_k \leftarrow \begin{cases} D_k & \text{(delay centers)} \\ D_k(1 + Q_k) & \text{(queueing centers)} \end{cases}$

  $X \leftarrow \dfrac{n}{Z + \sum_{k=1}^{K} R_k}$

  for $k \leftarrow 1$ to K do $Q_k \leftarrow XR_k$

end

Algorithm 6.2 — Exact MVA Solution Technique

Figure 6.7 — Single Class Solution Population Precedence

Figure 6.7 illustrates the precedence relations of the solutions required to apply the exact MVA solution technique. As just described, the solution of a closed model with N customers requires the solution with $N-1$ customers, which requires the solution with $N-2$ customers, etc. Thus, the full solution requires N applications of equations (6.1)-(6.4). Since each of the N applications of the equations requires looping (several times) over the K service centers, the computational expense of the solution grows as the product of N with K. The space requirement, in contrast, is about K locations, since the performance measures for the network with n customers can be discarded once they have been used to calculate the performance measures for $n+1$ customers. Note that all solutions between 1 customer and N customers are computed as by-products of the N customer solution. Thus, there is no additional expense involved in obtaining these intermediate solutions (although of course some additional space is required if all of them are to be retained). This is an important characteristic of the solution technique that will be exploited in Chapter 8 when we discuss flow equivalent service centers.

When Algorithm 6.2 terminates, the values of R_k, X, and Q_k (all for population N) are available immediately. Other model outputs are obtained by using Little's law. Here is a summary:

system throughput:	X
system response time:	$N/X - Z$
average number in system:	$N - XZ$
device k throughput:	XV_k
device k utilization:	XD_k
device k queue length:	Q_k
device k residence time:	R_k

Closed Model Example (Exact Solution)

Table 6.2 shows the computation of the solution of the network of Figure 6.6 with the transaction class replaced by a terminal class. There are three centers, with service demands $D_{CPU} = .605$ seconds, $D_{Disk1} = 2.1$ seconds, and $D_{Disk2} = 1.35$ seconds. The terminal class has three customers ($N=3$) and average think time of 15 seconds ($Z=15$). The algorithm begins with the known solution for the network with zero customers, and calculates the $R_k(n)$, $X(n)$, and $Q_k(n)$ for each successively larger population n, up to three.

In studying Table 6.2, note that the sum of the queue lengths at the three centers does not equal the customer population. This is the case because we are dealing with a class of terminal type, and some of the customers are "thinking". (Algorithm 6.2 accounts for this by the inclusion of the think time, Z, in one of its equations.) We can calculate the average number of "thinking" customers by subtracting the average number in system, $Q = N - XZ$, from the total customer population, N, yielding XZ (which equals zero for a batch class).

6.4. Solution Techniques

	k	n=0	n=1	n=2	n=3
	CPU	-	.605	.624	.644
R_k	Disk1	-	2.1	2.331	2.605
	Disk2	-	1.35	1.446	1.551
X		-	.0525	.1031	.1515
	CPU	0	.0318	.0643	.0976
Q_k	Disk1	0	.1102	.2403	.3947
	Disk2	0	.0708	.1490	.2350

Table 6.2 — Exact MVA Computation

Model outputs can be computed from the results for $N=3$:

$$X(3) = .152$$
$$R(3) = 3/.152 - 15.0 = 4.74$$
$$Q(3) = N - X(3)Z = 3 - (.152)(15) = .72$$
$$X_{CPU}(3) = X(3)V_{CPU} = (.152)(121) = 18.39$$
$$U_{CPU}(3) = X(3)D_{CPU} = (.152)(.605) = .092$$
$$Q_{CPU}(3) = .098$$
$$R_{CPU}(3) = .64$$

6.4.2.2. Approximate Solution Technique

The key to the exact MVA solution technique is equation (6.4), which computes the arrival instant queue length for population n based on the time averaged queue length with population $n-1$. The nature of the algorithm is a direct consequence of this relationship.

By replacing equation (6.4) with an approximation:

$$A_k(N) \approx h\big[Q_k(N)\big]$$

for some suitable function h, a more efficient, iterative algorithm can be obtained. (The function h actually might depend on values other than $Q_k(N)$. For instance, the approximation we will propose shortly also depends on N. However, we use this notation for simplicity, and to suggest that the key requirement is the value of $Q_k(N)$.) The accuracy of the algorithm depends, of course, on the accuracy of the function h that is used. (A particular choice for h will be presented shortly.)

This general approach is outlined in Algorithm 6.3. It is seen easily that the time and space requirements of this algorithm depend on the number of centers but are independent of the customer population of the network being evaluated (except indirectly; the number of iterations required for convergence may be affected by the population). This can be a substantial improvement over the exact MVA technique, which requires time proportional to the product of the number of centers and the number of customers.

---

1. Initialize: $Q_k(N) \leftarrow \dfrac{N}{K}$ for all centers k.
2. Approximate: $A_k(N) \leftarrow h\!\left[Q_k(N)\right]$ for all centers k.
   (The choice of an appropriate function h is discussed in the text.)
3. Use equations (6.3), (6.1), and (6.2) in succession to compute a new set of $Q_k(N)$.
4. If the $Q_k(N)$ resulting from Step 3 do not agree to within some tolerance (e.g., 0.1%) with those used as inputs in Step 2, return to Step 2 using the new $Q_k(N)$.

Algorithm 6.3 — Approximate MVA Solution Technique

Crucial to this faster solution technique is the function h. Unfortunately, no function h is known that is exact for all separable networks. Instead, an approximation must be used. A particularly simple and reasonably accurate approximation is:

$$\begin{aligned} A_k(N) &= Q_k(N-1) \\ &\approx h\!\left[Q_k(N)\right] \\ &\equiv \frac{N-1}{N} Q_k(N) \end{aligned} \qquad (6.5)$$

Equation (6.5) estimates the arrival instant queue length by approximating its exact value, the queue length with one fewer customer. This approximation is based on the assumption that the ratios $\dfrac{Q_k(N)}{N}$ and $\dfrac{Q_k(N-1)}{N-1}$ are equal for all k, i.e., that the amount that each queue length is diminished by the removal of a single customer is equal to the amount that customer contributes to the queue length. In general, this assumption is quite accurate. In particular, it is asymptotically correct for very large N, and trivially correct for models with only a single customer (since it predicts that arrival instant queue lengths are zero). Thus, the approximation is guaranteed to be good at the two extremes. Experience with the technique has demonstrated that it also gives remarkably good results for intermediate populations. Since this error is well within the bounds of other discrepancies inherent in the computer system analysis

6.5. Theoretical Foundations

process (e.g., the accuracy of parameter values), the approximate MVA technique is satisfactory as a general solution technique.

Closed Model Example (Approximate Solution)

Table 6.3 lists the successive approximations for the device queue lengths obtained by applying this approximate solution technique to the same example used previously with the exact solution technique. The stopping criterion used was agreement in successive queue lengths within .001. The exact solution of the model is listed in the table for comparison. (Note once again the apparent anomaly caused by the fact that the class in this model is of type terminal. We initialize by distributing the customers equally among the three centers. As the iteration progresses, customers "disappear" from the table. At the conclusion of the iteration, the difference between the full customer population and the sum of the queue lengths at the centers represents the average number of users "thinking".)

iteration	Q_{CPU}	Q_{Disk1}	Q_{Disk2}	X	R
0	1.00	1.00	1.00		
1	.1390	.4826	.3102	.1379	6.7583
2	.0988	.4150	.2436	.1495	5.0659
3	.0972	.4043	.2366	.1508	4.8950
4	.0972	.4024	.2359	.1510	4.8732
5	.0973	.4021	.2359	.1510	4.8700
exact solution	.0976	.3947	.2350	.1515	4.8020

Table 6.3 - Approximate MVA Computation

6.5. Theoretical Foundations

Separable queueing network models are a subset of the general class of queueing network models obtained by imposing restrictions on the behavior of the service centers and customers. The name "separable" comes from the fact that each service center can be separated from the rest of the network, and its solution evaluated in isolation. The solution of the entire network then can be formed by combining these separate solutions. In an intuitive sense, a separable network has the property that each service center acts (largely) independently of the others.

There are five assumptions about the behavior of a model that, if satisfied, guarantee that the model is separable. These are:

General Analytic Techniques: Models with One Job Class

- *service center flow balance* — Service center flow balance is the extension of the flow balance assumption (see Chapter 3) to each individual service center: the number of arrivals at each center is equal to the number of completions there.
- *one step behavior* — One step behavior asserts that no two jobs in the system "change state" (i.e., finish processing at some device or arrive to the system) at exactly the same time. Real systems almost certainly display one step behavior.

The remaining three assumptions are called *homogeneity assumptions.* This name is derived from the fact that in each case the assumption is that some quantity is the same (i.e., homogeneous) regardless of the current locations of some or all of the customers in the network.

- *routing homogeneity* — To this point we have characterized the behavior of customers in the model simply by their service demands. A more detailed characterization would include the routing patterns of the jobs, that is, the patterns of centers visited. Given this more detailed view, routing homogeneity is satisfied when the proportion of time that a job just completing service at center j proceeds directly to center k is independent of the current queue lengths at any of the centers, for all j and k. (A surprising aspect of separable models is that the routing patterns of jobs are irrelevant to the performance measures of the model. Thus, we will continue to ignore them.)
- *device homogeneity* — The rate of completions of jobs from a service center may vary with the number of jobs at that center, but otherwise may not depend on the number or placement of customers within the network.
- *homogeneous external arrivals* — The times at which arrivals from outside the network occur may not depend on the number or placement of customers within the network.

These assumptions are sufficient for the network to be separable, and thus to be evaluated efficiently. However, the specific solution algorithms we have presented thus far require one additional assumption, which is a stronger form of the device homogeneity assumption:

- *service time homogeneity* — The rate of completions of jobs from a service center, while it is busy, must be independent of the number of customers at that center, in addition to being independent of the number or placement of customers within the network.

The weaker of the two assumptions, device homogeneity, permits the rate of completions of jobs from a center to vary with the queue length there. Centers with this characteristic are called *load dependent* centers. A delay center is a simple example of a load dependent center, since the rate of completions increases in proportion to the number of customers at the center. Service time homogeneity asserts that the rate of completions is independent of the queue length. Centers with this characteristic are called *load independent*. The queueing centers we have described so far are examples of load independent centers. The particular versions of the MVA algorithms presented in this chapter are applicable only to networks consisting entirely of load independent and delay centers. In Chapters 8 and 20 we discuss the modifications necessary to accommodate general load dependent centers.

Although the assumptions above are necessary to prove mathematically that the solution obtained using Algorithm 6.2 is the exact solution of the model, they need not be satisfied exactly in practice for separable models to provide good results. Experience has shown that the accuracy of queueing network models is extremely robust with respect to violations of these assumptions. Thus, while no real computer system actually satisfies the homogeneity assumptions, it is rare that violations of these assumptions are a major source of inaccuracy in a modelling study. More typically, the problems encountered in validating a model result from an insufficiently accurate characterization by the model at the system level, usually because of inaccurate parameter values for service demands or workload intensities. The only important exceptions to this are cases in which the limitations on the structure of the model imposed by the assumptions required for separability prohibit representation of aspects of the computer system important to performance (for example, the modelling of memory constraints or priority scheduling). In these cases, we would like models that are as easy to construct and to evaluate as separable networks, but that also represent the "non-separable" aspects of the computer system. In Part III of this book we show that collections of separable models evaluated together (typically iteratively) provide just such tools. Thus, separable models not only are adequate simple models of computer systems, but also are the basic building blocks out of which more detailed models can be constructed.

6.6. Summary

In this chapter we have examined the construction and evaluation of single class, separable queueing network models. Separable models have the following desirable characteristics:

- *efficiency of evaluation* — Performance projections can be obtained from separable models with very little computation. General networks of queues require so much computation to evaluate that they are not practical tools.

- *accuracy of results* — Separable models provide sufficiently accurate performance projections for the majority of modelling studies. We have described a number of case studies to illustrate this point. For the most part, the inaccuracy inherent in establishing parameter values and in projecting workload growth dominates the inaccuracy inherent in separable models. Thus, there is little motivation to look for more accurate models.
- *direct correspondence with computer systems* — The parameters of separable models (service centers, workload types, workload intensities, and service demands) correspond directly to a high level characterization of a computer system. Thus, it is easy to parameterize these models from measurement data in constructing a baseline model, and it is relatively simple to alter the parameters in an intuitive way to reflect projected changes to the computer system in the model.
- *generality* — In cases where the restrictions required in the construction of separable models exclude an important aspect of a computer system from being represented in an individual separable model, collections of separable models can be used. Thus, separable models are the basic tool that we will use throughout the book as we extend our models to include increasingly detailed aspects of computer systems.

We have studied single class separable models in this chapter because they form a natural bridge between the bounding models of Chapter 5 and the more detailed multiple class models of Chapter 7. Important characteristics of single class models in this regard are:

- *ability to project performance* — Single class models contain sufficient detail that performance estimates, rather than performance bounds, can be projected.
- *simplicity* — Single class models are the simplest models for which this is true: the simplest to define, parameterize, evaluate, and manipulate. In light of this, they are the models of choice in situations where they are sufficiently detailed to answer the performance questions of interest.
- *pedagogic value* — The more detailed multiple class models presented in Chapter 7 are considerably more cumbersome notationally than single class models, but actually are very simple extensions of these models. Thus, an understanding of single class models aids in understanding the definition, parameterization, and use of multiple class models.

In the next chapter we extend our modelling capabilities to accommodate systems containing several distinct workload components, which we represent using multiple class, separable queueing network models.

6.7. References

Single class models originally were viewed in the stochastic setting. Jackson [1963] described networks of exponential queues and showed that their solution was separable. Gordon and Newell [1967] obtained similar results for closed networks, and showed that the state probabilities have a simple solution known as "product form".

Buzen [1973] introduced the first efficient evaluation algorithm for closed models. Reiser and Lavenberg [1980] developed the exact mean value analysis algorithm described here. The fact that $A_k(N) = Q_k(N-1)$ in separable queueing networks was established by Sevcik and Mitrani [1981], and independently by Lavenberg and Reiser [1980]. The approximate MVA algorithm is based on work by Bard [1979] and Schweitzer [1979]. Chandy and Neuse [1982] and others subsequently have developed related approximations.

The case studies of Sections 6.3.1, 6.3.2, and 6.3.3 were carried out by Scherr [1967], Lipsky and Church [1977], and Levy [1979], respectively. Scherr's monograph is the source of Figure 6.2, and Levy's paper the source of Figure 6.5.

Denning and Buzen [1978] discuss the homogeneity assumptions in greater detail than we have presented.

[Bard 1979]
Yonathan Bard. Some Extensions to Multiclass Queueing Network Analysis. In M. Arato, A. Butrimenko, and E. Gelenbe (eds.), *Performance of Computer Systems*. North-Holland, 1979.

[Buzen 1973]
Jeffrey P. Buzen. Computational Algorithms for Closed Queueing Networks with Exponential Servers. *CACM 16*,9 (September 1973), 527-531.

[Chandy & Neuse 1982]
K. Mani Chandy and Doug Neuse. Linearizer: A Heuristic Algorithm for Queueing Network Models of Computing Systems. *CACM 25*,2 (February 1982), 126-133.

[Denning & Buzen 1978]
Peter J. Denning and Jeffrey P. Buzen. The Operational Analysis of Queueing Network Models. *Computing Surveys 10*,3 (September 1978), 225-261.

[Gordon & Newell 1967]
W.J. Gordon and G.F. Newell. Closed Queueing Networks with Exponential Servers. *Operations Research 15* (1967), 244-265.

[Jackson 1963]
   J.R. Jackson. Jobshop-like Queueing Systems. *Management Science* *10* (1963), 131-142.
[Lavenberg & Reiser 1980]
   S.S. Lavenberg and M. Reiser. Stationary State Probabilities of Arrival Instants for Closed Queueing Networks with Multiple Types of Customers. *Journal of Applied Probability* (December 1980).
[Levy 1979]
   Allan I. Levy. Capacity Planning with Queueing Network Models: An IMS Case Study. *Proc. CMG X International Conference* (1979), 227-232.
[Lipsky & Church 1977]
   L. Lipsky and J.D. Church. Applications of a Queueing Network Model for a Computer System. *Computing Surveys* *9,3* (September 1977), 205-222. Copyright © 1977 by the Association for Computing Machinery.
[Reiser & Lavenberg 1980]
   M. Reiser and S.S. Lavenberg. Mean Value Analysis of Closed Multichain Queueing Networks. *JACM 27*,2 (April 1980), 313-322.
[Scherr 1967]
   Allan L. Scherr. *An Analysis of Time-Shared Computer Systems.* Research Monograph No. 36, MIT Press, 1967. Copyright © 1967 by the Massachusetts Institute of Technology.
[Schweitzer 1979]
   P. Schweitzer. Approximate Analysis of Multiclass Closed Networks of Queues. *Proc. International Conference on Stochastic Control and Optimization* (1979).
[Sevcik & Mitrani 1981]
   K.C. Sevcik and I. Mitrani. The Distribution of Queueing Network States at Input and Output Instants. *JACM 28*,2 (April 1981), 358-371.

6.8. Exercises

1. Suppose we wish to plot response time estimates obtained from a separable single class queueing network model for all populations from 50 to 75 online users:
   a. If the exact solution technique were used, how many applications of the algorithm would be required to compute performance measures for all 26 populations?

6.8. Exercises

b. Using the approximate solution technique, how many applications of the algorithm would be required?

Suppose that users of this system overlapped the preparation of each request with the processing of the previous request, so that effective think time varied with system response time, and thus with the user population. (For instance, average think time might be 10 seconds with 50 active users, and 8 seconds with 65 active users.)

c. Under this assumption how many applications of each algorithm would be required?

d. Why would it be incorrect simply to modify Algorithm 6.2 (the exact solution technique) so that the think time, Z, was a function of the user population?

2. Exercise 4 in Chapter 5 asked you to graph asymptotic and balanced system bounds for a simple model in two cases: batch and terminal workloads. Use Algorithm 6.2 to compute throughput and response time for these cases for values of N from 1 to 5. Use Algorithm 6.3 for $N=5$ and $N=10$. Compare these results with the bounds obtained previously.

   a. How much additional effort was required to parameterize the single class model in comparison with the bounding models?

   b. How do the techniques compare in terms of computational effort?

   c. How do the results of the techniques differ in terms of their usefulness for projecting performance? In terms of your confidence in the information that they provide?

3. Implement Algorithm 6.3, the approximate mean value analysis solution technique. Repeat Exercise 2 twice: once using this implementation, and once using the Fortran implementation of Algorithm 6.2 (exact mean value analysis) contained in Chapter 18. Compare the results.

4. Modify the program given in Chapter 18 to allow delay centers, and to allow classes of transaction type.

5. Use the modified program, as follows:

   a. Evaluate a model with three centers with service demands of 8, 5, and 4 seconds, and a transaction class with arrival rate .1 requests/second.

   b. Using the response time obtained in (a), calculate an appropriate think time for use in an equivalent model with the transaction class replaced by a terminal class with 10 users.

c. Evaluate the model constructed in (b).

d. Explain the differences between the performance measures obtained in (a) and (c).

6. Use the arrival instant theorem to show that in a balanced model (one in which the service demands at all centers are equal to $D_k = D/K$), system throughput is given by:

$$X = \frac{N}{N+K-1} \times \frac{1}{D_k}$$

(This result is the basis of balanced system bounds, as presented in Chapter 5.)

7. Both the exact and the approximate MVA algorithms involve four key equations (6.1 through 6.4).

a. For each of these four equations, provide an intuitive justification in a few words.

b. In a few sentences, describe how the exact MVA algorithm is constructed from these four components.

c. In a few sentences, describe how the approximate MVA algorithm is obtained from the exact algorithm.

Chapter 7

Models with Multiple Job Classes

7.1. Introduction

Multiple class models, like single class models, provide estimates for performance measures such as utilization, throughput, and response time. The advantages of multiple class models over single class models include:

- Outputs are given in terms of the individual customer classes. For example, in modelling a transaction processing system, response times for each of a number of transaction types could be obtained by including each type as a separate class. With a single class model, only a single estimate for response time representing the average over all transaction types could be obtained.
- For systems in which the jobs being modelled have significantly different behaviors, such as systems with a mixture of CPU and I/O bound jobs, a multiple class model can provide more accurate results. This means that some systems can be modelled adequately only by multiple class models, since the single class assumption that jobs are indistinguishable is unacceptable.

The disadvantages of multiple class models relative to single class models include:

- Since there are multiple customer classes in the model, multiple sets of input parameters (one set per class) are required. The data gathering portion of the modelling process therefore is more tedious.
- Most current measurement tools do not provide sufficient information to determine the input parameters appropriate to each customer class with the same accuracy as can be done for single class models. This not only complicates the process of parameterization, but also means that the potentially greater accuracy of a multiple class model can be offset by inaccurate inputs.
- Multiple class solution techniques are somewhat more difficult to implement, and require more machine resources, than single class techniques.

128   **General Analytic Techniques**: Models with Multiple Job Classes

For the most part, these disadvantages result from inadequate modelling support software, and thus should become less significant as queueing network modelling becomes more widespread. The first two disadvantages can be eliminated by measurement tools that are designed with knowledge of the information required to establish a model. The third disadvantage is significant only if one is developing queueing network modelling software. Commercially available software packages are capable of evaluating multiple class models. Thus, once the model inputs have been obtained, it is no more difficult to deal with a multiple class model than with a single class model.

7.2. Workload Representation

As illustrated in Chapter 4, the inputs of multiple class models largely correspond to those of single class models. The major additional consideration is the specification of scheduling disciplines. Since customers in single class models are indistinguishable, the scheduling disciplines at the various service centers are characterized entirely as being either delay or queueing. However, in multiple class models, customers are distinguishable, and so the choice of scheduling discipline can be important.

There are a large number of scheduling disciplines that can be represented in (separable) multiple class queueing network models. For practical purposes, however, the following disciplines have proven to be sufficient:

- *first-come-first-served (FCFS)* — Under FCFS scheduling, customers are served in the order in which they arrive. Although this is the simplest of scheduling disciplines to implement, it is difficult to model analytically. To do so, it is necessary to impose the restriction that all customer classes have the same service requirement at each visit to the service center in question ($S_{c,k}$). It is possible, however, for different customer classes to require different total numbers of visits to the service center ($V_{c,k}$), thus providing for distinct service demands there ($D_{c,k}$). A FCFS center might be appropriate to represent a disk containing user files for a number of classes. Since the basic operations performed at the device by the various classes are the same, it is reasonable to assume that the average service times across classes are nearly equal. The actual number of file accesses for a customer of each class can be represented in the model by appropriate values of the $V_{c,k}$ for each class c.

- *processor sharing (PS)* — Processor sharing is an idealization of round robin (RR) scheduling. Under RR, control of the processor circulates among all jobs in the queue. Each job receives a *quantum* of service before it must relinquish control to the next job in the queue, rejoining the queue at its tail. Under PS, the length of the quantum is effectively zero, so that control of the processor circulates infinitely rapidly among all jobs. The effect is that jobs are served simultaneously, but each of the n jobs in service receives only $1/n$-th of the full power of the processor. For example, each of three jobs at a processor shared, 3 MIPS (million instructions per second) CPU would receive service at a rate of 1 MIPS. PS often is appropriate to model CPU scheduling in systems where some form of RR scheduling actually is employed.
- *last-come-first-served preemptive-resume (LCFS)* — Under this discipline an arriving job preempts the job in service (if any) and immediately begins service itself. When a job completion occurs, the most recently preempted job resumes service at the point at which it was interrupted. LCFS might be used to model a CPU in a system where the frequency with which high priority system tasks are dispatched is high enough that LCFS is a reasonable approximation.
- *delay* — As in single class models, multiple class delay centers are used to represent devices at which residence time consists entirely of service (there is no queueing delay).

Although the first three disciplines seem quite different, the performance measures obtained from a model will be the same regardless of which is used. In most cases, we therefore distinguish only between *queueing* and *delay* disciplines, without being more specific.

7.3. Case Studies

In this section we present three simple case studies where multiple class separable queueing network models were used to obtain performance projections. The first examines the difference in performance projections provided by single and multiple class models. The second illustrates the principal advantage of multiple class models over single class models, namely the ability to specify inputs and outputs in terms of individual classes. The third demonstrates the successful use of a multiple class model to evaluate a loosely-coupled multiprocessor system.

7.3.1. Contrast with Single Class Models

In this case study we will construct single class and multiple class models of a hypothetical system, and will use these models to project the effects on response times of a CPU upgrade. Our purpose is to illustrate the qualitative differences between the projections that can be obtained from single and multiple class models.

The hypothetical system has two resources, a CPU and a disk. There are two workload components, one batch and the other interactive. Measurements provide the following information (times are in seconds):

$$B_{batch,CPU} = 600 \qquad B_{interactive,CPU} = 47.6$$
$$B_{batch,Disk} = 54 \qquad B_{interactive,Disk} = 428.4$$
$$C_{batch} = 600 \qquad C_{interactive} = 476$$
$$N_{batch} = 10 \qquad N_{interactive} = 25$$
$$Z_{batch} = 0 \qquad Z_{interactive} = 30$$

To construct a single class model of this system, we define a single "average" customer class, in essence by imagining that the measurement data did not distinguish on the basis of workload type. Our model will have two service centers (*CPU* and *Disk*) and a single, terminal class. This class will have 35 customers with think times of 13.271 seconds ($\frac{476}{600+476} \times 30$). Service demands will be .602 seconds at the CPU ($\frac{600+47.6}{1076}$) and .448 seconds at the disk ($\frac{54+428.4}{1076}$).

The multiple class model will have two service centers and two classes: a batch class of 10 customers, and a terminal class of 25 customers with think times of 30 seconds. Batch service demands will be 1.0 and .09 seconds at the CPU and disk, respectively. Interactive service demands will be .10 and .90 seconds at the CPU and disk, respectively.

Table 7.1 shows the outputs for the single class and multiple class models, for the base system and for an upgraded system in which the CPU speed is increased by a factor of five. The single and multiple class models agree well for the base system. They differ considerably for the system with the CPU upgrade, however, even when the projections of the single class model are compared to the "overall" projections of the multiple class model. For example, the multiple class model shows an overall throughput of 5.26 for the system with the upgraded CPU, compared with 2.11 for the single class model. Further, while the single class model projects a 60% improvement in average response time, the multiple class model projects an 80% improvement for batch jobs, but a 200% *degradation* for interactive users.

These differences can be accounted for by the nature of the workload. In the single class model, each "average" job requires a significant

7.3. Case Studies

	single class overall	
	base	upgrade
X	1.64	2.11
R	8.07	3.32
U_{CPU}	.985	.254
Q_{CPU}	10.70	.34
U_{Disk}	.733	.946
Q_{Disk}	2.58	6.63

	multiple class					
	overall		batch		interactive	
	base	upgrade	base	upgrade	base	upgrade
X	1.66	5.26	.93	4.64	.74	.62
R	7.52	3.16	10.79	2.16	3.40	10.57
U_{CPU}	1.000	.943	.926	.928	.074	.015
Q_{CPU}	10.57	5.28	9.72	5.20	.85	.08
U_{Disk}	.752	.979	.084	.418	.668	.561
Q_{Disk}	2.37	11.02	.28	4.80	2.09	6.22

Table 7.1 — Single and Multiple Class Results

amount of disk processing, and so the speedup of the CPU has a limited effect due to the performance constraint imposed by this secondary bottleneck. In the multiple class model, the batch class is heavily CPU bound, while the interactive class is heavily I/O bound. Thus, increasing the speed of the CPU greatly increases the batch throughput but is of little direct benefit to the interactive class. Further, because of the increased batch throughput, the interactive class suffers increased competition from the batch class at the disk center, and thus experiences a performance degradation.

In summary, this example illustrates two important points regarding the use of queueing network models:

- A model can project effects that intuition might not recognize. In this case, we have the counter-intuitive result that performance can degrade with a CPU upgrade.

- Single class models of systems with significantly heterogeneous workloads may give misleading results, both because the performance projections for the "average" job may be inaccurate, and because it is not possible to obtain projections for specific classes from the average results.

7.3.2. Modelling Workload Growth

The system studied here was a Digital Equipment Corporation PDP-10 running a special-purpose software package layered on the TOPS-10 operating system. The objective of the study was to project response times as the number of online users increased and as the number of users that simultaneously could be memory resident was altered. Although benchmarking using a remote terminal emulator (RTE) was possible, a queueing network modelling approach was chosen. This decision was motivated by the fact that projections for a large number of system configurations were required, and timely results with rough (say, 30%) accuracy were more desirable than the more accurate but considerably more time consuming results possible using benchmarking.

The system workload was divided into three components, primarily on the basis of similarity of resource usage. The first component consisted of users running jobs, the second of users executing system utility functions (such as printing or plotting), and the third of users editing. All classes were represented as terminal workloads. Service demands for these three classes were obtained by monitoring an RTE experiment involving a representative (although synthetic) jobstream. This base model was validated by comparing model outputs with measurements taken during the RTE experiment. Agreement was good, so the study proceeded to the projection phase.

	class	model		actual	
		R	U_{CPU}	R	U_{CPU}
Benchmark 1 **50 users**	running jobs utility editing total	9.97 122.8 63.4	 63.4	10.91 99.27 77.8	 77.8
Benchmark 2 **70 users**	running jobs utility editing total	9.2 63.6 1.83	 97.5	11.7 70.3 2.03	 100.0

Table 7.2 — Performance Projections

To assess the impact of workload growth on response times, the workload intensities of the three classes were increased to reflect various larger user populations. The model then was evaluated to obtain performance projections. For several specific user populations, additional RTE experiments were conducted to assess the accuracy of the model. Table 7.2 compares the model results with those obtained during RTE experiments for two user populations. The accuracy is reasonably good, despite the

7.3. Case Studies

extremely simple model used. (Response time improves as the user population increases because of an increase in main memory size that was represented in the model and implemented in the actual configuration. This additional memory resulted in reduced swapping. Techniques for modelling swapping are presented in Chapter 9.)

7.3.3. A Multiprocessing System

The configuration under consideration consisted of two Cyber 173 systems with private memories and disk subsystems, plus a set of shared disks supporting a Federated File System (FFS). The Cyber systems were used both to process local workloads and to process FFS requests from remote sites. The purpose of the study was to assess the impact of an expected growth in the batch components of the systems' workloads. Figure 7.1 shows the model that was employed.

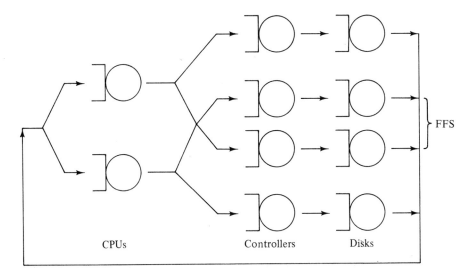

Figure 7.1 − The Multiprocessing System Model

Measurements obtained from software monitors were used to parameterize the model. Service demands were calculated for five workload components: system A interactive, system A batch, system B interactive, system B batch, and FFS accesses by remote systems. All workload components initially were represented using transaction classes, with the FFS arrivals split evenly between systems A and B. An attempt at validating this model showed reasonably accurate throughputs and utilizations, but poor estimates for queue lengths and response times. It was observed that the model projected that on average thirteen jobs would be active simultaneously in each system. However, it was known that system limitations permitted a maximum of five memory resident jobs. Because of this, the batch and interactive workload components of each system were converted to classes of batch type in the model, with workload intensities corresponding to the measured multiprogramming levels. This change resulted in nearly identical throughputs and utilizations and improved device residence time estimates, and so was adopted as the "validated" model. (This study points out the possible danger in using simple models with transaction classes when studying systems that have memory constraints. A more satisfactory model for memory constrained systems is presented in Chapter 9.)

The increase in the batch workloads was represented by increasing the workload intensities of the corresponding classes in the model, with all other parameters remaining unchanged. These were adjusted so that the estimated model throughput of batch jobs matched the anticipated offered workload. Response time estimates from this model were obtained as indications of the ability of the systems to handle the increased workload. It was projected that the systems would be able to handle the maximum expected batch volumes and still provide adequate interactive and FFS response times.

7.4. Solution Techniques

The solution techniques for multiple class models yield values for performance measures such as utilization, throughput, response time, and queue length, for each individual customer class. These techniques are natural extensions of the single class solution techniques. As in the single class case, the details of the solution technique depend on the types of the workloads (open or closed). This dictates the organization of our discussion.

7.4.1. Open Model Solution Technique

Let C be the number of classes in the model. Each class c is an open class with arrival rate λ_c. We denote the vector of arrival rates by $\vec{\lambda} \equiv (\lambda_1, \lambda_2, \ldots, \lambda_C)$. Because the throughputs of the classes in open models are part of the input specification, the solution technique for these models is quite simple. We list below the formulae to calculate performance measures of interest.

- *processing capacity*

    A system is said to have sufficient capacity to process a given offered load $\vec{\lambda}$ if it is capable of doing so when subjected to the workload over

7.4. Solution Techniques

a long period of time. For multiple class models, sufficient capacity exists if the following inequality is satisfied:

$$\max_k \left\{ \sum_{c=1}^{C} \lambda_c D_{c,k} \right\} < 1$$

This simply ensures that no service center is saturated as a result of the combined loads of all the classes. In the derivations that follow, we will assume that this inequality is satisfied.

- *throughput*

  By the forced flow law the throughput of class c at center k as a function of $\vec{\lambda}$ is:

  $$X_{c,k}(\vec{\lambda}) = \lambda_c V_{c,k}$$

- *utilization*

  From the utilization law:

  $$U_{c,k}(\vec{\lambda}) = X_{c,k}(\vec{\lambda}) S_{c,k} = \lambda_c D_{c,k}$$

- *residence time*

  As with single class models, residence time is given by:

  $$R_{c,k}(\vec{\lambda}) = \begin{cases} D_{c,k} & \text{(delay centers)} \\ D_{c,k}\left[1 + A_{c,k}(\vec{\lambda})\right] & \text{(queueing centers)} \end{cases}$$

where $A_{c,k}(\vec{\lambda})$ is the average number of customers seen at center k by an arriving class c customer. The intuition behind this formula is similar to that for single class models. For delay centers, a job's residence time consists entirely of its service demand there, $V_{c,k} S_{c,k}$. The explanation of the formula for queueing centers depends on the scheduling discipline used. For FCFS centers, the residence time is simply the sum of an arriving job's own service time, $V_{c,k} S_{c,k}$, and the service times of the jobs already present at the arrival instant, $V_{c,k}\left[A_{c,k}(\vec{\lambda}) S_{c,k}\right]$, since at FCFS centers all classes must have the same service time at each visit. For PS centers, the residence time is the basic service requirement, $V_{c,k} S_{c,k}$, "inflated" by a factor representing the service rate degradation due to other jobs competing in the same queue, $1 + A_{c,k}(\vec{\lambda})$. For LCFS centers the equation has no simple intuitive explanation, but nonetheless is valid.

An implication of the assumptions made in constructing separable networks is that the queue length seen on average by an arriving customer must be equal to the time averaged queue length. Thus, for queueing centers:

$$R_{c,k}(\vec{\lambda}) = D_{c,k}\left[1 + Q_k(\vec{\lambda})\right]$$

where $Q_k(\vec{\lambda})$ is the time averaged queue length at center k (the sum over all classes). Applying Little's law:

$$R_{c,k}(\vec{\lambda}) = D_{c,k}\left[1 + \sum_{j=1}^{C} \lambda_j R_{j,k}(\vec{\lambda})\right]$$

Notice now that the right hand side of the above equation depends on the particular class c only for the basic service demand $D_{c,k}$. Thus, $\dfrac{R_{c,k}(\vec{\lambda})}{R_{j,k}(\vec{\lambda})}$ must equal $\dfrac{D_{c,k}}{D_{j,k}}$, giving $R_{j,k}(\vec{\lambda}) = \dfrac{D_{j,k}}{D_{c,k}} R_{c,k}(\vec{\lambda})$. Substituting into the equation above and re-writing, we have:

$$R_{c,k}(\vec{\lambda}) = \frac{D_{c,k}}{1 - \sum_{j=1}^{C} U_{j,k}(\vec{\lambda})} \quad \text{(queueing centers)}$$

- *queue length*

  Applying Little's law to the residence time equation above, the queue length of class c at center k, $Q_{c,k}(\vec{\lambda})$, is:

  $$Q_{c,k}(\vec{\lambda}) = \lambda_c R_{c,k}(\vec{\lambda})$$

  $$= \begin{cases} U_{c,k}(\vec{\lambda}) & \text{(delay centers)} \\ \dfrac{U_{c,k}(\vec{\lambda})}{1 - \sum_{j=1}^{C} U_{j,k}(\vec{\lambda})} & \text{(queueing centers)} \end{cases}$$

- *system response time*

  The response time for a class c customer, $R_c(\vec{\lambda})$, is the sum of its residence times at all devices:

  $$R_c(\vec{\lambda}) = \sum_{k=1}^{K} R_{c,k}(\vec{\lambda})$$

7.4. Solution Techniques

- *average number in system*

    The average number of class c customers in system can be calculated using Little's law, or by summing the class c queue lengths at all centers:

$$Q_c(\vec{\lambda}) = \lambda_c R_c(\vec{\lambda}) = \sum_{k=1}^{K} Q_{c,k}(\vec{\lambda})$$

These formulae are summarized as Algorithm 7.1.

---

processing capacity : $\max_k \left\{ \sum_{c=1}^{C} \lambda_c D_{c,k} \right\} < 1$

throughput : $X_c(\vec{\lambda}) = \lambda_c$

utilization : $U_{c,k}(\vec{\lambda}) = \lambda_c D_{c,k}$

residence time : $R_{c,k}(\vec{\lambda}) = \begin{cases} D_{c,k} & \text{(delay)} \\ \dfrac{D_{c,k}}{1 - \sum_{j=1}^{C} U_{j,k}(\vec{\lambda})} & \text{(queueing)} \end{cases}$

queue length : $Q_{c,k}(\vec{\lambda}) = \lambda_c R_{c,k}(\vec{\lambda})$

$= \begin{cases} U_{c,k}(\vec{\lambda}) & \text{(delay)} \\ \dfrac{U_{c,k}(\vec{\lambda})}{1 - \sum_{j=1}^{C} U_{j,k}(\vec{\lambda})} & \text{(queueing)} \end{cases}$

system response time : $R_c(\vec{\lambda}) = \sum_{k=1}^{K} R_{c,k}(\vec{\lambda})$

average number in system : $Q_c(\vec{\lambda}) = \lambda_c R_c(\vec{\lambda}) = \sum_{k=1}^{K} Q_{c,k}(\vec{\lambda})$

Algorithm 7.1 − Open Model Solution Technique

Open Model Example

Figure 7.2 shows a simple open model with two customer classes and two service centers, and illustrates the calculation of various performance measures. (All times are in seconds.)

Model Inputs:

$$V_{A,CPU} = 10 \quad V_{A,Disk} = 9 \quad V_{B,CPU} = 5 \quad V_{B,Disk} = 4$$
$$S_{A,CPU} = 1/10 \quad S_{A,Disk} = 1/3 \quad S_{B,CPU} = 2/5 \quad S_{B,Disk} = 1$$
$$D_{A,CPU} = 1 \quad D_{A,Disk} = 3 \quad D_{B,CPU} = 2 \quad D_{B,Disk} = 4$$
$$\lambda_A = 3/19 \text{ jobs/sec.} \quad\quad \lambda_B = 2/19 \text{ jobs/sec.}$$

Model Structure:

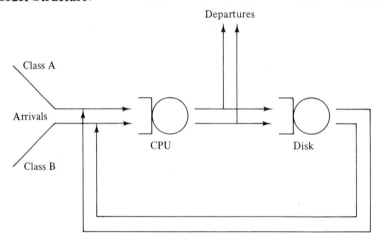

Selected Model Outputs:

$$X_{A,CPU}(\vec{\lambda}) = \lambda_A V_{A,CPU} = \frac{3}{19} \times 10 = 1.58 \text{ jobs/sec.}$$

$$U_{A,CPU}(\vec{\lambda}) = \lambda_A D_{A,CPU} = \frac{3}{19} \times 1 = .158$$

$$R_{A,CPU}(\vec{\lambda}) = \frac{D_{A,CPU}}{1 - \sum_{j=A}^{B} U_{j,CPU}(\vec{\lambda})} = \frac{1}{12/19} = 1.58 \text{ secs.}$$

$$Q_{A,CPU}(\vec{\lambda}) = \frac{U_{A,CPU}(\vec{\lambda})}{1 - \sum_{j=A}^{B} U_{j,CPU}(\vec{\lambda})} = \frac{3/19}{1 - (\frac{3}{19} + \frac{4}{19})} = .25 \text{ jobs}$$

$$R_A(\vec{\lambda}) = R_{A,CPU}(\vec{\lambda}) + R_{A,Disk}(\vec{\lambda}) = \frac{19}{12} + \frac{57}{2} = 30.08 \text{ secs.}$$

Figure 7.2 − Open Model Example

7.4.2. Closed Model Solution Techniques

A closed, multiple class model consists of C classes, each of which has a fixed population. We denote the workload intensity by $\vec{N} \equiv (N_1, \ldots, N_C)$, where N_c is the class c population size. Because the throughputs of closed classes are not provided as inputs, obtaining solutions for closed models is somewhat more complicated than for open models. The solution technique used is an extension of the single class mean value analysis (MVA) algorithm. Like its single class counterpart, multiple class MVA relies on three key equations:

- For each class, Little's law applied to the queueing network as a whole

$$X_c(\vec{N}) = \frac{N_c}{Z_c + \sum_{k=1}^{K} R_{c,k}(\vec{N})} \qquad (7.1)$$

- For each class, Little's law applied to the service centers individually

$$Q_{c,k}(\vec{N}) = X_c(\vec{N}) \, R_{c,k}(\vec{N}) \qquad (7.2)$$

It also is useful to consider the total queue length at center k:

$$Q_k(\vec{N}) = \sum_{c=1}^{C} Q_{c,k}(\vec{N})$$

- For each class, the service center residence time equations

$$R_{c,k}(\vec{N}) = \begin{cases} D_{c,k} & \text{(delay centers)} \\ D_{c,k}\left[1 + A_{c,k}(\vec{N})\right] & \text{(queueing centers)} \end{cases} \qquad (7.3)$$

where $A_{c,k}(\vec{N})$ is the arrival instant queue length at center k seen by an arriving class c customer.

We note that performance measures can be computed using the above equations once the $A_{c,k}(\vec{N})$ are known.

As with single class models, there are two approaches to the evaluation of closed models, exact and approximate. (We emphasize again that the word "exact" refers to how the solution relates to the model, not how the solution of the model relates to the system being modelled.) As with the single class MVA algorithms, the two methods differ in how the arrival instant queue lengths are computed.

7.4.2.1. Exact Solution Technique

To obtain an exact solution of a closed model, one must compute the values of the $A_{c,k}(\vec{N})$ exactly. Given these values, equations (7.1)-(7.3) can be applied to compute the full solution of the model. The key to the exact MVA solution technique is the multiple class generalization of the relationship used in the single class case:

$$A_{c,k}(\vec{N}) = Q_k(\overrightarrow{N-1_c}) \qquad (7.4)$$

where $\overrightarrow{N-1_c}$ is population $\vec{N}$ with one class c customer removed. Intuitively, the queue length seen upon arrival to a center is equal to the time averaged queue length at the center with the arriving customer removed from the network.

Beginning from the trivial solution of the network with the empty population $\vec{0}$ ($Q_k(\vec{0}) = 0$ for all centers k), equation (7.4) can be used, along with equations (7.1)-(7.3), to construct iteratively the solutions for increasing populations, culminating in performance measures for the population of interest, $\vec{N}$. Note that in general the solution for each population $\vec{n}$ requires as input C solutions, one for each population $\vec{n-1_c}$, $c = 1, ..., C$. Figure 7.3 illustrates this by showing the precedence relations of the solutions required to evaluate a network with 3 class A customers and 2 class B customers: the solution of the empty network is required to compute solutions with populations consisting of a single customer, (1A,0B) and (0A,1B), which then can be used to compute solutions for populations with two customers, etc. As a result of these complex dependencies, the time and space requirements of the multiple class algorithm are significantly greater than those of the single class algorithm. They are proportional to:

time: $\quad CK \prod_{c=1}^{C} (N_c+1) \quad$ arithmetic operations

space: $\quad K \prod_{\substack{c=1 \\ c \neq c_{max}}}^{C} (N_c+1) \quad$ storage locations

where c_{max} is the index of the class with the largest population. A significant implication of these time and space requirements is that it can be impractical to compute the exact solution of networks with more than a few customer classes. For example, the solution of a network with 10 centers and 5 classes of 10 customers each requires more than 8,000,000 arithmetic operations and 145,000 storage locations. (In contrast, a single class model with 10 centers and 50 customers requires roughly 500 arithmetic operations and 10 storage locations.) This is the motivation for the approximate solution technique to be described in the next section.

7.4. Solution Techniques

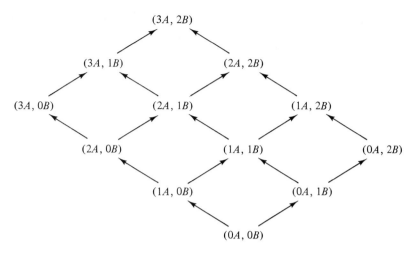

Figure 7.3 — Precedence of Intermediate Solutions

---

for $k \leftarrow 1$ to K do $Q_k(\vec{0}) \leftarrow 0$
for $n \leftarrow 1$ to $\sum_{c=1}^{C} N_c$ do
    for each feasible population $\vec{n} \equiv (n_1, \ldots, n_C)$ with n total customers do
    begin
        for $c \leftarrow 1$ to C do
            for $k \leftarrow 1$ to K do
$$R_{c,k} \leftarrow \begin{cases} D_{c,k} & \text{(delay)} \\ D_{c,k}\left[1 + Q_k(\overrightarrow{n-1_c})\right] & \text{(queueing)} \end{cases}$$
        for $c \leftarrow 1$ to C do $X_c \leftarrow \dfrac{n_c}{Z_c + \sum_{k=1}^{K} R_{c,k}}$
        for $k \leftarrow 1$ to K do $Q_k(\vec{n}) \leftarrow \sum_{c=1}^{C} X_c R_{c,k}$
    end

Algorithm 7.2 — Exact MVA Solution Technique (Closed Models)

142   General Analytic Techniques: Models with Multiple Job Classes

The exact MVA solution technique appears as Algorithm 7.2. When this algorithm terminates, the values of $R_{c,k}$, X_c, and Q_k (all for population $\vec{N}$) are available immediately. Other model outputs are obtained by using Little's law. Here is a summary:

class c system throughput:	X_c
class c system response time:	$N_c/X_c - Z_c$
average number of class c in system:	$N_c - X_c Z_c$
class c throughput at device k:	$X_c V_{c,k}$
class c utilization of device k:	$X_c D_{c,k}$
class c queue length at device k:	$X_c R_{c,k}$
class c residence time at device k:	$R_{c,k}$

Closed Model Example (Exact Solution)

Table 7.3 shows the computation required by the MVA solution of a closed model corresponding to the open model of Figure 7.2. The open classes have been replaced by batch classes, each with one customer. Other parameter values are the same.

	population vectors			
	(0A,0B)	(1A,0B)	(0A,1B)	(1A,1B)
$R_{A,CPU}$	-	1	-	4/3
$R_{A,Disk}$	-	3	-	5
$R_{B,CPU}$	-	-	2	5/2
$R_{B,Disk}$	-	-	4	7
X_A	-	1/4	-	3/19
X_B	-	-	1/6	2/19
$Q_{A,CPU}$	0	1/4	-	4/19
$Q_{A,Disk}$	0	3/4	-	15/19
$Q_{B,CPU}$	0	-	1/3	5/19
$Q_{B,Disk}$	0	-	2/3	14/19

Table 7.3 — Exact MVA Computation

7.4.2.2. Approximate Solution Technique

Because the exact solution technique can require excessive time and space for large numbers of classes, the approximate solution technique often is the only one that can be used in practice. Moreover, since the approximate technique is quite accurate, it is useful as a general technique, even for networks that could be solved exactly.

7.4. Solution Techniques

The multiple class approximate solution technique is a straightforward extension of the single class approximation. Equations (7.1)-(7.3) are employed, but the arrival instant queue lengths are estimated iteratively. The estimates are obtained based on the time averaged queue lengths at the service centers with the full customer population. Thus, the approximate solution technique does not require that one first compute solutions for all populations between the zero population and the full population, but instead iterates at the full population. An initial guess for time averaged queue lengths is made to start the iteration. The approximating function is applied to this guess, and the resulting approximate arrival instant queue lengths are used in equation (7.3). Applications of equations (7.2) and (7.1) result in new estimates for time averaged queue lengths, which then can be used to begin the next step of the iteration. The iteration continues until successive estimates of time averaged queue lengths are sufficiently close. The approximate solution technique is summarized as Algorithm 7.3.

---

1. Set $Q_{c,k}(\vec{N}) \leftarrow \dfrac{N_c}{K}$ for all c,k.
2. Approximate $A_{c,k}(\vec{N})$ by $h_c\left[Q_{1,k}(\vec{N}),\ldots,Q_{C,k}(\vec{N})\right]$, for all c,k. (The choice of h_c is discussed in the text.)
3. Apply equations (7.1)-(7.3) to compute a new set of $Q_{c,k}(\vec{N})$ for all c,k.
4. If the $Q_{c,k}(\vec{N})$ resulting from Step 3 do not agree to within some tolerance (e.g., 0.1%) with those used as inputs in Step 2, return to Step 2 using the new $Q_{c,k}(\vec{N})$.

Algorithm 7.3 — Approximate MVA Technique (Closed Models)

The significant advantage of this method over the exact technique is that it iterates on solutions of the network with the full customer population $\vec{N}$, rather than building up from the solution for the empty network. The approximation therefore requires much less storage than the exact technique, since it maintains the solution of the network for only one population ($\vec{N}$). In particular, the storage requirement is proportional to the product of C and K. The savings in time are harder to quantify because of the iterative nature of the approximate algorithm, although empirically these savings are considerable. The number of operations required per iteration is proportional to the product of C and K. (In other words, the populations of the classes are not a consideration.) Less than two dozen iterations typically are required for convergence to less than a 0.1% change in queue lengths. The accuracy of the technique typically is within a few percent of the exact solution for throughputs and utilizations, and within 10% for queue lengths and residence times.

As noted, the approximate solution technique is built upon estimates for the arrival instant queue lengths at each device for each class that depend only on information obtained from the solution of the network with the full population. A particular estimate for the function h_c that has been used successfully is:

$$\begin{aligned} A_{c,k}(\vec{N}) &= Q_k(\overrightarrow{N-1_c}) \\ &\approx h_c\left[Q_{1,k}(\vec{N}),\ldots,Q_{C,k}(\vec{N})\right] \\ &\equiv \left[\frac{N_c-1}{N_c}Q_{c,k}(\vec{N})\right] + \sum_{\substack{j=1 \\ j \neq c}}^{C} Q_{j,k}(\vec{N}) \end{aligned} \qquad (7.5)$$

Comparing equation (7.5) to the exact formula (7.4), it is evident that the assumption made in the approximation is that the removal of a customer from the network does not affect the placement of customers in other classes, and reduces queue lengths in its own class in proportion to their original size. Equation (7.5) has worked well in practice. More sophisticated estimates also have been used, although these are somewhat more difficult to implement and require more machine resources, in terms of both time and space.

An important benefit of the approximate technique is that non-integer multiprogramming levels easily are incorporated in the model. One simply sets N_c to the (non-integer) multiprogramming level and applies the approximation. No interpolation between separate integer solutions is required.

Closed Model Example (Approximate Solution)

Table 7.4 shows the intermediate and final values for the example given in Section 7.4.2.1 (Table 7.3), calculated using the approximate solution technique. The iteration was halted when the maximum change in all queue length estimates was less than .001.

7.4.3. Mixed Model Solution Technique

Mixed queueing network models are those in which some classes are open and some are closed. Such models may be constructed, for instance, to model a mixed batch and transaction processing system. We denote the workload intensity vector of the entire model by $\vec{I} \equiv (N_1 \text{ or } \lambda_1, N_2 \text{ or } \lambda_2, \ldots, N_C \text{ or } \lambda_C)$. Mixed models are evaluated using Algorithm 7.4.

7.4. Solution Techniques

iteration	class	performance measure			
		$Q_{c,CPU}$	$Q_{c,Disk}$	X_c	R_c
0	A	.500	.500		
	B	.500	.500		
1	A	.250	.750	.167	6.000
	B	.333	.667	.111	9.000
2	A	.211	.790	.158	6.333
	B	.263	.737	.105	9.500
3	A	.195	.805	.154	6.474
	B	.253	.747	.104	9.579
4	A	.193	.807	.154	6.495
	B	.249	.751	.104	9.610
5	A	.192	.808	.154	6.508
	B	.248	.752	.104	9.614
exact solution	A	.211	.789	.158	6.333
	B	.263	.737	.105	9.500

Table 7.4 – Approximate MVA Computation

An important aspect of queueing phenomena is illustrated by Step 2 of Algorithm 7.4. In that step, the performance measures of the closed classes of a mixed model are computed by creating a model that consists only of closed classes; the open classes have been eliminated. The effect of the open classes on closed class performance measures is represented by "inflating" the service demands of the closed classes at all devices. The "inflation factor" used is $1 - U_{\{O\},k}$, which is the percentage of time that the processor is not in use by the open classes. In essence, this factor indicates the effective speed of the processor as seen by the closed classes, given that some of its time is allocated to other (in this case open) classes. For example, if a 3 MIPS (million instructions per second) CPU is utilized 33% by transactions constituting an open class in the model, it appears to be a 2 MIPS CPU to the other classes. Dividing all service demands by .67 to create the closed model of Step 2 simply reflects the fact that more processing time is required on the effectively slower processor. This technique of inflating service times, which often is referred to as *load concealment*, will be used repeatedly in later chapters to reduce the complexity of models by eliminating customer classes while still incorporating their effects on performance.

Let $\{O\}$ be the set of open classes and $\{C\}$ the set of closed classes.

1. For each center k, obtain its utilization by each open class:
$$U_{c,k}(\vec{I}) = \lambda_c D_{c,k} \qquad c \in \{O\}$$
and its total utilization by all open classes:
$$U_{\{O\},k}(\vec{I}) = \sum_{c \in \{O\}} \lambda_c D_{c,k}$$
This simply is an application of the forced flow law and the utilization law to each open class.

2. Solve the closed model consisting of the K centers and the closed customer classes (but no open classes). The service demand $D^*_{c,k}$ of each class $c \in \{C\}$ at each center k in the closed model is set to:
$$D^*_{c,k} = \frac{D_{c,k}}{1 - U_{\{O\},k}(\vec{I})} \qquad c \in \{C\}$$
where $D_{c,k}$ is the service demand of class c at center k in the original mixed model. The throughputs, queue lengths, and residence times obtained from the solution of this model are the performance measures for the corresponding closed classes in the mixed model. Utilizations can be computed by applying the utilization law to the original set of service demands $D_{c,k}$.

3. Residence times and queue lengths for the open classes can be computed using the performance measures of the closed classes:
$$R_{c,k}(\vec{I}) = \frac{D_{c,k}\left[1 + Q_{\{C\},k}(\vec{I})\right]}{1 - U_{\{O\},k}(\vec{I})} \qquad c \in \{O\}$$
$$Q_{c,k}(\vec{I}) = \lambda_c R_{c,k}(\vec{I}) \qquad c \in \{O\}$$
where $Q_{\{C\},k}(\vec{I})$ is the total queue length of all closed classes at center k obtained from the solution of the closed model in Step 2.

Algorithm 7.4 — Exact MVA Solution Technique (Mixed Models)

Mixed Model Example

Figure 7.4 shows a mixed model with four classes and two centers. Classes A and B are open, while classes C and D are closed. As shown in the figure, the solution of the model is obtained in three steps corresponding to those of Algorithm 7.4.

7.5. Theoretical Foundations

As with single class models, certain assumptions about the behavior of a model are necessary to the mathematical proof that the solution obtained by the MVA procedure gives the exact performance measures for that model. With only one exception, the assumptions required in the multiple class case are straightforward extensions of those required in the single class case:

- *service center flow balance* — The number of arrivals of each class at each center is equal to the number of completions of that class there.
- *one step behavior* — Only a single customer can move (arrive to or depart from a service center) at a time.
- *routing homogeneity* — Given a more detailed view of customer behavior that includes the routing patterns of customers, routing homogeneity is satisfied if the proportion of time that a customer of class c leaving center j proceeds directly to center k depends only on c, j, and k, and is independent of the number of customers or their classes currently at any of the centers, for all c, j, and k.
- *device homogeneity* — This is the one assumption whose extension from the single class case is less than straightforward. In the single class case, we allowed the rate of completions of jobs from a center to vary in an arbitrary manner with the number of jobs at that center (although the rate could not otherwise be dependent on the number or placement of customers within the network). In the multiple class case, we do not allow completely arbitrary variation in completion rate as a function of population. Specifically, let n be the total number of customers at center k, n_c be the number of class c customers there, and $\mu_{c,k}(n,n_c)$ be the completion rate of class c customers at center k with those queue lengths. Device homogeneity is satisfied when:

$$\mu_{c,k}(n,n_c) = \frac{n_c}{n} \mu_{c,k}(1,1) \, a_k(n)$$

for all c and k, where $a_k(n)$ is a positive constant for fixed k and n. This assumption will be discussed further in Chapter 8.

Model Inputs:

$D_{A,CPU} = 1/4 \quad D_{B,CPU} = 1/2 \quad D_{C,CPU} = 1/2 \quad D_{D,CPU} = 1$
$D_{A,Disk} = 1/6 \quad D_{B,Disk} = 1 \quad D_{C,Disk} = 1 \quad D_{D,Disk} = 4/3$
$\lambda_A = 1 \quad\quad\quad \lambda_B = 1/2 \quad\quad N_C = 1 \quad\quad\quad N_D = 1$

Model Structure:

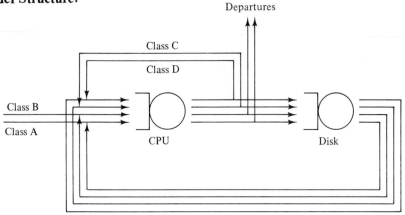

Evaluation:

1. Compute the total utilization of the devices by the open classes:

$$U_{\{O\},CPU}(\overline{I}) = \lambda_A D_{A,CPU} + \lambda_B D_{B,CPU} = .5$$
$$U_{\{O\},Disk}(\overline{I}) = \lambda_A D_{A,Disk} + \lambda_B D_{B,Disk} = .667$$

2. Solve the closed model obtained by deleting the open classes and inflating the service demands of the closed classes:

$$D^*_{C,CPU} = \frac{.5}{1-.5} = 1 \quad\quad D^*_{D,CPU} = \frac{1}{1-.5} = 2$$
$$D^*_{C,Disk} = \frac{1}{1-.667} = 3 \quad\quad D^*_{D,Disk} = \frac{1.333}{1-.667} = 4$$

This model is equivalent to the closed model solved in the example of Section 7.4.2, so the same performance measures will result, e.g., CPU queue lengths are .211 and .789 for classes C and D.

3. Using the queue lengths of the closed classes, compute the performance measures of the open classes. For example:

$$R_{A,CPU} = \frac{.25\,(1+1.0)}{1-.5} = 1.0 \quad\quad R_{B,CPU} = \frac{.5\,(1+1.0)}{1-.5} = 2.0$$

Figure 7.4 — Example Mixed Network

- *homogeneous external arrivals* — The rate of arrival of customers of each class is independent of the number and class of the customers currently in the system or the placement of those customers.

While these assumptions are sufficient for the model to be separable (and thus to be efficiently evaluated), the solution techniques that have been presented so far require one additional restriction:

- *service time homogeneity* — The completion rate of class c customers at center k times the ratio of the total number of customers at k to the number of class c customers at k is constant for all fixed c and k (i.e., when $a_k(n) = 1$ for all k, n).

This last assumption ensures that all service centers are *load independent*, which means that the rate of service is independent of the current state of the queue at the device. Somewhat more complicated models can be constructed using *load dependent* service centers, whose service rates depend on their queue lengths. These will be discussed in Chapter 8.

7.6. Summary

In this chapter we have focused on multiple class, separable queueing network models. We are interested in separable networks because they are reasonably accurate models of computer systems and can be solved efficiently; more general models require excessively large amounts of time and space. Exact solutions of separable models with a few customer classes, and accurate approximate solutions of models with many customer classes, can be obtained with modest machine resources.

The major advantage of multiple class models over single class models is also the main drawback. By identifying distinct workload components, output performance measures for each can be given separately. At the same time, input parameter values are required for each individual class. This typically requires considerable additional effort over that for a single class model, as measurement tools often do not provide sufficient information about resource consumption by classes.

While certain restrictive assumptions are required to construct separable models, it often is the case that separable models accurately project the behavior of complex computer systems despite these restrictions. In cases where aspects of a computer system important to its performance cannot be represented directly, variations on simple separable models must be used. These variations are the subject of Part III.

7.7. References

Almost all work with multiple class models has been conducted in the stochastic setting. The work of Baskett et al. [1975], which describes separable models for open, closed, and mixed workloads, is probably the most referenced paper in the area of queueing network models. Chandy et al. [1977] describe the stochastic properties required for a network to be separable.

The case studies in Sections 7.3.1, 7.3.2, and 7.3.3 were reported by Denning and Buzen [1978], Sanguinetti and Billington [1980], and Lindzey and Browne [1979], respectively.

[Baskett et al. 1975]
Forest Baskett, K. Mani Chandy, Richard R. Muntz, and Fernando G. Palacios. Open, Closed, and Mixed Networks of Queues with Different Classes of Customers. *JACM 22*,2 (April 1975), 248-260.

[Chandy et al. 1977]
K. Mani Chandy, John H. Howard, Jr., and Donald F. Towsley. Product Form and Local Balance in Queueing Networks. *JACM 24*,2 (April 1977), 250-263.

[Denning & Buzen 1978]
Peter J. Denning and Jeffrey P. Buzen. The Operational Analysis of Queueing Network Models. *Computing Surveys 10*,3 (September 1978), 225-261. Copyright © 1978 by the Association for Computing Machinery.

[Lindzey & Browne 1979]
G.E. Lindzey, Jr. and J.C. Browne. Response Analysis of a Multi-Function System. *Proc. ACM SIGMETRICS Conference on Simulation, Measurement and Modeling of Computer Systems* (1979), 19-26. Copyright © 1979 by the Association for Computing Machinery.

[Sanguinetti & Billington 1980]
John Sanguinetti and Richard Billington. A Multi-Class Queueing Network Model of an Interactive System. *Proc. CMG XI International Conference* (1980), 50-55.

7.8. Exercises

1. What is the principal advantage of multiple class models over single class models? The principal disadvantage?

7.8. Exercises

2. Evaluate the open model example of Figure 7.2 by hand with the following independent changes:
   a. Both arrival rates halved.
   b. $D_{A,CPU}$ doubled.
3. Extend the solution of the closed network shown in Table 7.3 to the case of two class A and two class B customers. Check your results against those obtained using the multiple class, exact MVA implementation in Chapter 19.
4. Construct an "equivalent" single class model to the model of Figure 7.2. Compare the performance measures of the single class model to the aggregate measures of the multiple class model.
5. In evaluating a model with a one transaction and one batch class, the solution technique involves the removal of the transaction class and the "service time inflation" of the batch class. This procedure yields an exact solution.

   Investigate the use of service time inflation to remove a batch class from a model. Consider a model with two batch classes and five centers. Class A has service demands 1, 2, 2, 2, 2 at the five centers, while class B has service demands 3, 1, 1, 1, 1.
   a. Use the software in Chapter 19 to obtain solutions to the model with populations (2 A, 2 B), (2 A, 8 B), and (2 A, 16 B).
   b. For each population $\vec{N}$, construct an approximate model with respect to class A by removing the class B customers from the model, and inflating the class A service demand at each center k by $1 - U_{B,k}(\vec{N})$. Compare the results for response time and system throughput with those obtained in (a). How do you derive sensible utilizations for class A from this approximate model?
   c. Give an intuitive explanation for the differences observed using the two class model of (a) and the single class approximation of (b).
6. Implement the approximate MVA solution technique (Algorithm 7.3) for models with two closed (batch or terminal) classes.
7. Argue that $O\left(KC\prod_{c=1}^{C}(N_c+1)\right)$ is the correct expression for the time complexity of Algorithm 7.2.
8. Argue that $O(KC)$ is the correct expression for the time complexity of Algorithm 7.3 (assuming that the number of iterations does not depend upon K or C).

Chapter 8

Flow Equivalence and Hierarchical Modelling

8.1. Introduction

The models studied in previous chapters were simple both in their construction and in the techniques required for their evaluation. Often it is useful to construct more sophisticated models so that additional details of the computer system may be represented. In this chapter we discuss a technique for doing so, *hierarchical modelling*. Hierarchical modelling is the process of partitioning a large model into a number of smaller submodels. Each of these submodels then is evaluated, and the individual solutions are combined to obtain the solution of the original model. The recombination is performed using a special type of service center called a *flow equivalent service center (FESC)*.

Consider the model shown in Figure 8.1, which represents two single-CPU systems with a shared I/O subsystem. In the general case, there is an arbitrarily defined subsystem, called the *aggregate*, which interacts with the other service centers in the network, called collectively the *complement* or *complementary network*. The aggregate itself may or may not be representable as a network of service centers. In the case of this example, the complement represents the CPUs, while the aggregate represents the complex I/O subsystem. A key step in the hierarchical approach is to replace the entire aggregate by a single service center that mimics its behavior, thus reducing the size of the network to be solved.

From the perspective of the service centers in the complement, the aggregate can be thought of as a black box whose behavior is characterized by the residence time there (i.e., the time interval from when a customer enters the aggregate until that customer departs the aggregate) and by the rate and pattern by which customers leave the aggregate to return to the complement (i.e., the departure process of the aggregate). As long as customers experience an appropriate delay at the aggregate, and the departure process of the aggregate is correct, the service centers in the complement are unaffected by the actual construction of the aggregate. Therefore, any representation of the aggregate that results in appropriate inter-departure times is sufficient to obtain the solution of the network

8.1. Introduction

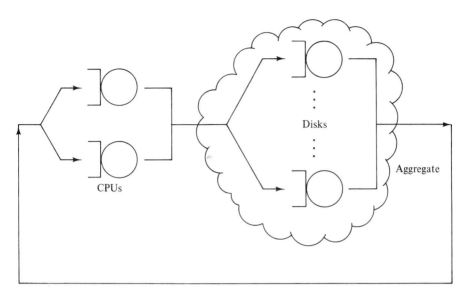

Figure 8.1 − Example Loosely-Coupled Multiprocessor Model

(with respect to the service centers in the complement). In particular, the performance measures obtained for the complementary network will be the same regardless of whether the aggregate is represented as a large number of service centers or as a single service center.

It is this realization that leads to the concept of flow equivalent service centers. An FESC is a single service center that, from the point of view of the complementary network, behaves identically to the aggregate itself. This means that the FESC must (minimally) cause the same average delay to customers passing through it as those customers would experience had they actually proceeded through the detailed representation of the aggregate. (In general, for an FESC to be exact, it must mimic the actual distribution of interdeparture times from the aggregate, not just the average. However, such detailed FESCs are too cumbersome to be of practical use, so we limit ourselves to FESCs that match only average residence time and throughput.) Since the FESC is a single service center, while the detailed representation of the aggregate presumably is much more complex, the use of FESCs is attractive because it leads to much simpler models.

FESCs are the keys to hierarchical modelling. Hierarchical modelling (often called *hierarchical decomposition*) is the process of modelling a system using multiple levels of models. The model at the highest level, level 0, consists of a number of FESCs, each of which represents some portion of the computer system being modelled. The level below that, level 1, consists of a number of models, each a more detailed representation of a subsystem represented in level 0 as an FESC. Each of the level 1 models itself may contain FESCs. In general, the characteristics of the FESCs at level l are determined by solving models at level $l+1$, until finally some level is reached at which all models are fully detailed, i.e., contain no FESCs. Figure 8.2 shows a possible decomposition scheme. (Notationally, FESCs are distinguished by an arrow through the server, suggesting variability.)

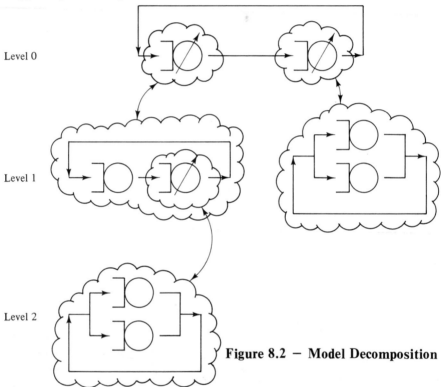

Figure 8.2 — Model Decomposition

Although the definition of the models normally proceeds from level 0 to level L, the evaluation of the models must occur in the opposite direction, i.e., from level L to level 0. Eventually the level 0 model is evaluated, and performance projections for the computer system being modelled are obtained from its solution.

There are two key requirements in hierarchical modelling beyond the original need to define the levels of models. The first is to find a suitable structure for FESCs. Our goal is to create a single service center that can replace an entire subsystem. Thus, we expect this center to be more complicated than the service centers we have seen so far, which represent only single resources. Intimately related to the problem of finding a suitable representation for the level l FESCs is the problem of obtaining

8.2. Creating Flow Equivalent Service Centers

parameter values for them from the submodels at level $l+1$. These issues are considered in Sections 8.2 and 8.3.

The second requirement of the hierarchical modelling process is to evaluate models containing FESCs. As mentioned above, we should expect FESCs to be more complicated than the types of centers we have seen so far. Correspondingly, we should expect the solution techniques required to evaluate models containing them to be more complicated. This issue is addressed in Section 8.4.

8.2. Creating Flow Equivalent Service Centers

In general, it is not possible to find FESCs that produce exact results for the complementary network. However, reasonably accurate approximations can be obtained. Figure 8.3 shows a typical situation in which an FESC might be used. The enclosed subsystem (the aggregate) would be replaced by the FESC.

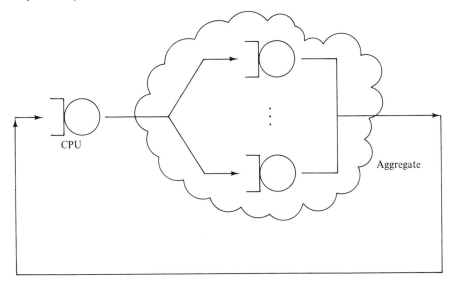

Figure 8.3 — Example Application of an FESC

The purpose of the FESC is to mimic the behavior of the aggregate. This behavior, as viewed by the complementary subnetwork, is the flow of customers out of the aggregate and into the complement. An approximation for this flow can be obtained by making the *decomposability assumption* that the average rate at which customers depart the aggregate depends only on the state of the aggregate, where the state is defined by the customer population within the aggregate. Thus, the state is independent of the placement of the customers at the various service centers. (For example, the state of an aggregate might be (2 class A, 1 class B). The total number of customers of each class is represented, but information about the location of each customer in the aggregate is ignored.) An aggregate therefore can be defined completely by a listing of its throughputs as a function of its possible customer populations.

The assumption that the output rate of the aggregate depends only on the customers in it implies the assumption that the aggregate achieves *local equilibrium* between successive interactions with the complement. Local equilibrium means that the behavior of the aggregate is independent of its starting condition. This situation occurs if, after an arrival to the aggregate, many transitions of customers between service centers in the aggregate occur before another arrival from the complement takes place. Local equilibrium is most likely achieved when the service centers in the aggregate all have service rates that are considerably faster than the service rates of the centers in the complement.

It is desirable that the aggregate achieve local equilibrium because in that case the average departure rate from the aggregate with a given population in it will be nearly the equilibrium throughput, regardless of the initial placement of those customers. This is exactly the assumption made in reducing the aggregate to a single service center whose state is described entirely by the number of customers present. If the aggregate did not achieve equilibrium, its output rate would depend on its initial configuration of customers, and so the single server representation would be deficient.

Flow equivalent service centers are represented in queueing network models using *load dependent service centers*. A load dependent service center can be thought of as a service center whose service rate (the reciprocal of its service time) is a function of the customer population in its queue. For instance, a delay center can be thought of as a load dependent service center that has service rate μ with one customer in the queue, and service rate $n\mu$ with n customers in the queue (in a single class model). In contrast, a queueing service center is *load independent*: it has service rate μ regardless of the number of customers in its queue.

An FESC for an aggregate is a load dependent service center with service rates $\mu_c(\vec{n})$ equal to the throughputs $X_c(\vec{n})$ of the aggregate for all populations $\vec{n}$ and classes c. (We will discuss methods for obtaining these rates in Section 8.3.) Because the FESC mimics the behavior of the aggregate, it can be used to replace the detailed description of the aggregate in the model with little effect on the performance measures obtained.

For single class models, a state of an aggregate is described simply by the number of customers anywhere within it, since customers are indistinguishable. A flow equivalent service center is formed by calculating

8.2. Creating Flow Equivalent Service Centers

throughputs $X(n)$ of the aggregate as a function of the number n of customers in the aggregate. These are used to create a load dependent service center with service rates $\mu(n) = X(n)$.

In the case where the workload is transaction type, a rather subtle problem can occur with the specification of the FESC. For these models, there is no limit to the number n of customers that might exist in the aggregate. Thus, an infinite number of throughput values seem to be required to specify the FESC. While this is the case in theory, in practice the situation is less bleak. Because real computer systems do not experience unbounded numbers of jobs in their queues, only a finite (and usually small) number of rates are required even for transaction type classes. Typically, distinct rates are specified for all n less than some given number n^* (which depends on the computer system being modelled). Rates for all larger n are then assumed to be equal to the rate with n^* customers. FESCs that have rates of this sort are said to have *limited load dependent* behavior. We will see specific applications of limited load dependence in Part III of this book.

In applying FESCs to multiple class models, the state of an aggregate is defined by a vector $\vec{n} \equiv (n_1, \ldots, n_C)$ giving the number of customers of each class present. Thus, the flow equivalent service center corresponding to a specific aggregate is the load dependent service center with output rate for class c, $\mu_c(\vec{n})$, equal to $X_c(\vec{n})$. Since the output rate of the FESC for each class must equal that of the aggregate, the "scheduling discipline" at multiple class FESCs cannot be a traditional one. (For example, if an FESC were scheduled FCFS, only the class currently in service at the FESC would exhibit the proper output rate, since all other classes would have output rates of zero.) Instead, an artificial scheduling discipline, called *composite queueing*, is used so that all classes receive service at once. One can think of the FESC as having C distinct queues, one for each customer class. These queues are served in parallel, with the class c queue being served at rate $\mu_c(\vec{n})$ when the population of the C queues is given by $\vec{n} \equiv (n_1, \ldots, n_C)$.

As with single class models, specifying rates for an FESC in a network that contains transaction type job classes can present problems in theory, because of the apparently unbounded number of rates required. In practice, though, FESCs with limited load dependent behavior are sufficient, and so models with transaction type classes pose no real problems.

A problem associated with multiple class FESCs that does not arise in the single class case is that the number of populations for which throughputs must be determined grows very quickly with the number of classes. In particular, $C\prod_{c=1}^{C}(N_c+1)$ throughputs are required for a network with a (closed) population of N_c class c customers (a throughput for each of the C classes, for each of the $\prod_{c=1}^{C}(N_c+1)$ possible aggregate populations). A network with five classes of ten customers each, for instance, requires nearly one million distinct throughputs. Fortunately, this problem can be dealt with in some cases by choosing an appropriate method for calculating the necessary load dependent throughputs (see Section 8.3).

It is important to keep in mind that while the hierarchical modelling process appears to give an exact representation of the model, in general it is only an approximation. The approximation arises in describing an entire subsystem by a single service center. In doing so, information regarding the placement of customers at the centers of the subsystem is lost, and so the FESC does not have sufficient information to mimic the subsystem exactly. In many situations, however, the resulting inaccuracy is negligible.

8.3. Obtaining the Parameters

The parameters required to specify an FESC are the load dependent service rates for each class as functions of the possible queue populations. As indicated previously, the rates for level l models generally are obtained from the solution of the corresponding level $l+1$ models. However, there are a number of different ways in which a level $l+1$ model can be evaluated:

- *measurements* — In some cases, it may be possible to observe the subsystem that is to be aggregated, and to obtain measurements of its throughput as a function of the number of customers present. For instance, one might measure the throughput of a channel/string pair as a function of the number of outstanding requests to that string. These measured throughputs then could be used directly to set the service rates of an FESC.

- *queueing network models* — The level l FESC might be representable at level $l+1$ as a queueing network consisting of load independent service centers (and possibly some FESCs with service rates set by solutions of lower-level models). This level $l+1$ model can be evaluated analytically, and the throughputs predicted from its solution used to set the service rates of the level l FESC.

- *simulation* — If some aspects of the aggregate make it difficult to evaluate analytically, a simulation of the aggregate can be performed to obtain the required load dependent throughputs.

- *special purpose analytic methods* — Models peculiar to a particular subsystem, such as a complex I/O subsystem, might be developed and solved analytically. The outputs of these models could be load dependent throughputs, which then would be used to define the FESC required in the next higher-level model.

In most cases we advocate the use of queueing network models for establishing the parameters of FESCs, for the same reasons that we advocate their use in general: a combination of reasonable accuracy and ease of use. Additionally, this approach has the overwhelming advantage of producing all $C\prod_{c=1}^{C}(N_c+1)$ rates required to parameterize the FESC with a single solution of the low-level model. (Remember that the exact MVA solution algorithm produces solutions for all populations from 0 to $\vec{N}$ as a by-product of obtaining the solution at population $\vec{N}$.)

Having obtained the parameters of the level l FESCs, we now must evaluate the level l model. As this model is simply one of the low-level models defining a level $l-1$ FESC, it is clear that we can use any of the preceding techniques to perform this analysis. However, for the reasons outlined above, it generally is the case that the second method (queueing network models) is used. In the next section we look in more detail at the process of applying this technique.

8.4. Solving the High-Level Models

The most obvious approach to evaluating high-level models is to apply the analytic techniques developed in previous chapters. In Chapter 20 we present extensions to the MVA solution technique that allow the efficient evaluation of networks containing load dependent service centers. Unfortunately, this approach is applicable only to separable queueing network models. *Non-separable* high-level models can arise when some non-separable aspect of the original model (such as a priority scheduled service center) is represented directly in the high-level model, or when the load dependent service centers have arbitrary service rate functions.

For the moment, let us assume that the original network to be analyzed is separable, so that the first of these two problems cannot arise. In this case, if we wish to evaluate the higher-level model using efficient analytic techniques, we require certain restrictions on the load dependent service rates of each FESC. In particular, it must be possible to describe the service rates of each FESC by a C dimensional matrix $g[0{:}N_1\,,\,0{:}N_2\,,\,\ldots\,,\,0{:}N_C]$, such that the service rate of class c with population $\vec{n}$, $\mu_c(\vec{n})$, is equal to $\dfrac{g[n_1\,,\,\ldots\,,\,n_c-1\,,\,\ldots\,,\,n_C]}{g[n_1\,,\,\ldots\,,\,n_C]}$, with the initial condition that $g[0, \ldots, 0] = 1$. A simple example of plausible throughput rates for a two-class aggregate that violate this condition is:

$$\mu_A(n_A=1, n_B=0) = 1/2$$
$$\mu_B(n_A=0, n_B=1) = 1/3$$
$$\mu_A(n_A=1, n_B=1) = 3/10$$
$$\mu_B(n_A=1, n_B=1) = 2/9$$

The first two rates require that $g[1,0]=2$ and $g[0,1]=3$ (remembering that $g[0,0]$ is equal to 1). The last two rates are incompatible, since the rate for class A requires that $g[1,1]$ be 10, while the rate for class B requires that it be 9.

While general techniques for estimating the service rates of FESCs do not lead to separable higher-level models, analyzing the lower-level models as separable networks (the second approach of Section 8.3) is guaranteed to do so. Based on this fact, an efficient strategy for use in the hierarchical modelling of separable networks is summarized as Algorithm 8.1. While the primary motivation for this strategy is its low computational requirement, it happens that when the original model is separable, this algorithm produces the exact solution.

In cases where the original model is not separable, Algorithm 8.1 must be modified slightly. If the non-separable aspect of the model is included in one of the lower-level models, then the step of the algorithm that solves that submodel must be modified, as the MVA solution technique is not applicable. Similarly, since the throughputs obtained from a non-separable submodel do not result in a separable FESC, the step of the algorithm dealing with the solution of the high-level model must be modified. If the non-separable aspects of the original model do not appear in any low-level models, but only in the higher-level model, only the step dealing with the solution of this model must be altered. An approach to solving non-separable models that can be used in place of MVA in applying Algorithm 8.1 is given in Section 8.5. That approach results in approximate solutions of the original model. However, experience has shown that such approximations usually are quite accurate.

8.5. An Application of Hierarchical Modelling

To this point we have been concerned with separable queueing network models. The principal advantage of separable networks over more general networks is that their solutions can be obtained very quickly. However, the conditions required for separability impose some restrictions that at times can result in insufficiently accurate models. There are three approaches that can be taken in such a case. One is to combine the solutions of a number of separable networks (possibly with some iteration

8.5. An Application of Hierarchical Modelling

> Given a closed, separable model with K centers and population $\vec{N}$, let centers 1 through A represent the aggregate, and centers $A+1$ through K the complement.
>
> 1. Create a low-level model by setting the service demands of centers $A+1$ through K to zero for all classes. This is equivalent to creating a model with centers 1 through A.
> 2. Evaluate this (separable) model with population $\vec{N}$, using the exact MVA solution technique. Obtain system throughputs $X_c(\vec{n})$ for all classes c and all populations from no customers to the full population $\vec{N}$.
> 3. Create a high-level model consisting of centers $A+1$ through K, an FESC representing centers 1 through A, and customer population $\vec{N}$. The service rate of the FESC for class c when the customer population in its queue is $\vec{n}$ should be $X_c(\vec{n})$.
> 4. Evaluate this high-level model using the extension to MVA described in Chapter 20. The solution of this model is an approximation to the solution of the original K center network. System performance measures for all customer classes, and performance measures for centers $A+1$ through K, are obtained as the results of this solution. Performance measures for centers 1 through A can be computed by combining information from the solutions of the high- and low-level models. For instance, the average queue length at center K in a single class model with population N can be estimated as:
>
> $$Q_K(N) = \sum_{n=1}^{N} \left[P[Q_{FESC}=n] \sum_{j=1}^{n} j \; P[Q_K=j \,|\, Q_{FESC}=n] \right]$$
>
> where $P[Q_{FESC}=n]$ is the probability that the queue length at the FESC is n (obtained from the high-level model), and $P[Q_K=j\,|\,Q_{FESC}=n]$ is the probability that center K has queue length j given that there are n customers in the aggregate (obtained from the low-level model).

Algorithm 8.1 — A Hierarchical Decomposition Solution Technique for Separable Models

to acquire necessary parameters) to obtain an estimate of the performance of the system. The second is to create a non-separable model. A modification of the MVA solution algorithm that reflects the non-separable aspects of the model then is used to obtain approximate performance measures. (Thus, we have an "exact" model but an approximate analysis technique.) Both of these approaches are used in Part III of this book. The final approach is to use a non-separable queueing network model and an analysis technique that yields the exact solution of the model. The price paid for this increased accuracy is that the solution requires a massive amount of computation.

In this section, we discuss the use of hierarchical modelling to decrease the cost of evaluating non-separable queueing network models. Our point of view is that we have determined that a non-separable queueing network model is required because of the need to represent a particular computer system characteristic, and are seeking a feasible means to evaluate this model. By judicious choices of aggregates, a large non-separable model can be replaced by a much smaller model, by substituting single FESCs for various subsystems of service centers. This (still non-separable) reduced model can be evaluated feasibly using one of the accurate but computationally expensive solution techniques for non-separable models. Thus, we have an approximate solution technique that allows explicit representation of very general features of computer systems and still is efficient enough to be practical.

In the next two subsections we examine two specific general solution techniques, one analytic and the other simulation.

8.5.1. Global Balance

The general analytic technique used to evaluate closed, non-separable networks is called *global balance*. The global balance solution technique involves creating and solving the large sets of linear equations that describe the behavior of these models. This technique is impractically expensive in most cases because of the enormous number of equations and unknowns involved. Global balance requires one equation per state of the network, where a state is (roughly) a placement of customers at the service centers. A model with K centers and C classes therefore has at least:

$$\prod_{c=1}^{C} \begin{bmatrix} N_c + K - 1 \\ K - 1 \end{bmatrix}$$

equations and unknowns, where $\begin{bmatrix} n \\ p \end{bmatrix}$ denotes the number of ways of choosing p objects from n. Systems of equations of this size are unmanageable even for apparently modest K, C, and $\vec{N}$. For instance, a network with 6 service centers, 5 classes, and 5 customers in each class has more than 10^{12} states, and so cannot be solved directly using global balance.

8.5. An Application of Hierarchical Modelling

The implication of the rapid growth in the size of the state space with the size of the model is that global balance can be applied only to very small models. Approximate solutions of large, general models can be obtained, however, by a combination of global balance and hierarchical decomposition. A large model is broken into pieces, each of which can be analyzed independently. These individual solutions then are combined into a single model using FESCs, and the solution of this much smaller model is obtained via global balance.

As an example, Figure 8.4 shows a model with three service centers (a CPU and two I/O devices) and two customer classes. Both I/O devices are queueing devices, while the CPU is scheduled with priority given to class A over class B. (An arriving class A customer goes into service immediately if there are no class A customers at the center, and queues behind those class A customers otherwise.) Because of the priority scheduling, the model is not separable, and thus cannot be evaluated using the MVA techniques of Chapter 7.

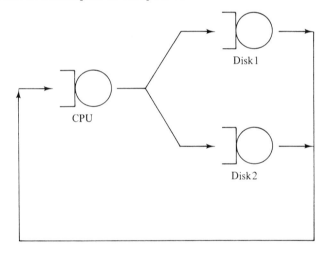

$N_A = 1$
$\quad V_{A,CPU} = 16 \quad V_{A,Disk1} = 15 \quad V_{A,Disk2} = 0$
$\quad S_{A,CPU} = 15 \quad S_{A,Disk1} = 20 \quad S_{A,Disk2} = -$
$\quad D_{A,CPU} = 240 \quad D_{A,Disk1} = 300 \quad D_{A,Disk2} = 0$

$N_B = 2$
$\quad V_{B,CPU} = 11 \quad V_{B,Disk1} = 4 \quad V_{B,Disk2} = 6$
$\quad S_{B,CPU} = 13 \quad S_{B,Disk1} = 20 \quad S_{B,Disk2} = 50$
$\quad D_{B,CPU} = 143 \quad D_{B,Disk1} = 80 \quad D_{B,Disk2} = 300$

Figure 8.4 — Global Balance Model

Recall that the service demand of class c at center k, $D_{c,k}$, is the product of the visit count, $V_{c,k}$, and the service requirement per visit, $S_{c,k}$. In separable models, we speak only of the $D_{c,k}$ because the performance measures are identical for all combinations of $V_{c,k}$ and $S_{c,k}$ that have the same product $D_{c,k}$. In non-separable models, different combinations of $V_{c,k}$ and $S_{c,k}$ with the same product $D_{c,k}$ will in general yield different results. Thus, in order to specify the non-separable model in Figure 8.4, we have had to provide the $V_{c,k}$ and $S_{c,k}$. We assume that each job begins and ends service at the CPU, so for each class the CPU visit count is one greater than the sum of the disk visit counts. This information will be used only in obtaining the exact solution to the model; our hierarchical approximation will consider the model at the level of service demands.

This example is small enough that global balance could be applied directly. In general, however, this will not be the case. Yet, since priority scheduling has an important influence on the performance of the system, it is necessary to represent it in the model. We do so here by applying global balance to the smaller model created by replacing all centers other than the CPU with an FESC. (Other techniques for modelling priority scheduling are presented in Chapter 11.) The resulting two center model (the priority CPU and the FESC) then can be evaluated using global balance, and this solution used as an estimate for the performance measures of the system. The entire process is outlined below:

- *isolate the I/O subsystem* — A model consisting of only the I/O subsystem is created (see Figure 8.5). Each class has a service demand at the CPU of zero, and a service demand at each disk as indicated in Figure 8.4.

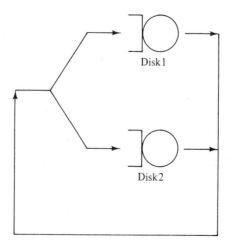

Figure 8.5 — Isolated I/O Subsystem Model

8.5. An Application of Hierarchical Modelling

- *evaluate the low-level model* — The low-level model just created is evaluated for every population that it could contain in the full network. Since this submodel is separable, the standard MVA technique can be applied. The performance measures of interest are the population dependent throughputs for each class:

\multicolumn{2}{c}{$\vec{n}$}		
A	B	$X_A(\vec{n})$	$X_B(\vec{n})$
0	1	0	.00263
0	2	0	.00316
1	0	.00333	0
1	1	.00275	.00217
1	2	.00255	.00293

These give the rate at which customers leave the aggregate and return to the CPU for each customer population in the aggregate, and thus are the parameters required to form an FESC.

- *create the high-level model* — The high-level model (Figure 8.6) consists of the original CPU service center and an FESC representing the I/O subsystem. At the CPU, each class has the service demand indicated in Figure 8.4. The FESC has the population-dependent service rates shown in the preceding table (e.g., .00275 for class A and .00217 for class B when one customer of each class is present). Remember that the FESC is scheduled using composite queueing, so that all customer classes are in service simultaneously and independently. Thus, service rates of .00275 for class A and .00217 for class B mean that a class A customer will leave (on average) in 363.6 (= 1/.00275) time units and a class B customer in 460.8 (= 1/.00217).

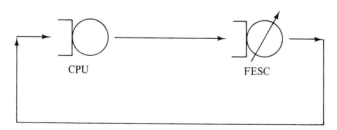

Figure 8.6 — The High-Level Model

- *evaluate the high-level model* — Since the high-level model contains a priority scheduled CPU service center, it cannot be solved using MVA (which pertains only to separable networks). However, the high-level model is small, and so can be solved by the global balance technique. We obtain:

$$X_A = .0016 \qquad X_B = .0020$$
$$Q_{A,CPU} = .396 \qquad Q_{B,CPU} = .838$$

The exact solution of the model of Figure 8.4, obtained by an expensive direct application of global balance, is:

$$X_A = .0016 \qquad X_B = .0020$$
$$Q_{A,CPU} = .373 \qquad Q_{B,CPU} = .790$$

Note that the performance measures obtained using the hierarchical approach are only approximations, although both the low- and high-level models were solved exactly. This is because the behavior of the I/O subsystem cannot be replicated exactly by the FESC, since information regarding the location of customers in the I/O subsystem is discarded.

The motivation for using an FESC in this example is that global balance can be applied to the resulting small high-level model, but (in more general cases) not to the original, large model. Use of the global balance technique was required because of the non-separable aspect of priority queueing in the model. In the following section we give a more detailed description of the global balance solution technique. The technique is described both in general terms, and more specifically as applied to the problem above. One should keep in mind that the global balance technique can be applied in many more situations than those involving priority scheduling. However, in all cases, the network to be solved must be quite small.

Details of Global Balance

The global balance solution technique can be used to compute the solutions of fairly general networks of queues. The technique is based on analyzing transitions of the system from one "state" to another.

We define a *state* of a *service center* in a queueing network model to be an ordering of customers in its queue. For example, the feasible states of a service center in a network with two class A customers and one class B customer are

$(AAB) \quad (ABA) \quad (BAA) \quad (AA) \quad (AB) \quad (BA) \quad (A) \quad (B) \quad (\,)$

The state of a service center provides information about which customers are in service and which are waiting. In some cases the state description need not contain information about the ordering of customers in the queue. For instance, if the queue above were scheduled with priority to class A over class B, there would be no need to list the order of customers since it is certain that class A will be served first.

We define a *state* of a *queueing network* to be a composite of the states of all of its service centers. Intuitively, the state of a queueing network

8.5. An Application of Hierarchical Modelling

contains all the information necessary to determine the behavior of the model at the moment.

We define the *state space* of a queueing network to be the set of feasible states. For instance, the state space of a model with two service centers and a single customer class of 3 customers is:

$$(3\ ;\ 0) \quad (2\ ;\ 1) \quad (1\ ;\ 2) \quad (0\ ;\ 3)$$

Here, the first number in each pair represents the number of customers at center one, and the second the number at center two. In general, the set of feasible states of a queueing model is determined by the number of customers of each class in the network, the service centers that each class visits, and the scheduling disciplines of the various centers.

We define a *state transition* to be the movement of the model from one of its states to another, caused by the movement of a customer within the model. For instance, if the model above were in state (3 ; 0), it would move to state (2 ; 1) when one of its customers completed service at center one and proceeded to center two. A common assumption made in analyzing queueing networks is that they exhibit *one step behavior:* each state transition involves the movement of exactly one customer. Thus, the network can move from state (3 ; 0) to state (2 ; 1), but not (directly) to state (1 ; 2). One step behavior is a reasonable assumption since it is very unlikely that any two jobs of the computer system can change locations at precisely the same time.

We define the *state transition rate* associated with a particular state transition to be the instantaneous rate at which that transition occurs, given that the network is in the starting state. For instance, if center one in the model above has a service time of 2 (a service rate of .5), and customers always alternate between centers 1 and 2, the rate associated with the transition from (3 ; 0) to (2 ; 1) is .5. In general, state transition rates depend on the service time of the moving customer at the center it departs, and the likelihood that a customer leaving this center proceeds immediately to another specific center. For single class models we have:

$$(n_1\ ;\ \ldots\ ;\ n_i+1\ ;\ \ldots\ ;\ n_j-1\ ;\ \ldots\ ;\ n_K) \rightarrow (n_1\ ;\ \ldots\ ;\ n_i\ ;\ \ldots\ ;\ n_j\ ;\ \ldots\ ;\ n_K)$$

with rate $\mu_i p_{i,j}$, where μ_i is the service rate of center i and $p_{i,j}$ is the proportion of time that a customer leaving center i proceeds directly to center j.

Given an arbitrary queueing network model, one can compute its state space, associated state transitions, and state transition rates from the model inputs. The solution of a model thus described can be obtained by making the *state space flow balance assumption* that the rate of flow of the network into any state must equal the rate of flow of the network out of that state. (This assumption is much like the flow balance assumption of

Chapter 3 applied to the network at the state space level.) The rate of flow out of a state S is the proportion of time spent in S multiplied by the sum of the state transition rates out of S. The rate of flow into a state S is the sum over every state of the network of the proportion of time spent in that state times the state transition rate from that state to S.

Finally, we define the *flow balance equations* to be the equations obtained by setting the total rate of flow into a state equal to the total rate of flow out of that state. The flow balance equations are a set of simultaneous linear equations in which the unknowns are the proportions of time spent in each possible network state. The global balance solution technique for queueing network models involves creating and solving these flow balance equations. Note that there is a single equation per state. Thus the complexity of global balance grows combinatorially with the size of the network, since the size of the state space does so.

As a particular example of the global balance technique we consider the solution of the high-level model of Figure 8.6.

- *create the state space* — Because the CPU uses priority scheduling, there is no need to include the order of customers in the queue there as part of the state description. Similarly, because the FESC uses composite queueing, the two customer classes act largely independently there and so queue ordering is not important. The model thus has six states. Using the notation (x;y) to indicate the state of the network with the CPU in state x and the FESC in state y, the state space of the model is:

    state1: $(ABB\ ;\ )$ state2: $(AB\ ;\ B)$ state3: $(BB\ ;\ A)$
    state4: $(A\ ;\ BB)$ state5: $(B\ ;\ AB)$ state6: $(\ ;\ ABB)$

- *calculate the state transition rates* — Each transition is caused by the movement of a customer from the CPU to the FESC or from the FESC to the CPU. The transition rate is equal to the rate at which this customer receives service at the origin center when in the origin state, multiplied by the proportion of time that this customer moves directly to the other (destination) center upon completion at the origin center.

    Because of the simple nature of the high-level model that we are considering, customers always move to the CPU upon completion at the FESC, and to the FESC upon completion at the CPU. Thus, $p_{A,CPU,FESC} = p_{B,CPU,FESC} = p_{A,FESC,CPU} = p_{B,FESC,CPU} = 1$. As a result, for example, the transition rate from state $(B;AB)$ to state $(AB;B)$, which involves the movement of a class A customer from the FESC to the CPU when one customer of each class is present at the FESC, is $.00275 \times 1 = .00275$. Figure 8.7 shows the state transition diagram for this model.

8.5. An Application of Hierarchical Modelling

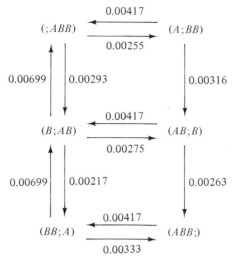

Figure 8.7 — The State Transition Diagram

- *create the flow balance equations* — The flow balance equations are obtained by setting flow in equal to flow out. The resulting set of equations do not determine a unique solution. Therefore, an arbitrary equation is discarded and replaced by an equation that ensures that the sum of the proportions of time spent in the states is one. In matrix notation, the balance equations for this example are:

$$\begin{vmatrix} -.00417 & .00263 & .00333 & 0 & 0 & 0 \\ 0 & -.00680 & 0 & .00316 & .00275 & 0 \\ .00417 & 0 & -.01032 & 0 & .00217 & 0 \\ 0 & 0 & 0 & -.00733 & 0 & .00255 \\ 0 & .00417 & .00699 & 0 & -.01191 & .00293 \\ 1 & 1 & 1 & 1 & 1 & 1 \end{vmatrix} \begin{vmatrix} P(\text{state 1}) \\ P(\text{state 2}) \\ P(\text{state 3}) \\ P(\text{state 4}) \\ P(\text{state 5}) \\ P(\text{state 6}) \end{vmatrix} = \begin{vmatrix} 0 \\ 0 \\ 0 \\ 0 \\ 0 \\ 1 \end{vmatrix}$$

- *solve the flow balance equations* — There are standard algorithms for solving sets of simultaneous linear equations. Gaussian elimination can be used on small systems. More sophisticated, iterative techniques may be required for larger models. The solution of the system of equations above gives the proportions of time spent in each state:

$$P(\text{state 1}) = .161 \quad P(\text{state 4}) = .110$$
$$P(\text{state 2}) = .125 \quad P(\text{state 5}) = .183$$
$$P(\text{state 3}) = .104 \quad P(\text{state 6}) = .317$$

- *compute performance measures* — Performance measures may be calculated from the proportions of time spent in the various states. For instance, class A's *CPU* utilization is given by:

$$U_{A,CPU} = P(\text{state 1}) + P(\text{state 2}) + P(\text{state 4}) = .396$$

8.5.2. Hybrid Modelling

Hybrid modelling is a joint simulation/analytic solution technique that attempts to combine the best aspects of each. Simulation is used so that aspects of the computer system leading to non-separable models can be represented. Analytic techniques are used for efficiency.

To understand the relationship of hybrid modelling to the analytic techniques that are the primary concern of this book, we first must present a brief examination of the simulation approach to modelling. We have chosen to describe a particular type of simulation, that of probabilistic, event driven simulation. While other approaches are possible, event driven simulation is the most useful in computer system performance analysis.

Simulation techniques are experimental in nature. However, rather than running a physical experiment with real hardware and workload components (i.e., a benchmark experiment), the functional operation of the physical system is represented in software. The software maintains a simulation clock, which keeps track of the simulated elapsed time of the experiment. The software also keeps track of the state of each simulated physical device. States typically include information about which simulated jobs are in service or queued at each device, and information about the completion time of the operation in progress at each device. The software drives the simulation by selecting the event that should occur soonest, updating the simulation clock to the time of that event, and changing the state of the simulation to correspond to the occurrence of the event. This change of state might include the scheduling of new events at future simulation times. For instance, suppose that at simulation time 104.35 seconds, the next event that should occur is the completion of the job in service at the CPU at time 104.50 seconds. The simulation would advance the clock to 104.50 seconds, remove the job from the CPU queue, and enqueue that job at the device where it would next require service. It would also place a new job in service at the CPU (assuming that there were waiting jobs), pick a service time for that job according to some probability distribution that was an input parameter of the model (say 0.23 seconds), and schedule the departure of that job for some future simulation time (in this case at 104.73 seconds). The final task of the simulation driver is to record performance statistics about the simulation experiment. For instance, the driver might maintain a count of the total number of simulated seconds during which the simulated CPU was busy. At the end of the experiment, the ratio of that quantity to the final value of the simulation clock would be the estimate for CPU utilization.

It should be clear from this description that a simulation is capable of representing nearly arbitrary amounts of detail of the operation of the real

8.5. *An Application of Hierarchical Modelling* 171

system. Of course, as more detail is incorporated, the size and expense of the simulation increase. Thus, to be useful, some amount of abstraction is required in forming the simulation model. For instance, a simulation model of a computer system might be identical to the queueing network models we have been examining (meaning that the input parameters of the simulation model and the queueing network model are the same). Alternatively, the simulation model might include more detail, such as a more accurate representation of a priority scheduling discipline used at the CPU. Finally, models with a large amount of detail (and very little abstraction) might include information about memory reference patterns (for use in determining page fault rates) or instruction mix (for use in determining effective CPU speed). Thus, simulation models are a superset of the queueing models with which we are concerned. Their advantage is their ability to incorporate detail. Their disadvantage is their expense: the computation required to obtain reliable performance estimates, the effort required to obtain the more detailed information needed to parameterize the more detailed models, and the effort required to gain insight into the critical parameters affecting performance in a model with a large number of inter-dependent parameters.

With this characterization of simulation in mind, we can proceed with the description of the basic hybrid modelling technique. Given a (non-separable) model of a system to be analyzed, isolate a subsystem (an aggregate of service centers) that can be solved conveniently in isolation. Create a flow equivalent service center to represent the submodel (by solving the submodel analytically to obtain the population dependent throughputs), and replace the subsystem by its FESC in the original model. Finally, solve this reduced model using simulation. Of course, it is possible to reverse the roles of simulation and queueing network modelling in this scheme (so that the low-level model is solved by simulation, and the high-level model analytically). This might be done, for instance, to model a complex I/O subsystem component of a large computer system, the remainder of which can be represented adequately as a separable queueing network.

In essence, this technique is identical to that of the previous subsection, with simulation substituted for global balance. Our motivation for proposing it also is the same: we have a powerful model solution technique (simulation) that we would like to employ, but the technique is too inefficient computationally for general use.

The inefficiency of simulation as a solution method is an effect of the statistical nature of the technique. Since simulation depends on observations of essentially random behavior sequences, many such sequences must be observed before we can have any confidence in the results (since any small number of sequences might be atypical). Thus, simulation is inherently expensive. This problem is compounded in cases where the events being simulated happen at significantly differing rates. For example, consider a model in which the I/O subsystem is represented in detail, and from which we would like to obtain system throughput. Suppose that for each I/O request, we simulate individually the I/O path selection, cylinder seek, rotational latency, path reconnect, and data transfer times. Further, suppose that the effect of data transmission errors is represented by simulating each transferred byte (so that errors can be inserted). In this case we have events occurring at rates varying from relatively slow (job completions in the system) to relatively fast (byte transfers). As mentioned before, to obtain any statistical confidence in the results for system throughput, many job completions must be observed (say 1000, as an example). Suppose each job performs 100 I/O operations on average. This means 100,000 I/O operations must be simulated. Now suppose each I/O operation transfers 4,000 bytes of information. This implies the simulation of 400,000,000 byte transfers. Obviously such a simulation will require immense machine resources.

Hybrid modelling can be used to best advantage in situations like the above where there are large time scale differences in the rates at which various events take place. Typically, the subsystem containing the events occurring the most frequently is modelled analytically, and the load dependent throughputs obtained from the solutions are used to create an FESC. This FESC replaces the subsystem, and the resulting model is simulated. Activity in the subsystem therefore is represented by the arrival and departure of customers from the FESC, which must occur at the same rate as events in the remainder of the model (since that is where the customers come from). Thus, this model can be simulated (relatively) efficiently.

Consider using a model to evaluate the performance of various long term scheduling policies (memory admission policies). Let the model consist of service centers representing the significant hardware resources (CPU, disks, etc.), a memory queue, and three customer classes. One class represents CPU bound jobs, one I/O bound jobs, and one balanced jobs. The scheduling policies to be evaluated use information about the current memory resident job mix to select a waiting job from one of the three classes, in an attempt to maximize system throughput.

Because of the memory queue and complicated memory admission policies to be considered, this model is not separable and so cannot be solved analytically (although perhaps the technique of the previous section could be applied successfully). A pure simulation approach would be very expensive, if not infeasible, because of the time scale difference between the rate at which long term scheduling decisions must be made and the rate at which events occur within the central subsystem. Thus, a hybrid approach is recommended. The central subsystem (CPU and I/O subsystem) model is isolated, yielding a separable model. This model is solved analytically for each feasible mix of customers of the three classes. Finally, a simulation of the memory admission policies is performed, with the time between job completions selected according to the rates of the FESC formed from the solutions of the central subsystem model solved previously. In essence, we use simulation to analyze a model consisting simply of the memory queue and an FESC representing the remainder of the computer system, with the parameters (service rates) of the FESC obtained by an analytic solution of the submodel the FESC replaces.

In an actual experiment with this technique applied to this problem, the maximum relative percentage difference between the hybrid technique and a simulation-only technique was 7%, while the simulation-only model took 56 times longer to execute. Given this combination of accuracy and efficiency, the hybrid technique is the approach of choice.

8.6. Summary

The key concept of this chapter is hierarchical decomposition, the process of splitting one model into a number of smaller submodels, each of which then can be analyzed in isolation. The solution of the original model is formed by combining the solutions of the submodels.

The submodels are combined using flow equivalent service centers. FESCs mimic the behavior of the submodels they represent by modelling the average output rates of these submodels as functions of their customer populations. Thus, FESCs are represented as load dependent service centers in the model.

The output rates of FESCs can be obtained in a number of ways, but by far the most important of these is the representation of the submodel as a queueing network model, which is solved by a single application of mean value analysis. Where this technique is applicable, it yields all the output rates for all populations of interest, and ensures that the FESC produced has analytically nice properties that allow efficient solutions of models that incorporate it. In some cases, however, this approach to solving the low-level model is not appropriate. (For instance, the parameter values of the low-level model might depend on the customer population. In this case the required load dependent rates cannot be obtained by a single application of MVA.) For these models, the load dependent rates used to parameterize the FESC generally will not lead to an efficiently analyzable higher-level model. We will deal with this problem in Part III of this book, when we use FESCs as tools in analyzing increasingly more sophisticated models of computer systems.

An important specific use of hierarchical modelling is the efficient approximate solution of non-separable queueing networks. There are two important approaches to solving these models: global balance, and simulation. Both techniques can require excessive computation for all but very small models. Thus, to employ these techniques (and so to use the modelling constructs they allow) one must restrict the model size. Hierarchical modelling is useful in this respect because the large models that naturally arise in modelling computer systems can be reduced using flow equivalent service centers to models of manageable size.

In Part III of this book we examine a number of specific components of computer systems that must be represented in a performance model. In many cases we are confronted with characteristics of computer systems that cannot be modelled directly using separable networks. Hierarchical modelling and flow equivalent servers are the keys to successful models in many of these cases.

8.7. References

Flow equivalent service centers were shown to yield exact solutions for single class separable networks by Chandy et al. [1975]. Sauer and Chandy [1975] first presented their use as an approximation.

The global balance solution technique is a classical approach to the solution of Markovian systems (see [Cox & Miller 1965], for example). Sauer and Chandy [1981] present this material in the computer system modelling context.

The utility of the hybrid modelling approach of Section 8.5.2 was pointed out by Schwetman [1978] and Tolopka and Schwetman [1979]. For other case studies employing hybrid modelling, see [Browne et al. 1975] and [Lindzey & Browne 1979].

[Browne et al. 1975]
J.C. Browne, K.M. Chandy, R.M. Brown, T.W. Keller, D.F. Towsley, and C.W. Dissley. Hierarchical Techniques for Development of Realistic Models of Complex Computer Systems. *Proc. IEEE 63*,6 (June 1975), 966-975.

[Chandy et al. 1975]
K.M. Chandy, U. Herzog, and L.S. Woo. Parametric Analysis of Queueing Networks. *IBM Journal of Research and Development 19*,1 (January 1975), 50-57.

[Cox & Miller 1965]
D.R. Cox and H.D. Miller. *The Theory of Stochastic Processes*. Wiley, 1965.

[Lindzey & Browne 1979]
  G.E. Lindzey, Jr. and J.C. Browne. Response Analysis of a Multi-Function System. *Proc. ACM SIGMETRICS Conference on Simulation, Measurement and Modeling of Computer Systems* (1979), 19-26.
[Sauer & Chandy 1975]
  C.H. Sauer and K.M. Chandy. Approximate Analysis of Central Server Models. *IBM Journal of Research and Development 19*,3 (May 1975), 301-313.
[Sauer & Chandy 1981]
  C.H. Sauer and K.M. Chandy. *Computer Systems Performance Modeling.* Prentice-Hall, 1981.
[Schwetman 1978]
  H.D. Schwetman. Hybrid Simulation Models of Computer Systems. *CACM 21*,9 (September 1978), 718-723.
[Tolopka & Schwetman 1979]
  S.J. Tolopka and H.D. Schwetman. Mix-Dependent Job Scheduling - An Application of Hybrid Simulation. *1979 National Computer Conference Proceedings, AFIPS Volume 48* (1979), AFIPS Press, 45-49.

8.8. Exercises

1. Modify the Fortran program of Chapter 18 to accommodate flow equivalent service centers. (The modifications required are described in Chapter 20.)

2. Use Algorithm 8.1 to evaluate a (separable) single class model consisting of a CPU center with service demand 10, and four disk centers with service demands 4, 3, 3, and 2. The customer class should be terminal type with 20 active users and 30 second think times. In applying the algorithm, treat the four disk centers as the aggregate, and the CPU center as the complementary network. Use the software created in answering Exercise 1 (extended to accommodate terminal classes) to analyze the high-level model that you construct. Compare the solution you obtain by applying hierarchical decomposition to that obtained by simply solving the full five-center network using MVA.

3. Use the global balance technique to solve the example model from Section 6.4.2.1. This exercise should illustrate dramatically the computational advantage of separable models (which can be solved using MVA) over general networks of queues (which require a global balance analysis to obtain the exact solution).

4. Figure 8.7 shows the state transition diagram for the model illustrated in Figure 8.6. There are two centers: a preemptive-priority-scheduled CPU, and an FESC representing the I/O subsystem. There are two classes: A, the high-priority class, with one customer, and B, the low-priority class, with two customers.
   a. Why is there no state $(BA;B)$?
   b. Why is there no transition from state $(BB;A)$ to state $(AB;B)$?
   c. Why is there no transition from state $(ABB;)$ to state $(AB;B)$?
   d. At what rate does class B depart the FESC when one class A and one class B customer are present there?

Part III

Representing Specific Subsystems

Many successful modelling studies are conducted without venturing beyond the techniques described in Part II. In other words, the system characteristics considered in these studies are restricted to those that can be represented directly using the parameters of separable queueing networks.

There are, of course, situations in which the analyst will wish to represent a specific subsystem in greater detail than is possible within the confines of separable networks. Techniques for doing so are the subject of Part III.

The efficient evaluation that is characteristic of separable networks is mandatory in analyzing contemporary computer systems. For this reason, *non-separable* networks typically are evaluated by "mapping" them onto (perhaps several) separable networks. Since this mapping necessarily is approximate, the techniques for doing so traditionally have been referred to as *approximate solution techniques*. This phrase is not really meaningful, though, since we use approximate techniques to evaluate even separable networks, and since any queueing network model is only an approximate representation of an actual system.

As yet there is no unifying theory underlying these techniques. There is, however, a small set of ideas on which they are based. Among these ideas are:

- *iteration* — making an initial guess at the value of a parameter, then iteratively refining this value, in a manner analogous to that of the MVA-based iterative approximate solution techniques for separable networks, described in Chapter 6;
- *load concealment* — representing the effect of a workload component or system characteristic indirectly, by "inflating" the service demands of those workload components that are represented explicitly, in a manner analogous to the calculation of performance measures for closed classes in mixed separable networks, described in Chapter 7;

- *decomposition* — evaluating a subsystem in isolation, perhaps using a heuristic, and incorporating the results of this analysis in a flow equivalent service center that can be included in a high-level model, as described in Chapter 8.

Not all of the techniques are fully general. We will see that *homogeneity assumptions* frequently are introduced in some aspect of a model to facilitate the detailed representation of a subsystem. As a specific example, in order to evaluate multiple class memory constrained queueing networks we will assume that the throughput of each class is dependent only on its own central subsystem population and the *average* central subsystem population of every other class.

We have organized our discussion into three chapters, which consider the representation of memory, disk I/O, and processors. Just as with the algorithms for evaluating separable queueing networks presented in Part II, the techniques presented in Part III generally will be incorporated in a queueing network analysis package at a level not visible to the analyst. While it is possible to use these techniques without understanding them, achieving such an understanding is important for two reasons: so that they can be used confidently and appropriately, and so that the analyst can devise related techniques when confronted with novel situations. Some examples of such novel applications will be given in Part V.

Chapter 9

Memory

9.1. Introduction

Memory and its management affect the performance of computer systems in two major ways. First, almost every system has a *memory constraint*: a limit on the number of "threads of control" that can be active simultaneously, imposed by the availability of memory. A memory constraint places an upper bound on the extent to which processing resources (CPUs, disks, etc.) can be utilized concurrently, and thus on the throughput of the system. Second, there is *overhead* associated with memory management. As an example, swapping a user between primary memory and secondary storage places service demands on the I/O subsystem (and the CPU, as well). To the extent that the operating system devotes processing resources to the management of memory, the progress of "useful" work is impeded.

Although memory seldom was mentioned explicitly in Parts I and II, specific implicit assumptions were made in each example:

- When we described the intensity of a workload by its population N (a *closed* model with a *batch* workload), we were assuming that the system had a memory constraint, that this constraint could be expressed in terms of a specific number of jobs (i.e., that all jobs required the same amount of memory), and that there was a sufficient backlog of work that the system was continuously operating at its maximum multiprogramming level.

- When we described the intensity of a workload by its population N and average think time Z (a *closed* model with a *terminal* workload), we were assuming that the system had a fixed number of interactive users, and that enough memory existed to accommodate as many of these users as might concurrently require it (i.e., that there was no memory constraint).

- When we described the intensity of a workload by its arrival rate λ (an *open* model with a *transaction* workload), we were again assuming that there was no memory constraint. The assumption in this case is in fact somewhat more extreme than in the case of a terminal workload, because there is no bound on the central subsystem population of a transaction workload.

In each case we either ignored overhead due to memory management or included an average value in the service demands of every customer.

These simple assumptions about system behavior are encountered frequently in modelling studies because they satisfy the conditions required for queueing network models to be *separable*, i.e., directly amenable to the efficient evaluation techniques described in Part II. The fact that these studies are successful indicates that the assumptions, if not strictly correct, are at least robust:

- In an actual computer system, the multiprogramming level of a batch workload may vary over time for many reasons: the amount of memory available to the batch workload may vary, or the memory requirements of individual batch jobs may differ, or the backlog of work may drop below the memory constraint. However, usually it is possible to validate a model using a single multiprogramming level that represents the time-weighted average of the observed multiprogramming levels. Projecting performance for a modified workload or configuration requires that the analyst estimate the effect of the modification on this average multiprogramming level.

- Although there are times in almost every interactive system when a user must wait for access to memory, these times may be so infrequent that the existence of the memory constraint can be ignored in constructing a model. A modification to the workload or configuration may affect the distribution of the number of users desiring memory, so the validity of the assumption must be checked in modelling such a modification. Doing so usually is not difficult.

- Although detailed paging behavior is difficult to model, many operating systems succeed in maintaining an average page transfer rate that is relatively insensitive to variations in configuration and workload. In such cases it is not difficult to characterize a customer's service demand at the paging device.

Of course, these simple assumptions are not always adequate. In this chapter we will extend the flexibility with which we represent memory and its management in queueing network models. The organization of the chapter reflects our belief that the throughput-limiting effect of a memory constraint is the *primary effect* of memory on performance, while the overhead associated with memory management is a significant *secondary effect*. The chapter has five principal sections. First, we explore some of the subtleties that can arise in the simple case of a system with a known average multiprogramming level. Next, we show how to represent the effect on system throughput of a memory constraint that is sometimes, but not continuously, reached. Then, we describe how to represent overhead due to swapping (Section 9.4) and paging (Section 9.5). Finally, we use case studies to relate these techniques to one another, supplementing the examples presented in each section.

9.2. Systems with Known Average Multiprogramming Level

This section serves to illustrate that subtleties can arise even in modelling the apparently straightforward case of a batch workload with a known average multiprogramming level.

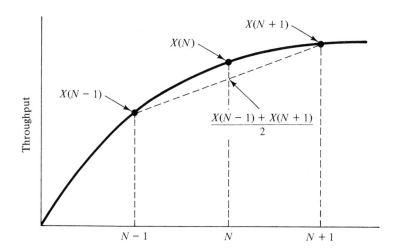

Figure 9.1 − **Throughput Versus Multiprogramming Level**

In all but the simplest of systems, the multiprogramming level of a workload (the number of active threads of control) is not constant, but varies over time due to factors such as competition for memory from other workloads, differences in the memory requirements of jobs, and the availability of jobs. As the multiprogramming level of a workload varies, so does its throughput. The relationship of throughput to multiprogramming level is illustrated qualitatively by the curve in Figure 9.1. At low multiprogramming levels, the marginal increase in throughput due to an additional active job is relatively large, since this job causes a relatively large increase in the concurrent activity of various processing resources.

As the multiprogramming level increases, the marginal increase in throughput becomes relatively small, because little additional concurrency is realized. (Figure 9.1 assumes that the overhead due to a job can be included as a component of its service demands, and is insensitive to multiprogramming level.)

Imagine that we observe such a workload for a period of time and measure its average multiprogramming level, N. For the sake of argument, let N be an integer. Now, consider two cases:

- If the system had operated at a constant multiprogramming level of N during the entire observation interval, then its throughput would have been $X(N)$, as indicated in the figure.
- If the system had operated at a constant multiprogramming level of $N-1$ during the first half of the interval and at a constant multiprogramming level of $N+1$ during the second half, then its throughput would have been $X(N-1)$ during the first half of the interval, $X(N+1)$ during the second half, and $\frac{X(N-1) + X(N+1)}{2}$ over all, which, as shown in the figure, is less than $X(N)$.

Clearly, if the system actually had operated as in the latter case but a queueing network model of the system is evaluated at the average multiprogramming level N, a discrepancy will result. This discrepancy often is small; systems almost inevitably are modelled successfully using an average multiprogramming level, which almost inevitably represents a time-weighted average of several different multiprogramming levels encountered during an observation interval. However, if greater accuracy is required, the model can be analyzed at each of the observed multiprogramming levels and a weighted average of the results taken. This approach can be applied to multiple class models as well as single class models. Naturally, though, the incentive to be satisfied with the results of an analysis at average workload intensities increases with the number of combinations that would have to be considered to do otherwise.

Here is an example based on actual data collected during a benchmark test of a system with three distinct workloads, each of batch type. As shown in Table 9.1, the multiprogramming levels of these workloads varied in a way that partitions the benchmark into three time periods. These periods are described by the first three lines of the table, which show the elapsed time (in seconds) at which the transitions between periods occurred, the duration of each period (again in seconds), and the proportion of the total observation interval due to each period.

In order to parameterize a queueing network model, we need not only the workload intensities, as shown in Table 9.1, but also the service demands. These service demands, calculated from measurements taken during the benchmark, are shown in Table 9.2.

9.2. Systems with Known Average Multiprogramming Level

quantity		period 1	period 2	period 3	average
time interval		0 - 1268	1268 - 1734	1734 - 2108	
duration		1268	466	374	
proportion of total		.602	.221	.177	
MPL	workload 1	2	2	3	2.18
	workload 2	1	0	0	0.60
	workload 3	2	3	0	1.87

Table 9.1 — Variation in Multiprogramming Level (MPL)

device	service demand, seconds/job		
	workload 1	workload 2	workload 3
CPU	12.906	1.315	0.632
disk 1	4.133	0.325	0.004
disk 2	8.580	0	0
disk 3	7.549	0.081	0.305
disk 4	0.424	0.001	0.181
disk 5	4.896	0.053	0.198
disk 6	6.437	0	0
disk 7	3.651	0	0
disk 8	0	0.082	0.888
disk 9	3.057	0.087	0.049
disk 10	4.980	0.141	0.080

Table 9.2 — Service Demands

First we consider a three class model of this system which we evaluate three times, using the three sets of multiprogramming levels corresponding to the three time periods of the benchmark. The results are shown in Table 9.3.

quantity		period 1	period 2	period 3	average
CPU utilization		.925	.782	.557	.825
throughput, jobs/minute	wkld. 1	1.343	1.475	2.498	1.58
	wkld. 2	14.71	0	0	8.86
	wkld. 3	29.71	44.14	0	27.6

Table 9.3 — Model Outputs for Three Time Periods

The alternative is to evaluate the same three class model once, using the average multiprogramming levels for each workload. Table 9.4 compares measurement data, the model using the average multiprogramming level, and the model representing the three time periods.

quantity		actual value	model results			
			average MPL		variable MPL	
			value	discrep.	value	discrep.
CPU utilization		.820	.819	0	.825	+ 1%
t'put., jobs/min.	wkld. 1	1.59	1.51	− 5%	1.58	− 1%
	wkld. 2	8.77	8.72	− 1%	8.86	+ 1%
	wkld. 3	27.0	28.9	+ 7%	27.6	+ 2%

Table 9.4 − Measurements Versus Two Modelling Approaches

Two summary comments, the first of which is technical, the second philosophical:

- As we have observed in other contexts (e.g., Chapter 4), average response time must be calculated in a different and less obvious way than average throughput, queue length, and utilization. These latter quantities are obtained by weighting the performance measure for each period by the relative length of that period. For example:

$$\overline{U} = \sum_{\substack{all \\ periods\ p}} (U\ during\ period\ p) \times \frac{duration\ of\ period\ p}{total\ duration\ of\ observation\ interval}$$

Average response time, on the other hand, is obtained by weighting the performance measure for each period by the relative number of jobs completed during that period:

$$\overline{R} = \sum_{\substack{all \\ periods\ p}} (R\ during\ p) \times \frac{(X\ during\ p) \times (duration\ of\ p)}{\sum_{\substack{all \\ periods\ p}} (X\ during\ p) \times (duration\ of\ p)}$$

- We observe frequently in queueing network modelling that significant increases in effort (both in data collection and in analysis) yield only small increases in accuracy. This is perhaps the most important point illustrated by this example.

9.3. Memory Constraints

Since the throughput-limiting effect of a memory constraint is the primary effect of memory on performance, its accurate representation can be important. We have noted that separable queueing network models allow the direct representation of certain extreme cases, such as a memory con-

9.3. Memory Constraints

straint that is continuously reached (batch workloads) and a memory constraint that is never reached (terminal or transaction workloads). Unfortunately, the interesting general case of a memory constraint that is sometimes, but not continuously, reached, is an instance of *simultaneous resource possession*, which violates the conditions required for separability. Fortunately, rather elegant techniques exist for the indirect representation of such a memory constraint in separable models. These techniques are the subject of the present section.

Our approach is based on the concepts of flow equivalence and hierarchical modelling, as described in Chapter 8. As shown in boxes 1 and 2 of Figure 9.2, we initially are confronted with a queueing network model that is non-separable because of the existence of a memory queue. First, we decompose the model into two parts: the *central subsystem* plus the *memory queue* (box 2) and the *external environment* (box 1). Next, we define a load dependent service center (shown in box 3) that is flow equivalent to 2 from the point of view of the external environment. We do this using a separable subsystem model, which can be evaluated efficiently. Finally, we analyze a high-level model consisting of this FESC and the external environment (1 and 3 taken together). The joint analysis of 1 and 3, which again can be carried out efficiently, will yield nearly the same results as the joint analysis of 1 and 2, which cannot.

This hierarchical analysis coincides nicely with the users' view of the system. Referring again to Figure 9.2, each customer can be in one of two principal *states*: *thinking* (i.e., at the terminals; equivalently, within box 1) or *ready* (i.e., desiring to compute; equivalently, within box 2). The primary concern of a user is the average time spent in the ready state (box 2), which corresponds to average response time. It happens that, because of the memory constraint, ready customers can be in one of two sub-states: *waiting* (i.e., in the memory queue; equivalently, above the dashed line in box 2) or *active* (i.e., memory resident and competing for the processing resources of the central subsystem; equivalently, below the dashed line in box 2). This influences the completion rate of customers — the rate at which customers flow from box 2 back to box 1 — and thus average response time. The objective of our analytic approach is to define an FESC that characterizes this completion rate as a function of the customer population within box 2. This characterization will account for competition within the central subsystem (i.e., below the dashed line in box 2), and also for the effect of the memory constraint on the actual population of the central subsystem.

We first discuss single class memory constrained systems, and then extend our discussion to the multiple class case.

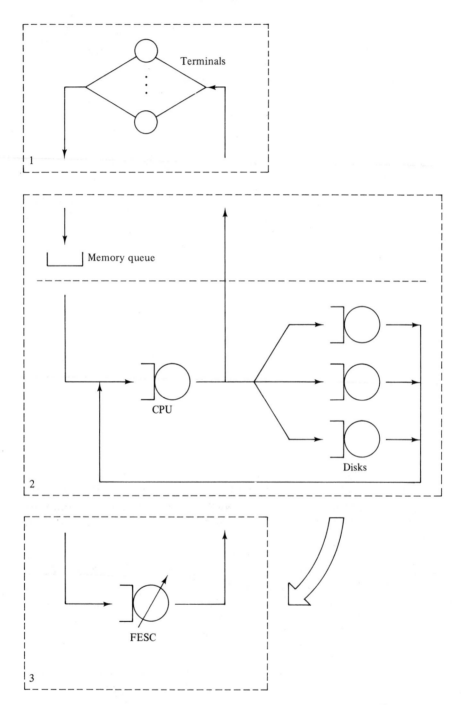

Figure 9.2 — Modelling a Memory Constrained System

9.3.1. The Single Class Case

We assume that customers have indistinguishable memory requirements, as well as service demands. We denote the memory constraint by M. If a customer becomes ready when there are fewer than M other ready customers (i.e., when there are $N-M$ or more thinking customers) then that customer becomes active immediately. If a customer becomes ready when there are M or more other ready customers (and thus M active customers fully occupying memory) then that customer must wait until memory becomes available.

Our task is to define an FESC for the central subsystem plus the memory queue. As noted in Chapter 8, a load dependent service center has a throughput that varies with its queue length. The queue length at the FESC in box 3 corresponds to the number of ready customers — the number of customers anywhere within box 2. In the actual system, how does throughput vary with the number of ready customers? The answer to this question is displayed qualitatively in Figure 9.3 both with the memory constraint (the solid curve) and without (the dashed curve). Once the memory constraint is reached (once there are M ready customers), no further increase in throughput results from an increase in the number of ready customers. Why is this the case? Because these additional ready customers are not active, but rather are waiting (for memory). This is made explicit by Table 9.5, in which $X(n)$ denotes the throughput of the central subsystem with a population of n customers.

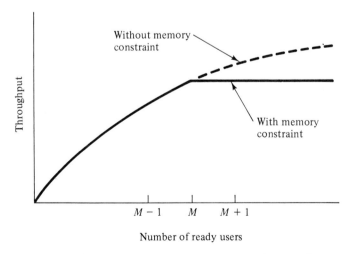

Figure 9.3 — **Throughput Versus Number of Ready Customers**

Representing Specific Subsystems: Memory

FESC queue length	ready customers	active customers	throughput
1	1	1	$X(1)$
2	2	2	$X(2)$
$\vdots$	$\vdots$	$\vdots$	$\vdots$
$M-1$	$M-1$	$M-1$	$X(M-1)$
M	M	M	$X(M)$
$M+1$	$M+1$	M	$X(M)$
$\vdots$	$\vdots$	$\vdots$	$\vdots$
N	N	M	$X(M)$

Table 9.5 − Throughput of a Memory Constrained System

It is a simple matter to determine $X(n)$. We define a *low-level model* consisting of the processing resources comprising the central subsystem. We evaluate this model for each feasible customer population n, i.e., for each number of active customers from 1 to M. For each population, we note the throughput. These throughputs are the $X(n)$ that are used to define the FESC used in the *high-level model*. This is stated more precisely in Algorithm 9.1.

---

1. Define a low-level model consisting of the service centers representing the processing resources that comprise the central subsystem.

2. Evaluate this model, which is separable, for each feasible population, $n = 1, \ldots, M$. Note the load dependent throughputs, $X(n)$.

3. Create a load dependent service center that is flow equivalent to the central subsystem plus the memory queue, by setting its throughput with queue length n, $\mu(n)$, to:

$$\mu(n) = \begin{cases} X(n) & n = 1, \ldots, M \\ X(M) & n > M \end{cases}$$

4. Define a high-level model consisting of this FESC and the external environment: if a terminal workload, then N customers with think time Z; if a transaction workload, then an external arrival rate λ. Evaluate this model, which is separable.

Algorithm 9.1 − Single Class Memory Constrained Systems

9.3. Memory Constraints

As an example application of this algorithm, consider a small timesharing system with a CPU, two disks, and 512K bytes of memory. An average interaction requires 3 seconds of CPU service, 4 seconds of service at one of the disks, 2 seconds of service at the other disk, and 100K bytes of memory. The operating system requires 150K bytes of memory, so that at most 3 users can be memory-resident simultaneously. There are 15 users, with average think times of 60 seconds. We wish to know:
- the average response time
- the average number of ready users
- the average number of active users
- the distribution of memory partition occupancy
- the average time spent queued awaiting access to memory
- the utilization of each processing resource
- the improvement in response time that would result if 256K of memory were added

We begin by analyzing the central subsystem for 1, 2, and 3 active users. This low-level model has three centers with service demands of 3, 4, and 2 seconds per interaction respectively. We obtain the load dependent throughputs shown below:

population	throughput, interactions/sec.
1	0.1111
2	0.1636
3	0.1930

Next we define a high-level model with $N = 15$ customers, $Z = 60$ seconds, and a load dependent center that is flow equivalent to the central subsystem plus the memory queue, defined as follows:

queue length	throughput
1	0.1111
2	0.1636
3	0.1930
4	0.1930
⋮	⋮
15	0.1930

We evaluate this model, obtaining the basic outputs shown in Table 9.6. Interactive response time is available directly: 25.7 seconds. So is the average number of ready customers: 4.5. From the queue length distribution at the FESC we see that 3.8% of the time the central subsystem is idle, 8.6% of the time there is a single active customer, 12.2% of the time there are two active customers, and 75.4% of the time there are three

> throughput: 0.175 interactions/second
> average residence time at the FESC: 25.7 seconds
> average queue length at the FESC: 4.5
> queue length distribution at the FESC:
>
> | queue length | probability |
> |---|---|
> | 0 | .038 |
> | 1 | .086 |
> | 2 | .122 |
> | 3 | .137 |
> | 4 | .142 |
> | 5 | .135 |
> | 6 | .117 |
> | >6 | .228 |

Table 9.6 — Basic Outputs

active customers (i.e., 3 or more ready customers). Thus the average number of active customers is 2.6. Substituting this into Little's law, $N = XR$, we find that the average time spent in the central subsystem once a memory partition has been obtained is $2.6/0.175 = 14.9$ seconds. Thus a customer spends $25.7 - 14.9 = 10.8$ seconds awaiting access to memory. To calculate device utilizations we employ the utilization law, $U_k = XD_k$. At the CPU, utilization must be $0.175 \times 3.0 = 52.5\%$. At the two disks, utilization must be 70% and 35%, respectively.

To assess the impact of additional memory we calculate FESC rates for 4, 5, and 6 customers in the central subsystem. (Three additional users can be accommodated by the new configuration.) The FESC now will have the characteristics shown below:

queue length	throughput
1	0.1111
2	0.1636
3	0.1930
4	0.2110
5	0.2226
6	0.2305
7	0.2305
⋮	⋮
15	0.2305

When we analyze a high-level model consisting of 15 users and this FESC, we obtain a response time of 20.7 seconds, a 20% improvement.

9.3. Memory Constraints

The utility of the technique described in Algorithm 9.1 arises both from its accuracy and from its efficiency. Its accuracy is due to the fact that the terminals and the central subsystem are *decomposable*: the rate at which customers interact in the central subsystem is much greater than the rate at which they flow between the thinking and ready states. Its efficiency is due to two factors:

- The load dependent throughputs used in defining the FESC can be obtained efficiently. In this case, the model of the central subsystem is a single class separable queueing network.
- The resulting high-level model can be analyzed efficiently. In this case, it also is a single class separable queueing network.

This approach to analyzing single class memory constrained systems epitomizes the use of flow equivalence and hierarchical modelling to evaluate non-separable queueing networks efficiently.

9.3.2. Multiple Classes with Independent Memory Constraints

Here we consider a system with C customer classes, $c = 1, ..., C$, having independent memory constraints M_c. (The classes may be thought of as differing not only in their workload intensities and service demands, but also possibly in their memory requirements.) There is an obvious generalization of Algorithm 9.1 to this case:

- Define a multiple class low-level model consisting of the service centers representing the processing resources that comprise the central subsystem.
- Evaluate this model for each feasible population vector, $\vec{n} = (n_1, n_2, ..., n_C)$, $0 \leq n_c \leq M_c$. Note the "population vector dependent" throughputs of each class, $X_c(\vec{n})$.
- In a manner analogous to Algorithm 9.1, use these throughputs to define a multiple class FESC.
- Define a multiple class high-level model consisting of this FESC and the external environment of each class. Evaluate this model.

Unfortunately, this generalization possesses neither of the efficiency properties of its single class counterpart:

- Obtaining the throughputs needed to parameterize the FESC requires evaluating the low-level model for every feasible population vector. The cost of this is proportional to:

$$CK \prod_{c=1}^{C} (M_c + 1)$$

- The resulting high-level model is not separable, so can be evaluated only by the global balance technique, which is prohibitively expensive unless there are few classes and the memory constraints are small.

To circumvent these difficulties we introduce two *homogeneity assumptions*:

- We assume that the throughput of class c when its own central subsystem population is n_c depends only on the average central subsystem populations of the other classes.
- We assume that each class sees the other classes as though their central subsystem populations were independent of one another.

The former assumption allows us to determine the load dependent throughputs of any class by analyzing a C class queueing network in which the populations of the other classes are fixed at their *average values*. These average values are determined from the high-level model; the high- and low-level models are solved iteratively, terminating when successive estimates are sufficiently close. The latter assumption allows us to define a separate FESC for each class. In essence, we analyze C separable single class high-level models, rather than a single non-separable C class high-level model.

The result is Algorithm 9.2. This algorithm is applicable to models in which some of the C classes are unconstrained. For ease of expression, we denote the number of constrained classes by $\hat{C} \leq C$ and order the classes so that the constrained classes have indices $c = 1, \ldots, \hat{C}$. The algorithm is a good example of the introduction of homogeneity assumptions in order to facilitate evaluation.

9.3.3. Multiple Classes with Shared Memory Constraints

Algorithm 9.2 assumed that each class was subject to a memory constraint that was independent of the behavior of the other classes. Here we generalize that algorithm to shared memory constraints: constraints on the total number of customers in memory (or in a region of memory), rather than on the populations of the individual classes. The only significant change to Algorithm 9.2 will be in the calculation of the $\mu_c(n)$ in Step 3.2.

Let there be F *domains*, or shared regions of memory. Each memory constrained class is assigned to a domain. To simplify the discussion we will assume that all domains are shared; dedicated domains are, of course, a special case of shared domains. Let M_f be the capacity of domain f, i.e., the number of customers that can reside in that domain. (We temporarily assume that the classes assigned to a particular domain have indistinguishable memory requirements.)

9.3. Memory Constraints

> 1. Obtain initial estimates of the average central subsystem customer population for each memory constrained class, $\bar{n}_c$ for $c = 1, \ldots, \hat{C}$. To do so, ignore all memory constraints in the original C class model, yielding a separable queueing network. Evaluate this network. For each memory constrained class c, set $\bar{n}_c$ to the minimum of M_c and the average class c central subsystem population in the unconstrained model.
>
> 2. In preparation for the iteration, modify the original model by changing each of the $\hat{C}$ memory constrained classes into a batch class with population equal to $\bar{n}_c$. Leave the unconstrained classes in their original form. The result is a C class separable queueing network. (The non-integer customer populations of the constrained classes are naturally suited to the MVA-based iterative approximate solution technique.)
>
> 3. For each memory constrained class $c = 1, \ldots, \hat{C}$:
>
>   3.1. Replace the $\bar{n}_c$ class c customers with each feasible population of class c, $n_c = 1, \ldots, M_c$. Evaluate the queueing network, obtaining the throughput of class c, $X_c(n_c)$.
>
>   3.2. Create an FESC, a single class load dependent service center whose throughput with queue length n, $\mu_c(n)$, is defined by:
>
> $$\mu_c(n) = \begin{cases} X_c(n) & n = 1, \ldots, M_c \\ X_c(M_c) & n > M_c \end{cases}$$
>
>   3.3. Define and evaluate a single class separable high-level model consisting of this FESC and the external environment of class c (N and Z, or λ). Obtain the queue length distribution at the FESC. (We let $P[Q_{FESC} = i]$ denote the probability that the queue length at the FESC is i.) Use this to calculate a new estimate for the average central subsystem population of class c:
>
>   .. continued ..

Algorithm 9.2 — Multiple Classes, Independent Memory Constraints

> .. continued ..
>
> $$\bar{n}_c = \sum_{i=1}^{M_c} i\, P[Q_{FESC}=i] + \left[1 - \sum_{i=0}^{M_c} P[Q_{FESC}=i]\right] M_c$$
>
> 4. Repeat Step 3 until successive estimates of the $\bar{n}_c$ for each constrained class are sufficiently close.
> 5. Obtain performance measures for the constrained classes from the $\hat{C}$ high-level models evaluated during the final iteration. Obtain performance measures for the unconstrained classes by solving the queueing network defined in Step 2 using the final estimates of the $\bar{n}_c$ for the constrained classes.

Algorithm 9.2 — **Multiple Classes, Independent Memory Constraints**

Our approach is to view a domain shared by several classes as several smaller domains, each used by a single class. The memory constraint on a specific class will be determined iteratively, by considering the average central subsystem populations of its *competitor classes*: all other classes sharing the domain, in the case of FCFS domain scheduling; all other classes of greater or equal priority sharing the domain, in the case of priority domain scheduling. This approach is embodied in Algorithm 9.3, parts of which are abbreviated because of their similarity to Algorithm 9.2.

Algorithm 9.3 can be used to evaluate models in which the classes sharing a specific domain have distinct memory requirements. This requires straightforward modifications to the functions M_f and δ_c, defined in the algorithm. Once modified in this way, the algorithm can also be used to evaluate single class memory constrained models in which customers differ in their memory requirements. This is accomplished by defining a single domain shared by several "artificial" classes. Each of these artificial classes corresponds to those customers with a specific memory requirement. Each has service demands identical to those of the "real" class, and a workload intensity adjusted to reflect the proportion of customers having the corresponding memory requirement.

9.3. Memory Constraints

1. Obtain initial estimates of $\bar{n}_c$ for $c = 1, \ldots, \hat{C}$. To do so, ignore all memory constraints in the original C class model. Evaluate the resulting separable network. For each memory constrained class c, set $\bar{n}_c$ to the *minimum* of the average class c central subsystem population in the unconstrained model and a "proportionate share" of its domain, calculated as:

$$M_{F(c)} \times \frac{\alpha_c}{\sum\limits_{\substack{i \in \text{c plus its com-}\\ \text{petitor classes}}} \alpha_i}$$

where $F(c)$ is a function that gives the domain to which class c is assigned ($M_{F(c)}$ is thus the capacity of the domain to which class c is assigned), and α_i is the average class i central subsystem population in the unconstrained model.

2. In preparation for the iteration, modify the original model by changing each of the $\hat{C}$ memory constrained classes into a batch class with population equal to $\bar{n}_c$.

3. For each memory constrained class $c = 1, \ldots, \hat{C}$:

   3.1. Replace the $\bar{n}_c$ class c customers with each feasible population of class c, n_c. Evaluate the queueing network obtaining the throughput of class c, $X_c(n_c)$. Feasible populations are integers from 1 to $\lfloor M_{F(c)} - \delta_c \rfloor$, where:

   $$\delta_c \equiv \sum\limits_{\substack{i \in \text{c's compe-}\\ \text{titor classes}}} \bar{n}_i$$

   Also evaluate the network at the non-integer population $M_{F(c)} - \delta_c$.

   3.2. Create an FESC, a single class load dependent service center whose throughput with queue length n, $\mu_c(n)$, is defined by:

   $$\mu_c(n) = \begin{cases} X_c(n) & n \leq M_{F(c)} - \delta_c \\ X_c(M_{F(c)} - \delta_c) & n > M_{F(c)} - \delta_c \end{cases}$$

   .. continued ..

Algorithm 9.3 — Multiple Classes, Shared Memory Constraints

> .. continued ..
>
> 3.3. Define and evaluate a single class separable high-level model consisting of this FESC and the external environment of class c (N and Z, or λ). Obtain the queue length distribution at the FESC. Use this to calculate a new estimate for the average central subsystem population of class c:
>
> $$\bar{n}_c = \sum_{i=1}^{\lfloor M_{F(c)} - \delta_c \rfloor} i\, P[Q_{FESC}=i] + \left[1 - \sum_{i=0}^{\lfloor M_{F(c)} - \delta_c \rfloor} P[Q_{FESC}=i] \right] (M_{F(c)} - \delta_c)$$
>
> 4. Repeat Step 3 until successive estimates of the $\bar{n}_c$ for each constrained class are sufficiently close.
> 5. Obtain performance measures as in Algorithm 9.2.

Algorithm 9.3 — **Multiple Classes, Shared Memory Constraints**

9.4. Swapping

In Section 9.3 we developed techniques for representing the throughput-limiting effect of a memory constraint. While concentrating on this primary effect of memory on performance, we allowed ourselves to ignore the problem of explicitly representing swapping.

On the one hand, swapping devices are no different than other I/O devices: they can be included in a model, and their service demands can be calculated by multiplying device utilization by the length of the measurement interval, then dividing this result by the number of interactions during that interval. In this sense, swapping activity has been included implicitly in all of the models we have constructed. On the other hand, we presently have no way of projecting changes to this service demand that might result from system or workload modifications. Service demand at the swapping device is not an intrinsic property of an interaction, like service demand at the CPU or at a file device. The analyst typically knows how to modify intrinsic parameters to reflect system changes. On the other hand, the influence of system modifications on the level of

9.4. Swapping

swapping activity is something we would like to learn from our model, rather than provide as an input. If the system modifications under consideration can be expected to influence significantly the level of swapping activity, then the modelling approach must include a procedure for estimating swapping device service demand.

The explicit representation of swapping is the subject of the present section. The techniques we develop will use the algorithms of Section 9.3 as a basis, since we wish to represent the effect of the memory constraint in addition to the overhead of memory management. For the sake of simplicity, the algorithms in this section will be expressed for the case of a single workload of terminal type (N customers with think time Z), and a single swapping device. Generalization to multiple workloads and multiple swapping devices is possible.

9.4.1. Swapping to a Dedicated Device

We first consider memory constrained systems with a single workload of terminal type, in which the swapping device is dedicated in the sense that activity there does not affect the throughput of the central subsystem. (The analytic simplicity resulting from this assumption will become apparent.) The basis of our approach is Algorithm 9.1. As shown in Figure 9.4, we modify the high-level model of that algorithm to include a center representing the swapping device, in addition to the FESC representing the central subsystem. The only new issue that we must confront is determining the service demand at the former center.

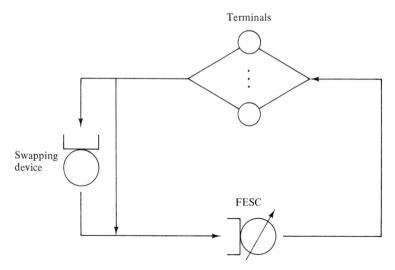

Figure 9.4 − **High-Level Model for Swapping to a Dedicated Device**

An interaction's service demand at the swapping device, D_{swap}, will be equal to the product of two terms: the probability that a swap precedes an interaction, $P[swap]$, and the service time for a swap in and subsequent swap out (both must occur), S_{swap}. S_{swap} is readily determined, but knowledge of the swapping policy of an operating system is necessary to estimate $P[swap]$. Here is an approach that can serve as a starting point. As in Algorithm 9.1, let there be N customers, M of whom can occupy memory simultaneously. We identify three cases:

- If $N \leqslant M$ then no swapping will occur. Thus $P[swap] = 0$.
- If $N > M$ then there will be some swapping. Let Q_{ready} be the average number of ready customers. If $Q_{ready} \geqslant M$ then a swap will precede every interaction. This is the case because we assume that only ready customers will be occupying memory, so a customer making a transition from the thinking state to the ready state will never be memory resident. Thus $P[swap] = 1$. (This clearly is an approximation, since we consider only the average number of ready customers.)
- If $N > M$ and $Q_{ready} < M$ then a swap will sometimes but not always precede an interaction. On the average there are $N - Q_{ready}$ thinking customers. Of these, $M - Q_{ready}$ are memory resident. So a customer leaving the thinking state requires a swap with probability:

$$P[swap] = 1 - \frac{M - Q_{ready}}{N - Q_{ready}} = \frac{N - M}{N - Q_{ready}}$$

The first of these three cases can be identified easily, since N and M are basic inputs. To distinguish between the second and third cases we need to know Q_{ready}, the average number of ready customers. This is an output of the model, not an input. Iteration is required, as described in Algorithm 9.4. (In the case that $N \leqslant M$, the swapping device can be ignored, and Algorithm 9.1 can be applied directly. For completeness, however, we include this case in Algorithm 9.4.)

From examination of the algorithm, our reliance on the assumption that the swapping device was dedicated should become evident. We constructed a flow equivalent representation of the central subsystem prior to iterating, and did not modify this representation subsequently. This requires that the load dependent throughputs of the central subsystem be independent of the level of swapping activity.

9.4.2. Swapping to a Shared Device

Especially in smaller systems, the swapping device also is apt to be used for other activities. To the extent that swap traffic impedes these activities (and vice versa), the analysis performed in the previous subsection will be invalid. Here, we will represent in our model this contention

9.4. Swapping

> 1. As in Algorithm 9.1, define a load dependent server that is flow equivalent to the central subsystem.
> 2. Define a high-level model consisting of the workload (N users with think time Z), the FESC from Step 1, and a center representing the swapping device. Initially, set the service demand at this last center, D_{swap}, to zero.
> 3. Evaluate this model. Obtain Q_{ready}, the average number of ready customers. This is equivalent to Q_{FESC}, the average queue length at the FESC. Use Q_{ready} to calculate a revised estimate for a customer's service demand at the swapping device, as follows:
>
> $$D_{swap} = S_{swap} \times P[swap]$$
>
> where:
>
> $$P[swap] = \begin{cases} 0 & N \leqslant M \\ 1 & N > M \text{ and } Q_{ready} \geqslant M \\ \dfrac{N-M}{N-Q_{ready}} & N > M \text{ and } Q_{ready} < M \end{cases}$$
>
> 4. Based on the discrepancy between the current and previous estimates for D_{swap}, decide whether to repeat Step 3 or to terminate.

Algorithm 9.4 — Swapping to a Dedicated Device

due to swapping. As before, an iterative analysis will be required. We will broaden the scope of the iteration to include the calculation of the load dependent throughputs, which now will vary with our estimate of swapping activity.

In generalizing Algorithm 9.4 a conceptual problem arises: Should the service center representing the swapping device appear in the high-level model (where swapping logically occurs) or in the low-level model (because by assumption this device also is used for file activity, which logically belongs in the low-level model). Fortunately this problem is not of practical concern, because only slight differences in results will occur. We choose to return to the high-level model used in Algorithm 9.1, and to represent all activity at the swapping device, both swapping activity and file activity, in the low-level model.

The low-level model, then, will consist of as many centers as there are processing resources. The service demand at most of these centers will be an intrinsic property of the workload, determined from measurement data. At the center representing the swapping device, however, the service demand will have two components: one due to file activity, determined from measurement data, and one due to swapping activity, determined iteratively as in Algorithm 9.4. The analysis is conducted as stated in Algorithm 9.5.

---

1. Define a low-level model consisting of the service centers representing the processing resources that comprise the central subsystem. At the center representing the swapping device, the service demand will have two components: one due to file activity, determined from measurement data, and one due to swapping activity, determined iteratively. Initially, assume that this latter component is equal to zero.
2. As in Algorithm 9.1, evaluate this low-level model for each feasible population, create an FESC, and define and evaluate a high-level model.
3. As in Algorithm 9.4, use the value of Q_{ready} obtained from the high-level model to calculate a revised estimate for the swapping activity component of the service demand at the swapping device. Based on the discrepancy between this estimate and the previous one, decide whether to repeat Steps 2 and 3 or to terminate.

Algorithm 9.5 − Swapping to a Shared Device

As an example, we return to the simple system considered in Section 9.3.1. Assume that the disk with an intrinsic service demand of 4 seconds also is used for swapping, and that the service time for a one-way swap of a 100K program is 150 msec.

On the first iteration we assume that no swapping occurs, so we evaluate the same low-level model used in Section 9.3.1, obtaining the same load dependent throughputs. We then construct and evaluate the same high-level model used in Section 9.3.1, obtaining the same value for the average number of ready users, 4.5. Now, we iterate. Since $Q_{ready} \geqslant M$ (the memory capacity was three customers in the example), we assume that a swap precedes each interaction. The service demand at the swapping device is equal to the sum of the intrinsic service demand there (4.0 seconds) and the service demand due to swapping. This latter service demand equals the product of the one-way swap service time (0.15

seconds), the probability that a swap precedes an interaction (1), and 2 (to account for the outswap that also must occur): 0.3 seconds. Total service demand at the swapping device is thus 4.3 seconds. We once again evaluate the low-level model for populations from 1 to 3, obtaining load dependent throughputs of 0.1075, 0.1577, and 0.1851, respectively. Using these rates to define a flow equivalent server, we again evaluate the high-level model, obtaining:

>throughput: 0.170 interactions/second
>average interactive response time: 28.0 seconds
>average number of ready users: 4.8

Since our revised estimate for Q_{ready} still is greater than the capacity of memory, we still estimate that a swap precedes every interaction, and further iteration is unnecessary. As we would expect, throughput and response time are slightly worse than in Section 9.3.1, where swapping activity was ignored.

9.5. Paging

Most computer programs exhibit *locality of reference*: although a program may have a large *address space*, only a small portion of that address space will be referenced during any short time interval. *Virtual memory systems* exploit this property by allocating to each program an amount of (physical) primary memory that is smaller than the program's (virtual) address space, then using a combination of hardware and software to translate virtual addresses into physical addresses and to transfer portions of the virtual address space between primary memory and disk.

There are two principal advantages to virtual memory: the system can accommodate programs whose virtual address spaces are larger than the amount of physical memory that is attached to the CPU, and the number of concurrently active programs can be larger than would otherwise be possible. There is also a disadvantage: CPU and I/O resources must be devoted to the management of the virtual memory.

Virtual memory systems may employ *paging*, or *segmentation*, or both. Our focus in this section will be on paging. We consider the system's physical memory to be divided into some number of fixed-size *page frames*, and the address space of each program to be divided into some number of *pages* of the same fixed size. The operating system must make decisions on both a system level (How many programs should be allowed to compete for memory resources? How many page frames should be allocated to each of these programs?) and on a program level (Which pages should occupy the page frames allocated to a program? Alternatively, which page should be removed from primary memory in order to accommodate a non-resident page that has just been referenced?) The I/O associated with moving pages between primary memory and disk in response to *page faults* is the aspect of system behavior whose modelling we will study in this section.

Modelling paging has much in common with modelling swapping. The fundamental issue is to determine the contribution of memory management activity to service demands. If it is not anticipated that the system modifications under consideration will have a significant effect on service demands at the paging devices, then these service demands can be taken from measurement data. As with swapping, though, the influence of system modifications on the level of paging activity is something we would like to learn from our model, rather than provide as an input. Paging activity is especially difficult to forecast because it is highly dependent on the characteristics of individual programs and on their interactions with each other through the memory management policies of the operating system.

Consider a simple example: a small multiprogrammed virtual memory system supporting a batch workload. Processing resources include a CPU at which jobs require an average of 3 seconds of service, two file disks at which jobs require an average of 8 and 2 seconds of service, respectively, and a paging disk.

Service demand at the paging disk is determined by considering in more detail the configuration of the system, the policies of the operating system, and the characteristics of the jobs. The system has 512 page frames of physical memory, 300 of which are available to user jobs. The operating system allocates memory on an *equipartition* basis: a multiprogramming level is selected and the available page frames are divided equally among the jobs. The memory reference characteristics of jobs and the page replacement policy of the operating system interact with one another in a manner that is reflected by the *program lifetime function*, shown in Figure 9.5. This function shows, for a single job, the average number of milliseconds of CPU service that elapse between page faults for various numbers of allocated page frames.

Suppose we are asked to model the performance of this system at multiprogramming levels of 2 through 8. A separate analysis must be conducted for each multiprogramming level. Each analysis must begin by determining the service demand at the paging disk. Consider a multiprogramming level of 5. Because 300 page frames are available for users, the equipartition policy will allocate $300/5 = 60$ page frames to each of the 5 jobs. The lifetime function tells us that at this memory allocation a job will experience an average of one page fault every 9 milliseconds of CPU processing. Since the average CPU service requirement of a job is 3 seconds, a job, on the average, will experience $3000/9 = 333$ page faults.

9.5. Paging

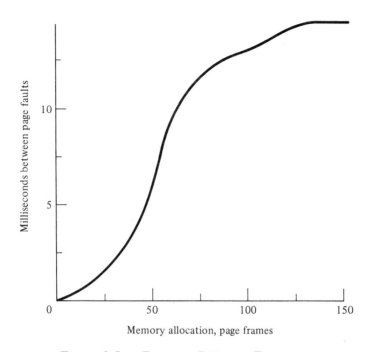

Figure 9.5 − Program Lifetime Function

Suppose we know that an average of 12.5 milliseconds of paging disk service is required to process a single page fault. Then on the average each job will place a service demand of $333 \times .0125 = 4.16$ seconds on the paging disk. The resulting queueing network model will have a population of 5 customers, and four service centers with service demands of 3, 8, 2, and 4.16 seconds.

Figures 9.6, 9.7, and 9.8 show respectively system throughput in jobs/minute, average job response time in seconds, and device utilizations, each as a function of multiprogramming level.

This example illustrates the techniques used to analyze paging systems. The difficulties that arise in such studies are related to the availability of data from which to parameterize the model. The example was very much simplified in this respect. For instance:

- It is extremely difficult to acquire paging lifetime data for a program. Doing so requires detailed tracing of the execution of the program in the context of the page replacement policy used by the operating system.
- The paging characteristics of a program are likely to vary as the program passes through different phases of execution, with each phase requiring a different lifetime function.

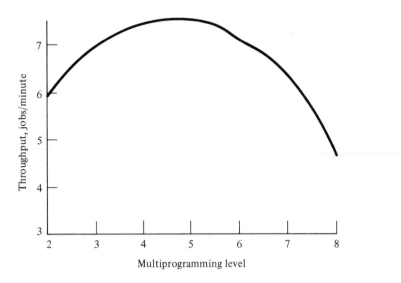

Figure 9.6 — Throughput Versus Multiprogramming Level

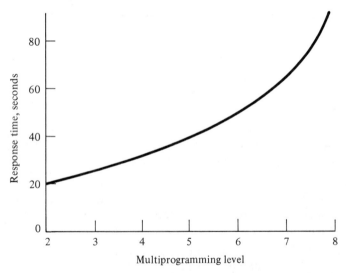

Figure 9.7 — Response Time Versus Multiprogramming Level

9.5. Paging

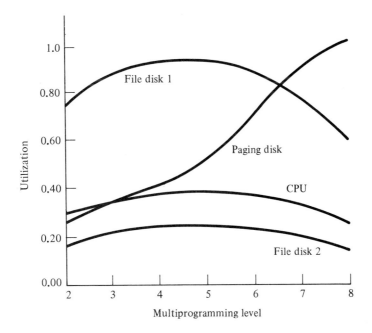

Figure 9.8 — Device Utilizations Versus Multiprogramming Level

- The paging characteristics of different programs will be dissimilar.
- Since different programs exhibit different paging characteristics, operating systems typically do not employ an equipartition strategy. At the very least, a different number of physical page frames will be allocated to each program.
- More likely, the operating system will change the number of physical page frames allocated to a program over the life of that program. Thus the number of programs that can be accommodated in memory simultaneously will vary with time.
- As the number of jobs that can be accommodated in memory varies, preemptive swapping may be employed. The swapping policy in a virtual memory system may be quite complex.

In practice, analysts using queueing network models to study virtual memory systems ignore many of these subtleties by making homogeneity assumptions similar to those we have encountered in other contexts. For example, it is common to consider only the average number of page frames allocated to a program, to assume that this average is the same for all programs belonging to the same class, and to assume that this average is largely independent of the load on the system. Studies incorporating such homogeneity assumptions generally are successful even in projecting the effect of modifications to the memory subsystem, e.g., the addition of memory. In the next section we will consider two such studies.

9.6. Case Studies

In this section we consider two successful case studies in which queueing network models were used to explore the effects of modifications to the memory subsystems of virtual memory systems. In the first study, a very simple model was used to evaluate the effects of increased paging device speed and of additional memory on the performance of an early IBM virtual memory system. In the second study, a more sophisticated model was used to evaluate workload and configuration changes to a Digital Equipment Corporation VAX/VMS system.

9.6.1. A Simple Model of an Early IBM Virtual Memory System

This study is from the early days of computer system analysis using queueing network models. At the time it was conducted, techniques for efficiently evaluating separable queueing networks (Chapters 6 and 7) and for representing memory subsystems using flow equivalence and hierarchical modelling (Chapters 8 and 9) were not widely known. This stimulated a number of clever "short cuts". The study serves to illustrate that useful results can be obtained for complex systems even in the presence of rather extreme simplifications. The system under consideration had the following characteristics:
- a small number of interactive users
- a CPU-intensive workload
- a large number of disks
- a low ratio of think time to response time (i.e., slow response)
- a paging virtual memory system
- a multiprogramming level limited to three to avoid thrashing

Figure 9.9 shows the model that was used in the study. It has one customer class. Each customer cycles through periods of thinking, (possibly) queueing for memory, and alternating bursts of CPU and I/O service. Because the multiprogramming level was limited to three and there were many possible paths to the I/O devices, little or no I/O queueing took place. This allowed the model to be simplified by representing the I/O subsystem as a single delay center. (The authors of the study probably evaluated the model by hand. Representing the large number of disks by a single delay center saved much tedious computation. Given a queueing network analysis package, it would be equally easy to represent all disks explicitly. This would be a "safer" procedure, since it would not rely on the assumption that no I/O queueing takes place.)

Because of the memory queue, the model is not separable. Even without the FESC approach described earlier in this chapter, though, it is possible to obtain accurate results in two extreme cases. The first is that

9.6. Case Studies

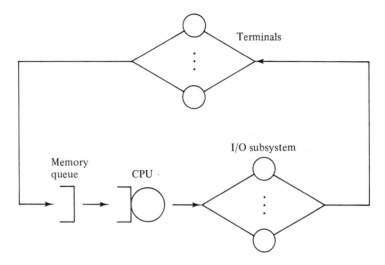

Figure 9.9 − The Model

memory utilization is low, so that little or no memory queueing takes place. This would occur, for instance, if response times were so short that most users spent the majority of their time thinking. Thus, the number of users in memory simultaneously would be small, and the chance that a user ever would need to queue for memory would be negligible. In this extreme case the memory queue could be ignored entirely, yielding a separable model.

The other extreme is that memory is utilized nearly 100%, so that the multiprogramming level of the system remains constant at its maximum. This was in fact the case in the system under consideration. This analytically fortunate situation allowed the model to be evaluated as follows:

- From the full model of Figure 9.9, extract the central subsystem (the queueing center representing the CPU and the delay center representing the I/O subsystem).
- Evaluate this central subsystem model with appropriate service demands and with a fixed population equal to the maximum multiprogramming level (in this case, three). Obtain throughput, X.
- Apply the response time law (N and Z must be provided).

For the system under consideration, evaluation of the central subsystem model gave a throughput of .395 interactions/second. From measurements, the number of interactive users was 10 and their average think time was 4 seconds. Applying the response time law:

$$R = \frac{N}{X} - Z = \frac{10}{.395} - 4 = 21.3 \text{ seconds}$$

The measured response time was 21.0 seconds.

Two changes to the configuration were being considered in an attempt to reduce the effect of the severe memory contention being experienced: upgrading the paging disks to drums, and adding memory. The upgrade to drums can be reflected in the model by adjusting the service demand at the delay center representing the I/O subsystem. The part of this service demand due to paging activity must be reduced to account for the elimination of the seek portion of data access (the drums have fixed heads) and for a decrease in the latency and data transfer portions (the drums have higher rotation speed than the disks). These adjustments can be estimated rather easily. Once a new service demand has been calculated, the evaluation can be carried out as before.

Representing the addition of memory is somewhat more challenging, since this modification affects paging activity (and thus the service demand at the delay center) in a manner that is not easily estimated. The addition of memory was studied for two cases: using the additional memory to increase the maximum multiprogramming level while maintaining the current number of page frames allocated to each active user, and using the additional memory to increase the number of page frames allocated to each active user while maintaining the current maximum multiprogramming level. To model the first case, it was determined that the additional memory would allow two more users to be active while maintaining the current memory allocation per user. Since the memory allocation per user would remain fixed, it was postulated that the page fault count of each user would be unaffected by the increase in the multiprogramming level. The memory addition was therefore modelled by increasing the number of customers in the central subsystem model from three to five and evaluating as before.

The other case, increasing the memory allocation to the three active users, can be expected to reduce the number of page faults per user. The service demand at the delay center in the model must be adjusted to reflect this. To estimate each user's service demand due to paging in the new environment, an experiment was conducted in which the maximum multiprogramming level of the existing system was reduced to two. (It had been determined that the number of page frames available to each of two active users on the existing configuration would be roughly the same as the number of page frames available to each of three active users on the proposed configuration.) I/O subsystem service demand was calculated from measurements during this experiment. The memory addition was modelled by using this value and a customer population of three, evaluating as before.

9.6. Case Studies

It is important to note a limitation arising from the fact that the evaluation technique assumes the central subsystem runs continuously at the maximum multiprogramming level. If response times improve significantly, this assumption may no longer be valid. Should this occur, the model may yield optimistic results. For any particular set of parameter values, the validity of the assumption can be checked by computing the average number of customers competing for memory (the average number of ready customers). If there are, on average, at least as many ready customers as can be accommodated in memory, the results of the model can be expected to be accurate. The average number of ready customers can be computed by applying Little's law to the central subsystem plus the memory queue. For the model of the original system:

$$N_{ready} = XR = .395 \times 21.3 = 8.4 \text{ customers}$$

The previous paragraph points out that proposed system modifications can have side effects that invalidate assumptions made by the particular evaluation technique in use. It is also possible for modifications to have side effects that invalidate measurements used to calculate model inputs. In the system described here, the user think time was measured as 4 seconds. This low value probably was due in part to the poor response time of the system: while one request was processing, users had time to prepare their next. If a system modification resulted in significantly improved response times, the think time would likely increase because of a reduction in this overlap.

Much of the success of a modelling study depends on the analyst's ability to anticipate significant side effects.

9.6.2. A Model of VAX/VMS

This section presents a queueing network model of Digital Equipment Corporation's VAX/VMS system. Memory management in VMS includes swapping, paging, and a shared cache of page frames. The questions addressed by this modelling study relate to workload and configuration changes that can be expected to affect paging and swapping behavior. The configuration is a small one, making homogeneity assumptions risky. For these reasons, the example serves to integrate a number of the techniques presented in this chapter, and we will examine it in considerable detail. The model is of an early release of VMS and does not reflect certain major changes in the system that have occurred since that time. The study predates the development of the algorithms for evaluating multiple class memory constrained queueing networks described in Section 9.3, so an alternative technique was employed.

9.6.2.1. Essentials of the System

As noted, memory management in VMS is accomplished through a combination of swapping, paging, and a shared cache of page frames.

A physical memory requirement, the *resident set size*, is associated with each process. An active process is guaranteed a number of page frames equal to its resident set size. Should a page fault occur in a process already using its entire allocation of page frames, a FIFO page replacement policy is used to select a page for removal from the resident set.

Since VMS makes no attempt to adjust processes' resident set sizes in response to observed behavior, an efficient allocation of page frames among active processes is unlikely. Since FIFO is a notoriously bad page replacement policy, an efficient choice of resident set membership is equally unlikely. To compensate for these shortcomings, VMS maintains a cache of page frames that is shared among the active processes. When a page is removed from a process' resident set it is added to this *shared page cache*. A fault on a page held in the cache can be resolved without disk I/O. Therefore we must distinguish between a page fault, which may not result in I/O, and a *paging transfer*, in which a page is retrieved from disk in response to a page fault. (Actually, pages are *clustered* for efficiency, and several pages are transferred in a single paging transfer.) The maximum and minimum sizes of the shared page cache are regulated by system parameters. If the cache exceeds its maximum size, pages are purged FIFO until the cache reaches its minimum size. Thus, as shown in Figure 9.10, physical memory can be divided logically into four parts: page frames permanently allocated to VMS, page frames containing processes' resident sets, page frames belonging to the shared page cache, and unallocated page frames.

Before a process that is swapped out can become active, it must be allocated sufficient page frames to accommodate its resident set. If enough unallocated page frames are not available, some other process must first be swapped out. Typically this process would correspond to an interactive user in the think state. The swapping rate at saturation is regulated by the *quantum*: a ready process is not eligible to be swapped out until it has acquired one quantum of CPU service.

One final detail. In point of fact, unallocated page frames are added to the shared page cache: the cache is allowed to grow until it reaches a size equal to the larger of its maximum size parameter and the number of page frames left over after VMS and the memory-resident processes have taken their toll. Cache pages that have been modified are written to disk when the maximum size parameter is reached, but the images of these pages are allowed to remain in memory and, if accessed, can be made available without disk I/O. The concept of an unallocated page frame principally is of use in understanding the swapping policy.

9.6. Case Studies

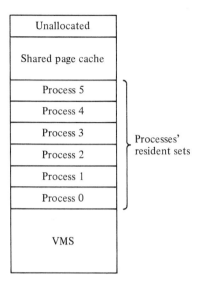

Figure 9.10 − A Logical View of Memory in VAX/VMS

9.6.2.2. The Queueing Network Model

The configuration under study is a small one: 512K bytes of memory and a single disk used for swapping, paging, and file activity. The workload is a benchmark consisting of one batch job (repeated compilation of a 10,000 line program) and 7 simulated interactive users each performing a specific task (compilation, execution, editing, trivial commands). The study involves validating a model of the base system, using this model to project the effect of specific modifications to the workload (eliminating the interactive users and running the batch job in isolation) and to the configuration (doubling the amount of physical memory), and finally making these modifications and comparing the results with the projections of the model. Four aspects of the system are of special interest in the context of the current chapter:

- There is a memory constraint.
- The proposed system modifications can be expected to affect the paging behavior of the system, which therefore must be modelled explicitly.
- The proposed system modifications also can be expected to affect the level of swapping activity, so this also must be modelled explicitly.

- The single disk means that swapping activity can be expected to interfere with the throughput of paging and file I/O.

The basis of the analysis is the familiar two-level hierarchical model: a low-level model that is evaluated at each feasible population in order to define an FESC for use in a high-level model. The low-level and high-level models are described in the following paragraphs.

The Low-Level Model

The low-level model has two service centers, representing the CPU and the disk, and two customer classes, representing the batch job and the interactive users. In the actual system there was a single batch jobstream that was locked into memory to reduce swapping activity, so in the low-level model the batch class has a constant multiprogramming level of one. In the actual system the seven interactive users had various resident set sizes, but the differences were small and on the average six interactive users could be accommodated in addition to the batch job. So in the low-level model there will be from zero to six customers in this class.

For each class, measurement data yields CPU service demand and the file activity component of disk service demand. Since we wish to explore system modifications that will affect paging and swapping behavior, we must develop techniques to estimate for each class the components of disk service demand due to these activities.

First, consider paging activity. Recall that each VMS process has a fixed allocation of page frames when it is memory resident. Because of this, the number of page faults sustained by a process will be insensitive to system load. However, the proportion of those page faults that result in disk I/O will vary with load, since this proportion is related to the number of page images in the shared page cache belonging to the process in question, which in turn depends on the number of processes actively using the cache. Thus the key to representing paging activity is estimating the effectiveness of the shared page cache.

We can measure the average number of page faults per interaction and we can calculate the average disk service time per paging transfer. We expect both of these quantities to be insensitive to the proposed modifications. The effectiveness of the shared page cache is reflected in the ratio of page faults to paging transfers. We can calculate this ratio for the benchmark measurement interval. In order to project performance under system modifications, we make the assumption that this ratio is linearly related to the average number of cache page frames available to each process actively using the cache. As an example, during the benchmark an average interaction caused 158 page faults and the ratio of page faults to paging transfers was 4:1. Thus an average interaction caused

158/4 = 39.5 paging transfers. The average number of active processes was eight: six interactive users, the batch job, and VMS (portions of which are pageable). Our assumption makes it possible to estimate that if the average number of active processes were three, the ratio of page faults to paging transfers would be $4 \times 8/3 = 10.7$, and an interaction would cause $158/10.7 = 14.8$ paging transfers. Our assumption also allows us to estimate that if the size of the shared page cache were doubled by the addition of memory (with three active processes), the ratio of page faults to paging transfers would become $10.7 \times 2 = 21.4$, and an interaction would cause $158/21.4 = 7.4$ paging transfers. Multiplying the average number of paging transfers per interaction by the average disk service time per paging transfer yields the paging activity component of disk service demand.

Next, consider swapping activity. The approach presented in Algorithm 9.5 is suitable except in the case that the average number of ready users exceeds the memory constraint. In this case, VMS will swap once per interaction plus once per quantum. The number of swaps per interaction due to the latter can be approximated by dividing the CPU service requirement per interaction by the quantum length.

The High-Level Model

We begin the analysis of the system by establishing initial values for the average numbers of ready and active interactive customers. These values allow us to estimate disk service demand due to paging (the average number of active customers is used for this) and swapping (the average number of ready customers is used for this). Given disk service demand, we can evaluate the low-level model. We do so for each feasible interactive population (the batch population is always one), obtaining load dependent throughputs which we use to construct an FESC.

The high-level model consists of this FESC and the workload (N customers with think time Z). Evaluation of this model yields revised estimates for the average numbers of ready and active interactive customers. If these revised estimates differ substantially from those used in the previous evaluation of the low-level model, we iterate using the new values.

Interactive response time and throughput, and thus the contribution of interactive users to device utilizations, can be determined directly from the high-level model. Batch throughput is calculated by taking the sum of the batch throughput at each interactive population (obtained from the low-level analysis) weighted by the proportion of time each of those interactive populations is encountered (obtained from the high-level analysis). Average batch response time and the batch contribution to device utilizations then can be determined by application of Little's law.

9.6.2.3. Use of the Model

In this section we illustrate the use of the model in some detail. Table 9.7 displays certain measured characteristics of the benchmark.

---

average interaction:

    0.74 CPU seconds
    158 page faults
    12.4 file I/O operations

batch job:

    330 CPU seconds
    101386 page faults
    918 file I/O operations

Table 9.7 — Measured Characteristics of the Benchmark Jobstream

Table 9.8 displays certain system parameters relating to paging activity that were measured during the benchmark.

63.4 page faults per second
55.8 pages transferred per second
15.9 paging transfers (physical I/Os) per second

Table 9.8 — Paging Activity Measures

Based upon knowledge of device characteristics, the average number of bytes transferred per swap and per file operation, and the page I/O clustering factors evident from Table 9.8, we calculate the I/O service times shown in Table 9.9.

.150 seconds per two way (in-out) swap
.037 seconds per file I/O
.039 seconds per paging transfer

Table 9.9 — I/O Operation Service Times

9.6. Case Studies

First we calculate service demands for interactive users. The CPU service demand is .74 seconds. The disk service demand due to file I/O is $12.4 \times .037 = .46$ seconds. From Table 9.8, the ratio of page faults to paging transfers is $63.4/15.9 = 3.99$. Thus an average interaction will cause $158/3.99 = 39.6$ paging transfers, with a resulting disk service demand of $39.6 \times .039 = 1.54$ seconds. In the benchmark, interactive think times were set to zero. (The system under study had extremely long response times, so users often typed ahead.) Thus there were always 7 ready and 6 active interactive users. We use the third component of the swapping approximation: each interaction requires $1 + .74/1 = 1.74$ swaps (the quantum length was 1 second), so interactive disk service demand due to swapping is $1.74 \times .150 = .26$ seconds. Total disk service demand is therefore $.46 + 1.56 + .26 = 2.26$ seconds.

Next we consider the batch job. CPU service demand is 330 seconds. Disk service demand due to file I/O is $918 \times .037 = 34$ seconds. Each batch job will cause $101386/3.99 = 25410$ paging transfers, with a resulting service demand of $25410 \times .039 = 991$ seconds. Since the batch job is not swapped, its total disk demand is $34 + 991 = 1025$ seconds.

Because there are always 7 ready and 6 active interactive users, we can take a short cut, analyzing the low-level model only one time, with a population of 1 batch job and 6 interactive users. With the exception of interactive response time, all interesting system performance measures can be obtained directly from the results of this analysis. Interactive response time is calculated as in the previous case study, by applying the response time law with $N=7$, $Z=0$, and X equal to the throughput obtained from the evaluation. Table 9.10 displays both observed and projected performance measures.

performance measure	observed	projected
total CPU utilization	.30	.32
swapping rate (swaps/sec.)	.72	.64
interactive		
throughput (int's./min.)	22.2	22.2
response time (secs.)	18.9	19.0
batch		
throughput (jobs/min.)	.0091	.0082

Table 9.10 − Original System

Next, we explore the effect of eliminating the interactive workload, running the batch job in isolation. The swapping rate will be zero. The cache will be shared by VMS and the batch job, rather than among 8 processes. It will expand to occupy the space vacated by the interactive users, increasing in size from 150 to 450 pages, a factor of 3. Our linear approximation to the effectiveness of the shared page cache estimates that the ratio of page faults to paging transfers will be $3.99 \times 3 \times 8/2 = 47.9$. We therefore calculate that the batch job's disk service demand due to paging will be $101386/47.9 \times .039 = 82.5$ seconds, and that its total disk service demand will be $34+82.5 = 116.5$ seconds. We evaluate the low-level model once, with a single batch job. Table 9.11 displays both observed and projected performance measures.

performance measure	observed	projected
total CPU utilization	.68	.73
batch throughput (jobs/min.)	.124	.133

Table 9.11 — Batch Only

Finally, we explore the effect on the original workload of doubling the size of memory. Once again, the swapping rate will be zero. All seven interactive users will be memory resident, so the page cache will be shared by 9 rather than 8 active processes. The size of the cache will increase from 150 to 1125 pages, a factor of 7.5. The linear approximation to the effectiveness of the shared page cache estimates that the ratio of page faults to paging transfers will be $3.99 \times 7.5 \times 9/8 = 33.7$. Interactive disk service demand due to paging will be $158/33.7 \times .039 = .183$ seconds, and total interactive disk service demand will be $.46+.183 = .643$ seconds. Batch disk service demand due to paging will be $101386/33.7 \times .039 = 117$ seconds, and total batch disk service demand will be $34+117 = 151$ seconds. We simply can evaluate the low-level model with a single batch job and 7 interactive customers. Table 9.12 displays both observed and projected performance.

performance measure	observed	projected
total CPU utilization	.89	.95
interactive throughput (int's./min.)	55.	65.7
response time (secs.)	7.6	6.38
batch throughput (jobs/min.)	.040	.026

Table 9.12 — Additional Memory

The projected performance measures shown in Tables 9.10 - 9.12 are sufficiently accurate to be useful. The discrepancies are reasonable when we consider the magnitude of the system modifications, the crudeness of the linear approximation to shared page cache effectiveness, and the absence of any consideration of the effect of paging and swapping rates on CPU overhead.

9.7. Summary

Memory and its management affect the performance of computer systems in two major ways. The existence of a memory constraint can impose a bound on the multiprogramming level, and thus the throughput, of a system. The overhead associated with memory management can impede the progress of "useful" work. In this chapter we have presented techniques for representing these effects, techniques which extend the flexibility of separable queueing network models.

It never is possible to represent every detail of an operating system's memory subsystem in a queueing network model. However, *nor is it necessary or desirable to do so*. This latter point is a philosophical cornerstone of computer system analysis using queueing network models, and cannot be overemphasized. In each particular modelling study − for each configuration, workload, and set of questions to be investigated − it is imperative to identify the *essential* characteristics of the system − those that can be expected to have primary effects on performance − and to represent these and only these in the model. A large body of case study literature testifies to the success of this approach.

In closing this chapter, we should mention two related points. First, the fact that we have organized Part III on a "subsystem" basis rather than on a "technique" basis means that the broad applicability of certain techniques is not emphasized. As an example, Algorithm 9.1 for evaluating single class memory constrained subsystems is applicable to any subsystem in which there is a population constraint. (See Exercise 2.)

The second related point is a brief mention of cache memory: relatively small, fast memory sometimes interposed between the CPU and primary memory, which is managed by hardware and firmware in a manner not unlike the paging that may occur one level removed in the memory hierarchy. The effect of cache memory is usually included in a queueing network model simply as an adjustment to the service demand at the CPU. This is consistent with the decomposition approach, since memory references occur extremely frequently relative to other events. The analyst must be aware that a statement about the instruction execution rate of a machine with a cache must necessarily rely on some assumption about the cache hit ratio, and that this assumption should be verified, probably by benchmark.

9.8. References

The implications of the fact that throughput is convex with respect to multiprogramming level were noted by Dowdy, Gordon, and Agre [Dowdy et al. 1979].

Brandwajn [1974] first analyzed single class memory constrained systems using a decomposition approach, although he did not couch the analysis in the simple terms of an FESC. Lazowska and Zahorjan [1982] and Brandwajn [1982] independently developed the extension to multiple classes. An interesting alternative for evaluating single class models with non-homogeneous memory requirements was suggested by Brown, Browne, and Chandy [Brown et al. 1977].

The iterative analysis of swapping behavior presented in Section 9.4 is due to Lazowska [1979]. The analysis of a paging system presented in Section 9.5 comes from Graham and Lazowska [1978].

The case study of the early IBM virtual memory system was conducted by Boyce and Warn [1975]. Lazowska [1979] performed the VAX/VMS case study. Hodges and Stewart [1982] use the same techniques to analyze a more recent version of VAX/VMS; this system is described in detail by Levy and Eckhouse [1980]. A good overview of memory management in general, and of paging and segmentation in particular, is provided by Denning and Graham [1975].

[Boyce & Warn 1975]
J.W. Boyce and David R. Warn. A Straightforward Model for Computer Performance Prediction. *Computing Surveys* 7,2 (June 1975), 73-93. Copyright © 1975 by the Association for Computing Machinery.

[Brandwajn 1974]
Alexandre Brandwajn. A Model of a Time-Sharing System Solved Using Equivalence and Decomposition Methods. *Acta Informatica* 4,1 (1974), 11-47.

[Brandwajn 1982]
Alexandre Brandwajn. Fast Approximate Solution of Multiprogramming Models. *Proc. ACM SIGMETRICS Conference on Measurement and Modeling of Computer Systems (1982), 141-149.*

[Brown et al. 1977]
R.M. Brown, J.C. Browne, and K.M. Chandy. Memory Management and Response Time. *CACM* 20,3 (March 1977), 153-165.

[Denning & Graham 1975]
Peter J. Denning and G. Scott Graham. Multiprogrammed Memory Management. *Proc. IEEE* 63,6 (June 1975), 924-939.

9.9. Exercises

[Dowdy et al. 1979]
 Lawrence W. Dowdy, Karen D. Gordon, and Jonathan R. Agre. On the Multiprogramming Level in Closed Queuing Networks. Technical Report TR-831, Department of Computer Science, University of Maryland, November 1979.

[Graham & Lazowska 1978]
 G. Scott Graham and Edward D. Lazowska. Quark: A Performance Evaluation Package for an Operating Systems Course. Technical Report 78-04-01, Department of Computer Science, University of Washington, April 1978.

[Hodges & Stewart 1982]
 Larry F. Hodges and William J. Stewart. Workload Characterization and Performance Evaluation in a Research Environment. *Proc. ACM SIGMETRICS Conference on Measurement and Modeling of Computer Systems* (1982), 39-50.

[Lazowska 1979]
 Edward D. Lazowska. The Benchmarking, Tuning and Analytic Modelling of VAX/VMS. *Proc. ACM SIGMETRICS Conference on Simulation, Measurement and Modeling of Computer Systems* (1979), 57-64. Copyright © 1979 by the Association for Computing Machinery.

[Lazowska & Zahorjan 1982]
 Edward D. Lazowska and John Zahorjan. Multiple Class Memory Constrained Queueing Networks. *Proc. ACM SIGMETRICS Conference on Measurement and Modeling of Computer Systems* (1982), 130-140.

[Levy & Eckhouse 1980]
 Henry M. Levy and Richard Eckhouse, Jr. *Computer Programming and Architecture: The VAX-11.* Digital Press, 1980.

9.9. Exercises

1. Suppose that in the example of Section 9.2 the observed average multiprogramming levels of the three classes had been 2.60, 0.40, and 1.75, but that no additional information was available (i.e., you did not know the actual distribution of multiprogramming mixes).
    a. How could you analyze this system using approximate MVA?
    b. How could you analyze this system using exact MVA?

2. Consider a Control Data 6000-series batch computer system consisting of a CPU, $K-1$ disks, and P peripheral processors, with a fixed multiprogramming level of N jobs. A job desiring disk service first must contend for access to any one of the PPs. Once allocated, the PP is held while the job contends for and uses the specific disk on which its data resides. At the conclusion of the I/O activity, both the disk and the PP are released, and the job enters the CPU queue. Thus, although there may be N jobs and $K-1$ disks, at most P jobs can be using disks simultaneously. The actual number may be less than P, either because fewer jobs desire disk service, or because several jobs desire access to the same disk.

   a. Draw an analogy between this modelling problem and the single class memory constraint problem discussed in Section 9.3.

   b. Analyze a system in which there are 10 jobs, a CPU at which each job has a service demand of 50 seconds, 3 PPs, and 5 disks at which each job has service demands of 20, 25, 30, 35, and 40 seconds, respectively. Report CPU utilization, disk utilizations, and average job response times. (Use the Fortran program in Chapter 18, extended to accommodate FESCs as described in Chapter 20.)

   c. Analyze the same system ignoring the PP constraint. (That is, represent the system using a separable single class model with 6 centers and 10 jobs.) What error in job response times results from this assumption? How about CPU utilization?

3. Re-work the example of Section 9.3.1 for the following values of think time:

   a. 10 seconds

   b. 180 seconds

   Simpler approaches to modelling memory constraints do not require the use of FESCs. The case study in 9.6.1 presents one such approach. Another approach is simply to ignore the memory constraint, which causes the model to be separable and thus amenable to the standard MVA algorithms.

   c. For think times of 10, 60, and 180 seconds in this example, how well do you think each of the simpler approaches will work?

   d. Test your intuition by applying both approaches in these three cases, and comparing the results to those obtained using the more accurate flow equivalent technique.

9.9. Exercises

4. Some computer systems do not impose a fixed limit on the number of jobs that can be loaded in memory, but instead load jobs in a FCFS manner until either there are no jobs left to be loaded or no memory in which to load them.
   a. In the case where all jobs can be thought of as belonging to a single class, how can Algorithm 9.2 be used to model such systems?
   b. If jobs in the system have widely differing memory requirements (e.g., many small jobs but occasional very large jobs), we may wish to model the system using multiple job classes. In this case, how can Algorithm 9.3 be used?
5. In Section 9.5 a technique was described for modelling the primary effect of the change in page fault rate with system load (or equivalently with main memory allocation per job): the change in the service demand at the paging device. An important secondary effect is a change in CPU overhead per job due to page fault handling.
   a. How would you reflect this secondary effect in the model (i.e., what parameters would you change)?
   b. How would you determine appropriate parameter values for a specific system?
6. Suppose that a system contains a number of disks dedicated to swapping, and a number dedicated to paging.
   a. What modifications to the techniques of this chapter need to be made for such systems?
   b. What additional measurement information would be required to parameterize such models?
   c. In the absence of such measurements, what reasonable guesses could you make to allow you to analyze the model?

Chapter 10

Disk I/O

10.1. Introduction

Processor and primary memory technology has moved forward rapidly in recent years. Comparable advances have not occurred in the design of I/O subsystems. As a result, I/O subsystems are playing an increasingly critical role in computer system performance. Queueing network models of disk I/O subsystems are the subject of the present chapter.

In any study involving queueing network models, the analyst must begin by determining which system devices should be represented as service centers in the model, and what the service demands at these centers should be. With these parameters as input, the computational algorithms described in Part II use Little's law to calculate the effect of resource contention, yielding performance measures such as utilizations, throughputs, residence times, and queue lengths. Most postulated modifications to the system or to the workload are represented in the model as modifications to the service demands.

The "canonical" queueing network model that we have used throughout the book consists of service centers representing the CPU and the individual disk devices. Such a model is a very abstract representation of the contemporary IBM disk I/O subsystem configuration illustrated in Figure 10.1. The architectural complexity of this subsystem results from difficult compromises between cost and performance. At one extreme, requiring the CPU to monitor directly all phases of I/O activity would lead to poor performance (although low cost). At the other extreme, endowing each disk with sufficient intelligence to transfer data in a fully independent manner would lead to high cost (although good performance). The obvious approach is to introduce some number of shared devices of varying intelligence (channels, controllers, string heads, etc.) on the *path* between the CPU and the disks.

How is it that a simple model, which does not represent explicitly the many I/O path elements, can validate? The answer is that, typically, the effects of these "details" are captured in the disk service demands

10.1. Introduction

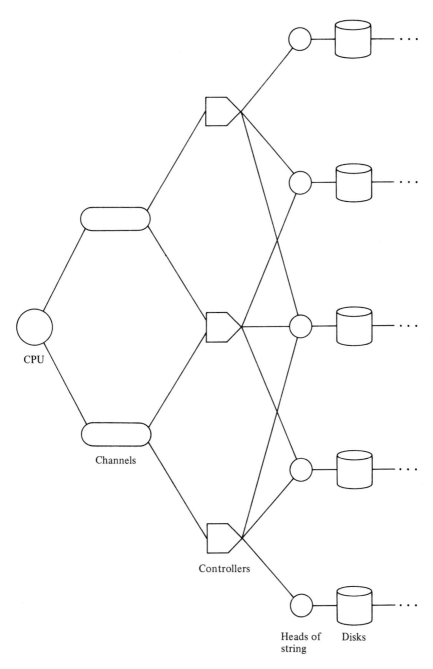

Figure 10.1 − **A Contemporary IBM I/O Subsystem**

obtained from measurement data. There are three intrinsic components of disk service time: *seek* (the time required to position the arm to the correct cylinder), *latency* (the time required for the start of the data record of interest to rotate under the heads) and *transfer* (the time required for the actual transfer of data). In addition, though, a disk is "held" by a customer during a *contention* period when data cannot be transferred due to the absence of a path back to the CPU. Thus, the result of I/O path contention is an *effective* disk service time (the sum of seek, latency, transfer, and contention times) that is longer than the *intrinsic* disk service time (the sum of seek, latency, and transfer times). Disk busy times increase correspondingly, and so the effect of I/O path contention is reflected in the disk service demand parameter of the queueing network model, which is calculated as $D_{disk} = B_{disk}/C$ (C here is the number of system completions).

How can our canonical model be used to project performance for modified environments? The answer to this question is at once very simple and very complex. On the one hand, many postulated system and workload modifications can be represented by appropriate adjustments to the service demand parameters of the model. For example, the primary effect of a 50% CPU upgrade can be represented by dividing all CPU service demands by 1.5: a customer that required six seconds of service on a 2 MIPS (million instructions per second) CPU will require four seconds of service on a 3 MIPS CPU. Similarly, the primary effect of adding I/O paths and reallocating disks can be represented by reducing the disk service demands, because I/O path contention can be expected to decrease. Unfortunately, it is difficult to quantify the amount of this reduction. The purpose of the I/O modelling techniques to be discussed in this chapter is to allow the analyst to deal with parameters that are meaningful: channels, controllers, strings, paths, disks, intrinsic I/O service requirements, etc. These techniques serve to translate a modification expressed in terms of these parameters into an appropriate modification of the disk service demands.

Our study will progress by introducing ever greater levels of detail into our models. Before proceeding, two remarks:

- For concreteness we will use terminology derived from IBM systems in this chapter. The architectural characteristics that we address and the modelling techniques that we develop, however, are equally applicable to systems of other manufacturers.

- The fact that the computer system under study has a complex I/O subsystem, such as that illustrated in Figure 10.1, does *not* mean that sophisticated I/O subsystem modelling techniques are required. In undertaking any study, the analyst must think carefully about the questions under consideration. If the primary effects of the postulated modifications can be represented by straightforward adjustments of disk service demands (or by no adjustment, as might be the case for a CPU upgrade), then sophisticated I/O subsystem modelling techniques are not called for.

10.2. Channel Contention in Non-RPS I/O Subsystems

In this section we develop a technique to represent the effect of channel contention in an I/O subsystem with disks that do not perform rotational position sensing (RPS). (RPS will be explained in the next section.) Customers cycle through such a system (illustrated in Figure 10.2) as follows:

- queue for the CPU
- when the CPU is available, use it
- queue for access to a specific disk
- when that disk is available, seek
- still holding the disk, queue for access to the channel (contention)
- when the channel is available, use both it and the disk to search for (latency) and transfer data.

Two preliminary remarks:

- In fact, momentary access to elements of the I/O path is required to initiate a disk seek. It is customary (and justified, based on experience) to ignore this in modelling disk I/O subsystems; we will do so throughout this chapter.
- Recall that topology is irrelevant in separable queueing networks; the crucial issue is our choice of service demands, not our placement of the channel relative to the disks in our figures.

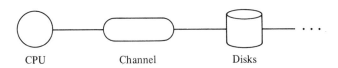

CPU　　　　　　Channel　　　　　Disks

Figure 10.2 − A Highly Simplified I/O Subsystem

As noted in the previous section, it is a straightforward matter to construct a queueing network model of a non-RPS disk I/O subsystem that, given parameters derived from measurements over a specific interval, accurately reproduces the performance observed during that same interval. Each disk should be represented individually, with service demand equal to measured disk busy time divided by measured system completions (in the single class case). The relative contributions of seek, latency, transfer, and contention times are unimportant.

In using the model to project performance for modified environments, it may be necessary to adjust not only the intrinsic service demands at the disks (for example, the substitution of a disk with a higher data transfer rate would result in a smaller transfer time component), but also the channel contention component (this same substitution would result in a decrease in channel holding times, and thus in channel contention). Note that conducting such a modification analysis imposes two requirements beyond those imposed by validating a baseline model:

- It may be necessary to deduce the relative contributions of seek, latency, transfer, and contention times in the measured disk busy times.
- It may be necessary to estimate the changes in each of these components that will result from the proposed modifications.

The emphasis in this section, and in the chapter as a whole, is on the most interesting aspect of these requirements: we will develop techniques that, given information about the intrinsic service requirements of requests at each disk (the seek, latency, and transfer times), will estimate the contention times experienced by requests associated with the various disks, and thus the effective service demands at the disks. In developing our techniques, we will assume that seek, latency, and transfer times are known. (A later section will discuss how to deduce these values from typical measurement data.) In using these techniques to project performance for a modified environment, the analyst would adjust the intrinsic service demands (e.g., transfer times) directly, relying on the algorithms to estimate revised contention components, and thus revised effective service demands. (Chapter 13 discusses modification analysis in more detail.)

Although it is the effective service demand at each disk k, D_k, that we require, it will be convenient to think of D_k as the product of V_k, the number of visits to disk k made by a customer, and S_k, the effective service requirement per visit. S_k, in turn, can be thought of as the sum of $seek_k$, $latency_k$, $transfer_k$, and $contention_k$, each of which are expressed on a per-visit basis. In other words:

10.2. Channel Contention in Non-RPS I/O Subsystems

$$D_k = V_k S_k$$
$$= V_k \left[seek_k + latency_k + transfer_k + contention_k \right]$$

We assume that all of these quantities except for $contention_k$ are known.

We must estimate $contention_k$, the time spent awaiting access to the channel by a request associated with disk k. In the spirit of mean value analysis, this can be viewed as the product of the channel holding time of a request associated with disk k and the number of requests encountered by a disk k request upon arrival at the channel. The channel holding time of a request associated with disk k is simply $latency_k + transfer_k$. To estimate the arrival instant channel queue length, we (falsely) view the channel as a center in an open system. Recall from Chapter 6 that the arrival instant queue length at any center in an open system is equal to $\frac{U}{1-U}$ where U is the utilization of the center. In the present case, we know that any requests ahead of a disk k request at the the channel must be associated with some disk other than k, so we modify this equation to be:

$$\frac{U_{ch} - U_{ch}(k)}{1 - U_{ch}}$$

where U_{ch} is the utilization of the channel, and $U_{ch}(k)$ is the contribution to this utilization of requests associated with disk k. Thus, if we knew U_{ch} and $U_{ch}(k)$ we could estimate the effective service demand of disk k as:

$$D_k = V_k \left[seek_k + latency_k + transfer_k + contention_k \right]$$
$$= V_k \left[seek_k + latency_k + transfer_k + \left[latency_k + transfer_k \right] \times \frac{U_{ch} - U_{ch}(k)}{1 - U_{ch}} \right]$$
$$= V_k \left[seek_k + \left[latency_k + transfer_k \right] \times \left[1 + \frac{U_{ch} - U_{ch}(k)}{1 - U_{ch}} \right] \right]$$
$$= V_k \left[seek_k + \frac{(latency_k + transfer_k)(1 - U_{ch}(k))}{1 - U_{ch}} \right]$$

Unfortunately, the various $U_{ch}(k)$ required to parameterize the model are known only after the model has been evaluated. This suggests the iterative scheme shown as Algorithm 10.1.

As an example, consider a batch computer system with an average multiprogramming level of 10, a CPU at which jobs have an average total

> 1. Define a queueing network model of the system in which the I/O subsystem is represented only by the disks. Initially, assume that system throughput, X, is zero. (This will cause the contention component of the disks' effective service demands to be set to zero during the first iteration.)
> 2. Iterate as follows:
>   2.1. For each disk k, estimate the contribution to channel utilization of requests associated with that disk as:
>
> $$U_{ch}(k) = X\, V_k \left[latency_k + transfer_k \right]$$
>
>   where X is obtained from the previous iteration.
>
>   2.2. Estimate channel utilization: $U_{ch} = \sum_{all\ disks\ k} U_{ch}(k)$
>
>   2.3. For each disk k, estimate its effective service demand as:
>
> $$D_k = V_k \left[seek_k + \frac{(latency_k + transfer_k)(1 - U_{ch}(k))}{1 - U_{ch}} \right]$$
>
>   2.4. Evaluate the queueing network model using MVA.
>
>   Repeat Step 2 until successive estimates of system throughput, X, are sufficiently close.
> 3. Obtain performance measures from the final iteration.

Algorithm 10.1 − Non-RPS Disks

service requirement of 15 seconds, a single channel, and five equally loaded non-RPS disks at each of which jobs have average total service requirements of 8 seconds seeking (i.e., $V_k seek_k = 8$), 1 second searching (latency), and 2 seconds transferring data. (Note that it is not necessary to descend to the "visit" level in order to apply Algorithm 10.1; we did so in our development for consistency with forthcoming sections.) We analyze this system using a queueing network with 10 customers and six service centers, corresponding to the CPU and the five disks. The service demand at the CPU is 15 seconds. The initial service demand at each disk is 11 seconds. (The equally loaded disks are not essential, but are used to simplify the example; they allow single calculations of $U_{ch}(k)$ and D_k to be used for all disks.) Table 10.1 displays the iteration.

The parameter values used in the first iteration correspond to an analysis in which channel contention is ignored. The results (throughput of .056, channel utilization of 84%) differ considerably from those

10.2. Channel Contention in Non-RPS I/O Subsystems

iter.	input	calculations			output
	X	$U_{ch}(k)$	U_{ch}	D_k	X
1	.0000	.000	.000	11.00	.0557
2	.0557	.167	.836	23.24	.0299
3	.0299	.090	.449	12.96	.0499
4	.0499	.150	.749	18.16	.0376
5	.0376	.113	.564	14.10	.0467
6	.0467	.140	.701	16.63	.0408
7	.0408	.122	.611	14.77	.0449
8	.0449	.135	.674	15.96	.0423
9	.0423	.127	.635	15.18	.0439
10	.0439	.132	.659	15.63	.0430
11	.0430	.129	.645	15.36	.0434
12	.0434	.130	.651	15.48	.0434

Table 10.1 — Execution of Algorithm 10.1

obtained at the end of the iteration (throughput of .044, channel utilization of 65%), when channel contention has been accounted for.

Algorithm 10.1 can be applied to computer systems with multiple channels, each connecting the CPU to a specific set of disks. Each channel subsystem must be considered separately in the algorithm. In Steps 2.1 and 2.2, a separate utilization is calculated for each channel. In Step 2.3, the effective service demand of each disk is estimated using the utilization of the channel to which it is attached.

Two simple modifications are required to generalize the algorithm to multiple class queueing networks. In Step 2.1 the channel utilization due to requests associated with disk k must be estimated as:

$$U_{ch}(k) = \sum_{c=1}^{C} \left[X_c V_{c,k} \left[latency_{c,k} + transfer_{c,k} \right] \right]$$

In Step 2.3 revised effective service demands must be estimated on a per-class basis as:

$$D_{c,k} = V_{c,k} \left[seek_{c,k} + \frac{(latency_{c,k} + transfer_{c,k})(1 - U_{ch}(k))}{1 - U_{ch}} \right]$$

Algorithm 10.1 is simple, efficient, and sufficiently accurate. Further, the situations in which its accuracy might be questioned are easily identified: those in which the utilization of the channel is high — certainly greater than 50%, a higher utilization than would be encountered in most applications. The source of this error is our view of the channel as a center in an open system, which we used in calculating the expected queue length encountered at the channel by arriving requests from disk

k. In reality, the number of requests queued at the channel is bounded, rather than unbounded as implied by the open system approximation. For a given utilization, a service center in an open queueing network will have a greater queue length than a service center in a closed network. The open system approximation therefore will tend to overestimate the queue length at the channel, and thus to overestimate channel residence times.

10.3. Channel Contention in RPS I/O Subsystems

Rotational position sensing (*RPS*) increases concurrency in the I/O subsystem by allowing disks to search for data (the latency period) independently of each other and of the channel. When the data record of interest rotates under the heads, the disk attempts to *reconnect*. (*Re*connect rather than *connect* because momentary access to elements of the I/O path was required to initiate the seek and the search; we shall continue to ignore this in our models.) If the path is free, this reconnect succeeds and the data transfer takes place. If not, another reconnect is attempted when the data next rotates under the heads, one disk revolution later. Reconnect attempts are continued in this manner until success is achieved. We refer to all reconnect attempts after the first as *retries*.

As in the previous section, we wish to estimate the effective service demand for each disk k:

$$D_k = V_k \left[seek_k + latency_k + transfer_k + contention_k \right]$$

We assume that all of these quantities except for $contention_k$ are known. In the case of RPS disks, we have:

$$contention_k = retries_k \times rotation_k$$

where $retries_k$ is the number of retries required by disk k before a successful reconnect, on average, and $rotation_k$ is the rotation time of disk k. The latter quantity is known from device characteristics; our objective thus is to estimate $retries_k$.

We assume that for any particular disk k, the probabilities of failure on various reconnect attempts are independent. (This assumption is not strictly correct, but at most a small error is introduced.) We let $P_k[reconnect\ fails]$ denote this probability of failure. Then:

10.3. Channel Contention in RPS I/O Subsystems

$$retries_k = 0 \times (1 - P_k[reconnect\ fails]) +$$
$$1 \times (1 - P_k[reconnect\ fails]) \times P_k[reconnect\ fails] +$$
$$2 \times (1 - P_k[reconnect\ fails]) \times (P_k[reconnect\ fails])^2 +$$
$$\vdots$$
$$= \sum_{i=1}^{\infty} \left[i\ (1 - P_k[reconnect\ fails]) \times (P_k[reconnect\ fails])^i \right]$$
$$= \frac{P_k[reconnect\ fails]}{1 - P_k[reconnect\ fails]} \quad \text{(a standard transformation)}$$

A reconnect attempt succeeds if the path back to the CPU is free, and fails otherwise. In other words, $P_k[reconnect\ fails]$ is equal to $P_k[path\ busy]$, the probability that disk k finds the path busy when it attempts to reconnect. Presently the channel is the only path element that we are considering, so $P_k[path\ busy]$ is equal to $P_k[channel\ busy]$, the probability that disk k finds the channel busy when it attempts to reconnect. At first glance, we might guess that $P_k[channel\ busy]$ is equal to U_{ch}. In fact, though, disk k will not "see" its own contribution to channel utilization. Thus:

$P_k[reconnect\ fails]$

$$= P_k[path\ busy]$$
$$= P_k[channel\ busy]$$
$$= P[channel\ busy\ |\ disk\ k\ not\ transferring]$$
$$= \frac{P[channel\ busy\ \&\ disk\ k\ not\ transferring]}{P[disk\ k\ not\ transferring]} \quad \text{(Bayes's rule)}$$
$$= \frac{U_{ch} - U_{ch}(k)}{1 - U_k(transfer)}$$

where $U_k(transfer)$ is the utilization of disk k due to data transfers. This quantity is equal to the utilization of the channel due to requests associated with disk k, $U_{ch}(k)$. Making this substitution and using the result in the expression for $retries_k$, we obtain:

$$retries_k = \frac{U_{ch} - U_{ch}(k)}{1 - U_{ch}}$$

Since these utilizations are known only once the model has been evaluated, we employ an iterative scheme, shown in Algorithm 10.2.

1. Define a queueing network model of the system in which the I/O subsystem is represented only by the disks. Initially, assume that system throughput, X, is zero. (This will cause the contention component of the disks' effective service demands to be set to zero during the first iteration.)
2. Iterate as follows:
   2.1. For each disk k, estimate the contribution to channel utilization of requests associated with that disk as:
   $$U_{ch}(k) = X \, V_k \, transfer_k$$
   where X is obtained from the previous iteration.
   2.2. Estimate channel utilization: $U_{ch} = \sum_{all\ disks\ k} U_{ch}(k)$
   2.3. For each disk k:
   - Estimate the average number of retries required before a successful reconnect as:
   $$retries_k = \frac{U_{ch} - U_{ch}(k)}{1 - U_{ch}}$$
   - Estimate an effective service demand as:
   $$D_k = V_k \Big[seek_k + latency_k + transfer_k + (retries_k \times rotation_k) \Big]$$
   2.4. Evaluate the queueing network model.
   Repeat Step 2 until successive estimates of system throughput, X, are sufficiently close.
3. Obtain performance measures from the final iteration.

Algorithm 10.2 — RPS Disks

As an example we return to the system considered in Section 10.2, but assume that the disks are capable of rotational position sensing. Let the rotation time of each disk be 17 msec., and let the number of operations per disk be 120. Table 10.2 displays the iteration. In comparison to the non-RPS case, we note that system throughput has increased by 17% while channel utilization has decreased by 23%.

10.4. Additional Path Elements

iter.	input	calculations				output
	X	$U_{ch}(k)$	U_{ch}	$retries_k$	D_k	X
1	.0000	.000	.000	.000	11.00	.0557
2	.0557	.111	.557	1.006	13.05	.0496
3	.0496	.099	.496	.788	12.61	.0509
4	.0509	.102	.509	.830	12.69	.0507
5	.0507	.101	.507	.822	12.68	.0507

Table 10.2 — Execution of Algorithm 10.2

Like its non-RPS predecessor, this algorithm can be applied to computer systems with multiple channels each connecting the CPU to a specific set of disks, by considering each channel subsystem separately in Steps 2.1 to 2.3. It also can be generalized to multiple classes by means of two simple modifications. The equation in Step 2.1 becomes:

$$U_{ch}(k) = \sum_{c=1}^{C} \left[X_c V_{c,k} \, transfer_{c,k} \right]$$

and the second equation in Step 2.3 becomes:

$$D_{c,k} = V_{c,k} \left[seek_{c,k} + latency_{c,k} + transfer_{c,k} + (retries_k \times rotation_k) \right]$$

(The rotation time of the disk and the average number of retries required before a successful reconnect are independent of the customer class.)

10.4. Additional Path Elements

The path between the CPU and a disk in a contemporary I/O subsystem contains several elements in addition to a channel. The contention component of the effective disk service demands is influenced by each of these path elements. Algorithm 10.2 estimates only the channel's contribution to the contention component. This algorithm can be used in modelling I/O subsystems with additional path elements, provided that a change in the channel's contribution will be the primary effect on the contention component of any contemplated modification. If this is not the case — if significant variations in the contributions to the contention component of other path elements are anticipated — then the algorithm must be extended to estimate these contributions. Such extensions are the subject of the present section.

10.4.1. Controllers

Figure 10.3 illustrates the interposition of a *controller* on the path between the CPU and a disk. Several controllers are attached to a channel, and several disks are attached to a controller. A controller is occupied when any of its associated disks are transferring data.

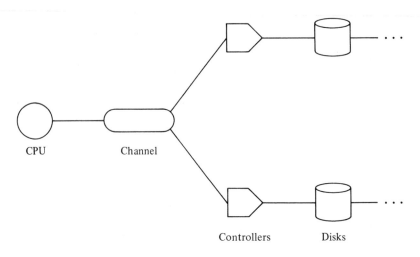

Figure 10.3 − Controllers

As in Section 10.3, our objective is to estimate D_k for each disk k. This requires that we estimate $contention_k$. To do so, we must estimate $retries_k$. This, in turn, requires that we estimate $P_k[reconnect\ fails]$, which is equal to $P_k[path\ busy]$. This quantity can be expressed as:

$$P_k[path\ busy] = P_k[controller\ busy] +$$
$$P_k[controller\ free\ \&\ channel\ busy]$$

By analogy to the derivation in the previous section, the probability that disk k finds its controller busy when attempting to reconnect is:

$$P_k[controller\ busy] = \frac{U_{ctlr} - U_{ctlr}(k)}{1 - U_k(transfer)}$$

The probability that disk k finds its controller free and its channel busy when attempting to reconnect is:

10.4. Additional Path Elements

$P_k[\text{controller free \& channel busy}]$
$= P[\text{controller free \& channel busy} \mid \text{disk } k \text{ not transferring}]$
$= \dfrac{P[\text{controller free \& channel busy \& disk } k \text{ not transferring}]}{P[\text{disk } k \text{ not transferring}]}$
$= \dfrac{U_{ch} - U_{ch}(ctlr)}{1 - U_k(transfer)}$

(In a generalization of our earlier notation, $U_{ch}(ctlr)$ is the utilization of the channel by requests associated with the controller to which disk k is attached.) To make our notation more compact, we replace $U_{ctlr} - U_{ctlr}(k)$, which is the utilization of the controller due to requests associated with disks other than k, by $U_{ctlr}(\bar{k})$. Similarly, we replace $U_{ch} - U_{ch}(ctlr)$, which is the utilization of the channel due to requests routed through controllers other than the one of interest, with $U_{ch}(\overline{ctlr})$. We obtain:

$$P_k[\text{path busy}] = \dfrac{U_{ctlr}(\bar{k}) + U_{ch}(\overline{ctlr})}{1 - U_k(transfer)}$$

and:

$$retries_k = \dfrac{U_{ctlr}(\bar{k}) + U_{ch}(\overline{ctlr})}{1 - U_{ch}}$$

An iterative solution can be obtained, in a manner analogous to Algorithm 10.2.

10.4.2. Heads of String

Some architectures introduce one further path element: a collection of disks constitutes a *string*, which is connected to a controller through a *head of string* (*hos*). Figure 10.4 illustrates this situation.

Like the controller and the channel, the head of string is occupied when any of its associated disks are transferring data. Thus:

$P_k[\text{path busy}] = P_k[\text{hos busy}] +$
$\qquad P_k[\text{hos free \& controller busy}] +$
$\qquad P_k[\text{hos free \& controller free \& channel busy}]$

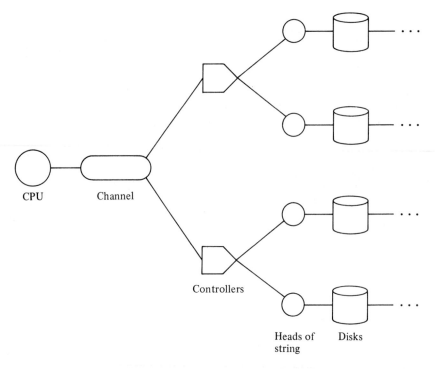

Figure 10.4 — Heads of String

Evaluating these terms yields:

$$P_k[hos\ busy] = \frac{U_{hos}(\bar{k})}{1 - U_k(transfer)}$$

$$P_k[hos\ free\ \&\ controller\ busy] = \frac{U_{ctlr}(\overline{hos})}{1 - U_k(transfer)}$$

$$P_k[hos\ free\ \&\ controller\ free\ \&\ channel\ busy] = \frac{U_{ch}(\overline{ctlr})}{1 - U_k(transfer)}$$

As a result:

$$P_k[path\ busy] = \frac{U_{hos}(\bar{k}) + U_{ctlr}(\overline{hos}) + U_{ch}(\overline{ctlr})}{1 - U_k(transfer)}$$

and:

$$retries_k = \frac{U_{hos}(\bar{k}) + U_{ctlr}(\overline{hos}) + U_{ch}(\overline{ctlr})}{1 - U_{ch}}$$

10.5. Multipathing

The architectures just described are *single path* architectures: each disk is connected to a single head of string, each head of string to a single controller, and each controller to a single channel, with the result that there is only one path from the CPU to any disk — a particular channel, controller, and head of string must be used. This imposes limitations in several respects:

- *reliability* — The failure of any path element will cause all disks "beneath it" to become inaccessible.

- *performance* — A disk may be unable to transfer data because, for example, although its head of string and its controller are free, its channel is busy transferring data for another disk associated with a different controller. There is no way to utilize another channel that may be free at the time.

- *sharing* — In a single path architecture it is not possible to organize several CPUs as a *loosely-coupled multiprocessor* coordinated by means of shared I/O devices.

Multipathing attempts to overcome these limitations. Figure 10.1 in the introduction to this chapter illustrates a multipathing I/O subsystem. In general, a disk may be connected to several heads of string, a head of string to several controllers, and a controller to several channels, perhaps attached to different CPUs. Each different combination of {channel, controller, head of string} that can be used to access a particular disk constitutes a unique path. The system includes an algorithm that selects a path for each data transfer. Existing algorithms fall into two general classes. In *static reconnection* algorithms, any free path is used to initiate an I/O sequence, but the disk must reconnect over this same path to transfer data. In *dynamic reconnection* algorithms, the reconnect may occur over any free path. (Interestingly, multipathing with static reconnection typically results in a performance *degradation* relative to the single path case, which is tolerated for the sake of reliability and sharing.)

In modelling multipathing, our basic approach remains unchanged, but the process of estimating the probabilities of reconnect failure for the various disks (the $P_k[\textit{reconnect fails}]$) becomes more involved. Three factors contribute to this complexity:

- To estimate the utilizations of the various path elements, the path selection algorithm must be considered, because at any "level" of the I/O subsystem hierarchy (i.e., at the level of the channels, the controllers, or the heads of string) the utilization due to requests associated with a particular disk is divided among several path elements in a manner determined by this algorithm. This problem is discussed in Section 10.5.1.

- Once the utilizations of the various path elements are known, it still is not straightforward to estimate the probability of reconnect failure for a particular disk. This is the case because several paths are available to each disk. The probability that each of these paths is found busy must be estimated. Then, the path selection algorithm must be considered to determine probability of reconnect failure given these path busy probabilities. This problem is discussed in Section 10.5.2.
- In the expression for the probability that a particular disk finds a particular path busy, additional terms must be introduced due to multipathing. This problem is discussed in Section 10.5.3.

Algorithm 10.3 shows the general structure of a technique for representing multipathing in queueing network models.

10.5.1. Estimating the Utilizations of Path Elements

In a single path architecture, the utilization of any particular path element (any channel, controller, or head of string) is equal to the sum of the data transfer utilizations of all disks "beneath it". In the case of multipathing, though, it may be possible to route the data transfers of any particular disk through several different {channel, controller, head of string} paths. Thus, the utilization of any path element is the sum of portions of the data transfer utilizations of a number of disks.

Even if we "know" the utilization of each disk due to data transfer (an improved estimate is obtained each time we iterate through all of Step 2 of Algorithm 10.3), the proportion routed through each path element can be estimated only once we have represented the behavior of the path selection algorithm. And, in order to represent the behavior of the path selection algorithm, we must know the utilizations of the path elements, because the path selection algorithm is driven by the probabilities that the various paths are found busy. In other words, estimating the utilizations of path elements, Step 2.2 of Algorithm 10.3, itself is an iterative process.

This iterative process would be relatively straightforward if I/O subsystems were fully interconnected — if every disk could use every head of string, controller, and channel. Unfortunately this is not the case. Both physical and logical constraints exist. These constraints could turn the estimation of the utilizations of path elements into a nasty combinatorial problem. Fortunately, though, interconnection structures tend to be quite limited and quite regular in practice, and various simplifying approximations can be introduced without significant loss of accuracy.

One possible approach (there are several) is suggested by the fact that in handling I/O operations for any particular disk k, the path selection

10.5. Multipathing

1. Define a queueing network model of the system in which the I/O subsystem is represented only by the disks. Make an initial estimate of system throughput, X.
2. Iterate as follows:
   2.1. Estimate the utilization of each disk k due to data transfer:
   $$U_k(transfer) = X\ V_k\ transfer_k$$
   2.2. Estimate the utilizations of the various path elements, by apportioning the data transfer utilizations of the disks among these path elements in a way that is consistent with the system's path structure and with the path selection algorithm. (See Section 10.5.1.)
   2.3. Estimate the effective service demand of each disk k:
   - For each path that can be used by disk k, estimate $P_k[path\ busy]$, the probability that disk k finds this path busy when it attempts to reconnect. (See Section 10.5.2.)
   - Considering these probabilities along with the system's path selection algorithm, estimate $P_k[reconnect\ fails]$, the probability that disk k fails to reconnect. (See Section 10.5.3.)
   - Given this probability, estimate $retries_k$ and D_k in the usual manner:
   $$retries_k = \frac{P_k[reconnect\ fails]}{1 - P_k[reconnect\ fails]}$$
   $$D_k = V_k\left[seek_k + latency_k + transfer_k + (retries_k \times rotation_k)\right]$$
   2.4. Evaluate the queueing network model.

   Repeat Step 2 until successive estimates of system throughput, X, are sufficiently close.
3. Obtain performance measures from the final iteration.

Algorithm 10.3 — Multipathing in the Rough

algorithm will choose among the possible paths in proportion to the probability that it finds them free. Thus:
- Establish initial estimates (say, zero) for the utilization of each path element.
- Iterate as follows:
    - Treat each disk k in turn:
        - For each path i to disk k, let $P_k[\text{path } i \text{ selected}]$ denote the proportion of disk k's transfers that use path i. Set the $P_k[\text{path } i \text{ selected}]$ to be proportional to the probabilities that disk k finds each path i free: $(1 - P_k[\text{path } i \text{ busy}])$, where $P_k[\text{path } i \text{ busy}]$ is calculated as in Section 10.5.2, using the current estimates for path element utilizations.
        - Update the estimates of the utilizations of the various path elements to include the new assignment of disk k's transfers.
    Once each disk has been considered, iterate, modifying previous values.

This procedure will not reproduce exactly the behavior of the path selection algorithm, but will provide a reasonable approximation.

10.5.2. Estimating the Path Busy Probabilities

As in the case of single path architectures, the probability that disk k finds any particular path busy when it attempts to reconnect is:

$$P_k[\text{path busy}] = P_k[\text{hos busy}] +$$
$$P_k[\text{hos free \& controller busy}] +$$
$$P_k[\text{hos free \& controller free \& channel busy}]$$

where *hos*, *controller*, and *channel* refer to the particular head of string, controller, and channel of interest — those that constitute the path in question.

In the multipathing case, additional terms are involved in expressing these probabilities in terms of the utilizations of path elements. The probability that disk k finds the path's head of string busy is unchanged:

$$P_k[\text{hos busy}] = \frac{U_{hos} - U_{hos}(k)}{1 - U_k(\text{transfer})}$$

The probability that disk k finds the path's head of string free but its controller busy has one additional term:

$$P_k[\text{hos free \& controller busy}] = \frac{U_{ctlr} - U_{ctlr}(\text{hos}) - U_{ctlr}(k \rightarrow \overline{\text{hos}})}{1 - U_k(\text{transfer})}$$

10.5. Multipathing

where $U_{ctlr}(k \rightarrow \overline{hos})$ is the utilization of the controller of interest due to requests associated with disk k routed through heads of string other than the one of interest. The probability that disk k finds the path's head of string and controller free but its channel busy has two additional terms:

$$P_k[\text{hos free \& controller free \& channel busy}]$$
$$= \frac{U_{ch} - U_{ch}(ctlr) - U_{ch}(hos \rightarrow \overline{ctlr}) - U_{ch}(k \rightarrow \overline{hos} \rightarrow \overline{ctlr})}{1 - U_k(transfer)}$$

where $U_{ch}(hos \rightarrow \overline{ctlr})$ is the utilization of the channel of interest due to requests routed through the head of string of interest but through controllers other than the one of interest, and $U_{ch}(k \rightarrow \overline{hos} \rightarrow \overline{ctlr})$ is the utilization of the channel of interest due to requests associated with disk k routed through heads of string and controllers other than the ones of interest.

10.5.3. Estimating the Probability of Reconnect Failure

In a single path architecture, $P_k[\text{reconnect fails}]$ is equal to $P_k[\text{path busy}]$. This simple relationship does not hold in the case of multipathing. Each disk k now has a number of paths to choose from. In determining the probability of reconnect failure, the busy probabilities of each possible path must be considered, along with the strategy used by the path selection algorithm.

With a static reconnection algorithm, the reconnection is attempted over whichever path was chosen for the initiation of the I/O sequence. Thus:

$$P_k[\text{reconnect fails}] = \sum_{\substack{i \in \text{possible} \\ \text{paths}}} P_k[\text{path } i \text{ selected}] \times P_k[\text{path } i \text{ busy}]$$

where $P_k[\text{path } i \text{ selected}]$ is the proportion of disk k transfers that use path i (from Section 10.5.1) and $P_k[\text{path } i \text{ busy}]$ is the probability that disk k finds path i busy when attempting to reconnect (from Section 10.5.2).

With a dynamic reconnection algorithm, the reconnection can take place over any free path. Thus:

$$P_k[\text{reconnect fails}] = P_k[\text{all possible paths busy}]$$
$$\approx \prod_{\substack{i \in \text{possible} \\ \text{paths}}} P_k[\text{path } i \text{ busy}]$$

(This equation assumes that the probabilities of various paths being busy are independent of one another. This assumption is not strictly correct, but any error introduced is apt not to be substantial.)

10.6. Other Architectural Characteristics

In this section we provide brief treatments of two additional architectural characteristics: shared disks and cached devices.

10.6.1. Shared Disks

As noted in Section 10.5, one virtue of multipathing is that it allows disks to be shared among several systems. Such a configuration often is referred to as a *loosely-coupled multiprocessor*. In principle the systems could be joined at any level in the I/O subsystem hierarchy. Figure 10.5 illustrates a typical case, in which a single controller is attached to two channels connected to different CPUs.

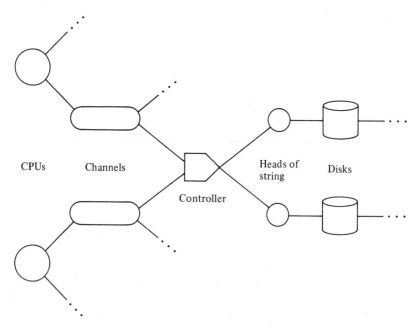

Figure 10.5 − Shared Disks

A loosely-coupled multiprocessor can be viewed in two ways: as a single system that happens to have multiple CPUs, or as a collection of separate systems that happen to share disks. The distinction is important, for the two views lead to different modelling approaches. The choice of view depends upon the way in which a particular processing complex actually is used, and the nature of the performance questions under consideration.

10.6. Other Architectural Characteristics

The first view, that of a single system that happens to have multiple CPUs, leads to a single large queueing network model that includes all devices and all workload components. The advantage of this modelling approach is its conceptual simplicity: no new ideas are involved. For this reason we will discuss this view no further.

The second view, that of a collection of separate systems that happen to share disks, leads to a collection of small queueing network models, one corresponding to each system. The advantage of this modelling approach is its modularity: a modification whose primary effect will be felt by one system can be investigated by defining, parameterizing, and evaluating one relatively small model. Conducting such an analysis is the subject of the remainder of this subsection.

Consider the queueing network model of any of the systems. The I/O subsystem component of this model will include service centers corresponding to all disks used by customers on that system, whether those disks are dedicated or shared. Certainly, contention in the I/O subsystem due to requests associated with other systems must be represented. If not, throughput of requests associated with the system of interest would be over-estimated. We will represent this contention in our model, but will do so in a way that is determined from measurement data. In modifying the model for purposes of performance projection, we will assume that the utilizations of disks and path elements due to requests associated with other systems remain unchanged.

In estimating the effective service demand at each disk in the model, we represent the effect of requests associated with other systems in two ways:

- *accounting for additional reconnect delay experienced because of path contention due to "foreign" requests* − In evaluating the expressions for the probabilities that various paths are found busy, the measured utilizations due to requests associated with other systems are added to the calculated utilizations due to customers in the model, for each shared path element and shared disk. This adjustment results in a realistic estimate for the contention component of effective service demand.

- *accounting for delay in acquiring the disk due to its use by "foreign" requests* − For each disk, the contention component calculated above is added to the seek, latency, and transfer components. This total is divided by one minus the measured utilization of the disk due to requests associated with other systems. The rationale is the same used in estimating channel contention for non-RPS disks (Section 10.2).

We recommend this approach whenever it is possible to assume relative stability in the utilizations of disks and path elements due to requests associated with other systems, in the presence of postulated modifications to the system of interest.

10.6.2. Cached Devices

A *cache memory* is a relatively small, relatively high speed memory that is used as a staging area for data. For many years cache memories have been interposed between processors and their primary memories. Very recently they have been introduced into I/O subsystems, typically by augmenting controllers with storage capacity (on the order of millions of bytes) and processing capacity. In this subsection we will take a brief look at modelling cached devices.

The cache contains duplicate copies of some of the disk-resident data. If the cache is well managed, the vast majority of the data that is referenced by I/O operations will be resident in the cache. Two parameters are crucial in determining the effectiveness of the cache. The first is the *hit ratio*: the proportion of I/O operations that refer to data residing in the cache. The second is the *read ratio*: the proportion of I/O operations that are reads rather than writes. These parameters are crucial because a *read hit* (a read operation referencing data resident in the cache) can be serviced without accessing the disk. Thus, it has a service time roughly equal to the data transfer time, with no seek or latency components. On the other hand, a *read miss*, a *write hit*, and a *write miss* each require that the disk be accessed. Furthermore, because of the overhead involved in managing the cache, a disk access in a cached environment is somewhat slower than a disk access in a conventional environment. Thus, a performance *degradation* can result from conversion to a cached I/O subsystem if a low read ratio exists, regardless of the hit ratio. A performance improvement will result if high hit and read ratios exist.

Let us consider a modelling study whose objective is to estimate the effect of converting an existing system to a cached I/O subsystem. We adhere to the basic model structure and evaluation techniques used in previous sections, and assume that a validated baseline model exists.

- We can reflect any changes in the seek, latency, and transfer times due to device characteristics in a straightforward manner.
- To account for the fact that a read hit can be serviced with no disk access, we adjust the effective service demands of the disks in the obvious way:

$$D_k = V_k \bigg[(1 - (hit\ ratio \times read\ ratio)) \times (seek_k + latency_k) + transfer_k + contention_k \bigg]$$

The hit ratio is not apt to be site-dependent in a significant way, so typical values can be obtained from manufacturer's data. The read ratio is not apt to change as a result of the conversion, so measurement data from the existing system can be used.

- The overhead of managing the cache may cause the utilizations to increase at various path elements, especially controllers. These increased utilizations should be represented, because they will affect path contention. Manufacturer's data is available that provides multiplicative factors to be used in estimating this overhead, given the basic transfer time. These factors can be used within the model in calculating the path busy probabilities.

10.7. Practical Considerations

Two practical considerations immediately arise in contemplating the application of the techniques we have described:
- How can the relatively detailed parameters required by these techniques be inferred from the measurement data that typically is encountered?
- How can these techniques be embedded in queueing network modelling software?

These related concerns are the subjects of the present section.

10.7.1. Inferring Parameter Values from Measurement Data

The techniques we have presented require that the following information be provided as input:
- a specification of the path structure of the I/O subsystem
- for each disk:
  - the visit count
  - the average seek, latency, and transfer times per visit
  - the average rotation time

Given this information, these techniques iteratively estimate the average contention time per visit at each disk, and thus the effective service time per visit, S_k, and the effective service demand, D_k.

In this section we consider the common situation in which the values of some of these parameters are not available directly, so must be inferred before our techniques can be applied. Inevitably, the visit counts and utilizations of the disks are known from measurement data. From these, the *actual* effective service times per visit and effective service demands can be calculated. We know that the actual effective service demands, if used to parameterize a model, would yield excellent results without the use of the techniques described in this chapter. (These techniques are required to conduct a modification analysis in which a change to the contention component of the effective service demands is anticipated to be a primary effect.) A fruitful way to view our task is that we must partition the actual effective service times per visit into seek, latency, transfer, and contention components, in such a way that when the seek, latency and transfer components are provided as inputs to the model (along with path structure, visit counts, and rotation times), the techniques that we have developed will calculate effective service times per visit and effective service demands that are roughly the same as the actual values. Once this has been achieved, we will consider the baseline model to be validated and will be prepared to use it for performance projection.

We denote the actual effective service time per visit at disk k by S_k^*, and the actual effective service demand by D_k^*. We proceed as follows:

- To estimate $latency_k$, we refer to the device characteristics.
- To estimate $transfer_k$ we employ the utilizations and visit counts of the channels, which are available readily from measurement data. From these, the service time per visit to each channel can be obtained. In the single path case, we set $transfer_k$ to this value (for the appropriate channel, of course). In the multipathing case, we take an average of the values of the channels accessible from disk k. Estimating $transfer_k$ on the basis of measured channel service times is important. The various path elements are processors rather than wires, and overhead is associated with each transfer. Estimating $transfer_k$ by considering block sizes and transfer rates would ignore this overhead, yielding an optimistic value. In stating our approach, we have made the homogeneity assumption that the data transfer service requirements of all disks on a particular channel are the same. Adjustments are possible if block size information is available.
- To estimate $seek_k$ it is tempting to refer to the device characteristics. Unfortunately, this approach is notoriously unreliable. We know that:

$$seek_k + contention_k = S_k^* - latency_k - transfer_k$$

where each of the quantities on the right hand side is known. In order to obtain consistent estimates for the two quantities on the left hand side, we will evaluate the queueing network, using either Algorithm 10.2 (for the single path case, augmented as in Section 10.4) or Algorithm 10.3 (for the multipathing case), and let the results determine the estimates. More specifically:

  — In Step 2.1 of either algorithm, we use the values of $transfer_k$ estimated above.
  — In each iteration of Step 2 in either algorithm, we use D_k^* as the effective service demand of disk k. (Fixing this value does not entirely eliminate iteration, because the throughput of the model will differ slightly from the throughput of the system.)

10.8. Summary

— When the algorithm terminates, it will have estimated $P_k[\text{reconnect fails}]$ and retries_k for each disk. Since rotation_k is known (from the device characteristics), this means that an estimate for contention_k has been obtained. We set our estimate for seek_k to:

$$\text{seek}_k = S_k^* - \text{latency}_k - \text{transfer}_k - \text{contention}_k$$

We now are prepared to use the model for performance projection.

10.7.2. Incorporation in Queueing Network Modelling Software

The preceding discussion provides a number of insights concerning the support that a queueing network analysis software package might provide for modelling complex I/O subsystems.

The package might provide a convenient syntax for specifying the path structure of the I/O subsystem. As input, the analyst would provide this path structure, plus the effective service demands and visit counts at each disk, and the service demands and visit counts at each channel. The package might make use of internal information concerning various device types to provide quantities such as average latency and rotation times.

The analyst would indicate when the model has been specified fully. At this point, the package would evaluate the model, inferring the detailed parameter values and storing them internally.

At this point, it is possible to undertake modification analyses. The package might support this process in a number of ways. For example, the path structure might be modifiable using the same syntax in which it was specified, with the package adjusting the detailed parameter values.

Chapter 16 contains a more extensive discussion of software support for queueing network modelling.

10.8. Summary

In this chapter we have presented a single model structure that can be used to represent complex contemporary I/O subsystems at varying levels of detail. In this model structure, the I/O subsystem is represented by service centers corresponding to the various disks, each with an *effective service demand*, D_k, equal to:

$$V_k \left[\text{seek}_k + \text{latency}_k + \text{transfer}_k + \text{contention}_k \right]$$

We have developed algorithms for estimating the contention component of the effective service demand under a number of different assumptions about the structure of the I/O subsystem and the level of detail of the model. We have discussed various practical considerations, such as obtaining the necessary parameters for these algorithms from typical measurement data and incorporating these algorithms in queueing network modelling software.

For a variety of reasons the material in this chapter should not be viewed as definitive: the I/O subsystem architectures of various vendors differ substantially in their details, these architectures are evolving rapidly, and techniques for representing these architectures in queueing network models are an area of current research activity. Our algorithms should be viewed as an indication of what can be done, and as a set of techniques that can be used directly and also can be tailored as necessary to the requirements of specific systems.

In closing this chapter, we reiterate an important point made in its introduction. The fact that the computer system under study has a complex I/O subsystem does *not* mean that sophisticated I/O subsystem modelling techniques are required. If the primary effects of the postulated modifications can be represented by straightforward adjustments of disk service demands, then sophisticated I/O subsystem modelling techniques are not called for. The benefits of omitting sophistication include a simpler parameterization and fewer assumptions.

10.9. References

In this chapter we have developed models in which service centers of the load-independent queueing type are used to represent each disk, iteratively estimating the effective service demands at these centers. Two equally reasonable alternate approaches exist. The first of these can be described as follows:
- Define a queueing network model of the system in which the I/O subsystem is represented only by the disks, and each disk is represented by a service center of the delay type.
- Iterate as follows:
  - For each disk:
    - Estimate the effective service demand.
    - Use this value in a formula from queueing theory to estimate the average residence time at the disk.
    - Substitute this value into the corresponding delay center.
  - Evaluate the queueing network model.

  Repeat until successive estimates of system throughput are sufficiently close.

10.9. References

This approach is common in practice. Although its origins are unknown, it has been used by Bard [1980, 1982], by Wilhelm [1977], and by Zahorjan, Hume, and Sevcik [Zahorjan et al. 1978].

The second alternate approach, due to Brandwajn [1981], involves multiple applications of the principles of flow equivalence and hierarchical modelling described in Chapter 8:

- Consider each string (the disks attached to a particular head of string) in turn. Define an FESC by evaluating (for each feasible population) a submodel in which each disk on the string is represented by a service center of load-independent queueing type.
- Consider each controller subsystem (the heads of string and disks attached to a particular controller) in turn. Define an FESC by evaluating (for each feasible population) a submodel in which each string is represented by the FESC defined in the previous step.
- Consider each channel subsystem (the controllers, heads of string, and disks attached to a particular channel) in turn. Define an FESC by evaluating (for each feasible population) a submodel in which each controller subsystem is represented by the FESC defined in the previous step.
- Evaluate a high-level model consisting of the CPU and the channel subsystem FESCs defined in the previous step.

Two of the three model structures described above, including the one adopted in this chapter, require that effective service demands be estimated for each disk. The treatment of non-RPS disks (Section 10.2) belongs to the folklore of queueing network modelling. The treatment of RPS disks (Section 10.3) also is difficult to attribute. Wilhelm [1977] and Zahorjan, Hume, and Sevcik [Zahorjan et al. 1978] are responsible for two accessible renditions. The latter analysis incorporates the fact that the probabilities of failure on successive reconnect attempts are not independent.

Bard is responsible for the original work on multipathing, both in the case of static reconnection algorithms [Bard 1980] and in the case of dynamic reconnection algorithms [Bard 1982]. Bard's approach relies on a *maximum entropy* formulation of the problem.

Buzen and von Mayrhauser [1982] present an interesting analysis of various considerations affecting the modelling and the performance of the IBM 3880-13 cached storage controller. The discussion in Section 10.6.2 is based partially on their work.

Hunter [1982] explores the process of parameterizing queueing network models of I/O subsystems from typical measurement data, in the context of IBM's MVS operating system. The discussion in Section 10.7.1 is based partially on his work.

[Bard 1980]
Yonathan Bard. A Model of Shared DASD and Multipathing. *CACM* 23,10 (October 1980), 564-572.

[Bard 1982]
Yonathan Bard. Modeling I/O Systems with Dynamic Path Selection, and General Transmission Networks. *Proc. ACM SIGMETRICS Conference on Measurement and Modeling of Computer Systems* (1982), 118-129.

[Brandwajn 1981]
Alexandre Brandwajn. Models of DASD Subsystems: Basic Model of Reconnection. *Performance Evaluation 1*,3 (November 1981), 263-281.

[Buzen & von Mayrhauser 1982]
Jeffrey P. Buzen and Anneliese von Mayrhauser. BEST/1 Analysis of the IBM 3880-13 Cached Storage Controller. *Proc. CMG XIII International Conference* (1982), 156-173.

[Hunter 1982]
David Hunter. Modelling Real DASD Configurations. In R.L. Disney and T.J. Ott (eds.), *Applied Probability - Computer Science: The Interface, Vol. I.* Birkhauser, 1982, 451-468. Also appears as IBM T.J. Watson Research Center Report RC-8606.

[Wilhelm 1977]
Neil C. Wilhelm. A General Model for the Performance of Disk Systems. *JACM 24*,1 (January 1977), 14-31.

[Zahorjan et al. 1978]
J. Zahorjan, J.N.P. Hume, and K.C. Sevcik. A Queueing Model of a Rotational Position Sensing Disk System. *INFOR 16*,3 (October 1978), 199-216.

10.10. Exercises

1. The example of Section 10.2 involves a CPU, five equally loaded disk devices, and a channel utilized roughly 65%. Clearly the channel represents a performance problem. Suppose a second channel were added to the system, and two of the five disks moved to it.

    a. Use the iterative technique of Section 10.2 to estimate system throughput under the assumption that the disks do not have rotational position sensing capability. Compare the channel contention component of effective disk service demand with the new configuration to that shown in Table 10.1 for the single channel configuration.

10.10. Exercises

  b. Perform the same calculations under the assumption of RPS disks. Compare your results to those shown in Table 10.2.

2. Consider the simple models of channel contention discussed in Section 10.2 (Algorithm 10.1 for non-RPS disks) and Section 10.3 (Algorithm 10.2 for RPS disks). Show that for fixed seek times, rotation times, data transfer times, and visit counts, the "effective service demand" will be lower with rotational position sensing than without it, for any disk throughput that does not saturate the channel. (Assume a single transaction workload, and a latency equal to one half of a rotation.)

3. Consider a new disk technology in which each disk contains a one track buffer. Assuming a simple channel/disk view of the I/O subsystem (i.e., ignoring other path elements), the disk would operate as follows. When performing a read operation, seek and initial latency would be performed independently of the channel. If the channel was idle when the data to be read rotated under the heads, the disk would gain control of the channel and perform the data transfer. If the channel was busy when the data became available, the entire track would be copied into the disk's buffer, and the disk would queue in a FCFS manner for the channel. When the channel became available, the data would be transferred from the buffer. When performing a write operation, the buffer would not be used (i.e., the disk would operate as a standard RPS device).

  a. Give an expression for the effective disk service time. What input parameters are required?

  b. Describe an (iterative) approximation technique for modelling this disk technology.

4. In deriving the expression for $retries_k$ (the average number of retries required by device k), we have assumed that the probability that a reconnect attempt fails is independent of the number of attempts made so far. However, it appears that in practice the probability that the second and subsequent attempts fail is slightly larger than the probability that the first attempt fails.

  a. What does this indicate about the tendency of the procedures described in this chapter to under- or over-estimate system response time?

  b. Suppose you knew that the probability of a reconnect attempt failing was 10% higher on the second and subsequent attempts than on the first attempt. Give an expression for the average number of retries required.

c. In practice, an unlimited number of reconnect failures is not possible. After some fixed number of failures, the disk queues for the channel, and reconnects as soon as possible regardless of the position of the desired data relative to the heads. What does this indicate about the tendency of the procedures described in this chapter to under- or over-estimate system response time? Does this amplify or diminish the effect indicated by your answer to (a)?

5. The complex approach to modelling multi-element I/O paths taken in this chapter was necessary for two reasons. First, a single job may use more than one path element at a time. Such simultaneous resource possession cannot be modelled directly by separable queueing networks. Secondly, measurement tools frequently do not provide sufficient information about the usage of the I/O path elements.

   a. What sorts of measurement information would be useful in modelling complex I/O subsystems?

   b. How could you modify the procedures given in this chapter to take advantage of such information?

Chapter 11

Processors

11.1. Introduction

Thus far we have considered only single CPU systems. We also have ignored the effects of the scheduling discipline that determines the order in which customers are served. In this chapter we will consider the representation of multiprocessors and scheduling disciplines.

In the realm of multiprocessor systems, an important distinction exists between *loosely-coupled multiprocessors* and *tightly-coupled multiprocessors*. In a loosely-coupled multiprocessor, the processors interact primarily through shared direct access storage devices. Since the processors operate essentially independently, they can be represented as separate service centers in a queueing network model, with different customer classes used to distinguish I/O operations originating from different CPUs. This approach was discussed in Chapter 10. In a tightly-coupled multiprocessor, the processors share main memory, and typically are under the control of a single operating system. Special techniques are required in building queueing network models of tightly-coupled multiprocessors; these techniques are the subject of Section 11.2.

Scheduling disciplines were ignored in the case of single class models (Chapter 6) because of two assumptions made there: that customers are indistinguishable (or "statistically identical") in their service demands, and that the expected *remaining* service time of a customer in service at a center does not depend on how much service the customer already has received. (The implication of this second assumption is that the expected time until the next customer completion at any particular center is not changed by removing one customer from service in order to serve another.) Given these two assumptions, system performance measures do not depend on the scheduling discipline used, as long as the processor is not idle when there is work to be done. The second assumption is violated, however, if the bursts of service required by a customer on successive visits to a processor vary widely in duration. Section 11.6 discusses an approach to modelling first-come-first-served (FCFS) scheduling when service bursts are highly variable.

In multiple class models, the situation is more complex. In Chapter 7 the following restrictions were placed on the scheduling disciplines used at queueing centers:

- The scheduling discipline cannot discriminate among customers based on class identity.
- If the scheduling discipline is FCFS, then the average time required to complete a customer in service must be independent not only of the amount of service it has acquired, but also of its class.
- If the scheduling discipline is *not* FCFS, then it must be one of a special group of disciplines that includes processor sharing (PS) and last come first served (LCFS). One important property of this group of disciplines is that each customer receives service immediately upon arrival at a center.

Under these restrictions, the performance estimates of a multiple class model are identical regardless of which of FCFS, PS, and LCFS scheduling is used at any center. Unfortunately, these scheduling disciplines do not adequately represent those used in many operating systems. In particular, class identity and the amount of acquired service often are used in making scheduling decisions. Separable models of such systems may not accurately reflect the relative performance of various workload components (classes). In Section 11.3 we suggest a way to model systems in which scheduling is done according to strict priorities among classes. In Section 11.4 we consider the more difficult case in which priorities are not based purely on class identity. Finally, in Section 11.5 we treat the case of FCFS scheduling when the service requirement per visit to the FCFS center differs from class to class.

11.2. Tightly-Coupled Multiprocessors

Tightly-coupled multiprocessor systems are in widespread use. These systems have two or more processors cooperating to complete work from a single shared queue.

It is easiest to view a tightly-coupled multiprocessor as a single service center, since in the system there is a single queue of jobs for all processors. The service rate of this center (i.e., the number of instructions delivered per time unit) is ideally the sum of the service rates of the individual processors. Consequently, the straightforward approach to modelling n tightly-coupled processors is to create a single center representing them in the model, and to divide the service demands of all customers at that center by n.

11.2. Tightly-Coupled Multiprocessors

This technique provides a simple, first-cut modelling approach, but it ignores two important aspects of multiprocessors. The first aspect is that the total service rate of n processors can be significantly less than n times the rate of a single processor because of competition for software locks (such as those controlling access to the shared queue of jobs) and interference in accessing main memory. Thus, we need a more realistic assessment of the total processing power actually delivered by the multiprocessor. The second aspect is that the effective service rate of a multiprocessor is not constant, but depends on the number of jobs queued at the center. Consider a four processor system. Ideally, if four (or more) jobs desire service at the center, all four processors can be kept busy, and the effective service rate of the center is its maximum rate. However, if less than four jobs are queued at the center, some of the processors will be idle, and so the effective service rate will be reduced correspondingly.

The first of these problems, that of accounting for the interference of the processors with one another in estimating effective service rates, is best solved by using the results of benchmark studies of the configurations under consideration, such as those typically provided by trade journals and vendors. For example, such figures might indicate that an IBM 3033MP (a tightly-coupled dual processor) is roughly 1.7 times as powerful as a single 3033 processor when running a mixed TSO and batch workload under the MVS operating system. Since the power of a multiprocessor can vary significantly depending on the operating system run on it and the nature of the workload to be processed, standard estimates are not likely to be highly reliable. As in all cases where the input parameters are not known with high confidence, it is good practice to evaluate the model for several effective service rates representing a reasonable range, thereby assessing the sensitivity of the results to the parameter whose value is in question.

The second of these problems, that of accounting for variability in the effective service rate of the multiprocessor as a function of the number of jobs needing processor service, is solved easily using a flow equivalent service center. Figure 11.1 graphs effective service rate as a function of the queue length for a four processor system. Service rates increase with queue length until all four processors are busy, after which increasing the number of jobs contending for the processors does not result in any increase in effective service rate. The dashed line illustrates the ideal growth in service rate, and the solid curve represents the effect of contention. The flow equivalent service center used to represent the multiprocessor is parameterized by giving the effective service rates for each possible customer population that could be seen there. This set of population and service rate pairs is essentially a tabular representation of the curve shown in Figure 11.1.

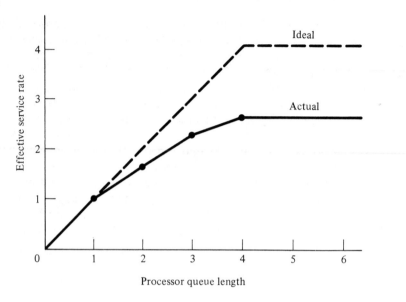

Figure 11.1 − **Service Rate Function of a Four Processor System**

11.3. Priority Scheduling Disciplines

In most current operating systems, processor scheduling disciplines are based on priorities. These priorities may be static (giving consistent preference to one workload component over another) or they may be dynamic (reflecting changing estimates of workload characteristics). Priority scheduling disciplines are not compatible with separable models. Since these disciplines can have a substantial effect on performance, it is important to be able to represent them. A number of approaches have been devised.

One approach was described as an example in Chapter 8. First, the I/O subsystem, which by itself was separable, was analyzed in isolation, and a multiple class flow equivalent service center was constructed. Then, a high-level model was defined that consisted of two centers: this FESC, and the priority scheduled CPU. Finally, the global balance technique was used to evaluate this model. This approach is quite accurate. Its drawbacks are, first, that it requires special purpose global balance software, and second, that because of the complexity of a global balance analysis it becomes infeasible for models with more than a few classes or customers, and for models with multiple priority scheduled centers.

Because of the difficulties in using the technique described in the last paragraph, another approach is required. The one we present here is

11.3. Priority Scheduling Disciplines

based on the mean value analysis technique. In practice, it has been found to be acceptably accurate, and is applicable even to very large models. Consider a model with C customer classes, each of which has a distinct priority at the CPU. (The generalization to several classes with equal priorities is straightforward.) For notational simplicity, assume that the classes are ordered so that higher numbered classes have priority over lower numbered classes. We develop an approximation to the residence time of class c customers at the CPU by considering successively the effects of jobs with lower, equal, and higher priorities than class c:

- *lower priority customers* (classes 1 through $c-1$)

  Because class c has preemptive priority over classes 1 through $c-1$, customers in these classes do not interfere with class c customers. Considering only these lower priority classes we obtain the following approximation to the CPU residence time of class c:

  $$R_{c,CPU}(\vec{I}) \approx D_{c,CPU}$$

- *equal priority customers* (class c)

  Each class c customer arriving at the CPU must queue behind any other class c customers already there. Class c customers that arrive subsequently do not cause further delay. Accounting for both lower and equal priority classes we have:

  $$R_{c,CPU}(\vec{I}) \approx D_{c,CPU}\left[1 + Q_{c,CPU}(\overrightarrow{I-1_c})\right]$$

  where $\overrightarrow{I-1_c}$ is the vector of workload intensities with one class c customer removed if class c is not transaction type (i.e., if class c is closed), and is the full workload intensity vector otherwise (i.e., if class c is open).

- *higher priority customers* (classes $c+1$ through C)

  An arriving class c customer must wait for all higher priority customers already in the queue. It also must wait for all higher priority customers that arrive while it is at the CPU. Because of this complication, it is not possible to estimate accurately the number of higher priority customers for which the class c customer must wait. Instead, we consider the servicing of higher priority customers to be "breakdowns" of the processor with respect to delivering service to the class c customers. Because of these breakdowns, more than $D_{c,CPU}$ time units are required for the class c customer in service to accumulate $D_{c,CPU}$ time units of service. In particular, since the CPU is busy $\sum_{j=c+1}^{C} U_{j,CPU}(\vec{I})$ of the time with higher priority customers, it takes

  $$\frac{D_{c,CPU}}{1 - \sum_{j=c+1}^{C} U_{j,CPU}(\vec{I})}$$

  time units for the currently selected class c

customer to complete. (For instance, once service is begun, it takes twice as long to complete on a processor 50% busy with higher priority customers than on a FCFS processor.) The final approximation for the residence time of class c, accounting for lower, equal, and higher priority classes, is thus:

$$R_{c,CPU}(\vec{I}) \approx \frac{D_{c,CPU}\left[1 + Q_{c,CPU}(\overrightarrow{I-1_c})\right]}{1 - \sum_{j=c+1}^{C} U_{j,CPU}(\vec{I})} \qquad (11.1)$$

A solution technique could be constructed from the mean value analysis technique by substituting equation (11.1) for the standard residence time equation in Algorithms 7.1 or 7.2. However, rather than further complicating these basic algorithms each time we extend our modelling techniques, we prefer to build upon them, using the basic algorithms as subroutines in our extended algorithms. (We return to this concept of layered implementation in Chapter 16.)

In the case of priority scheduling, we can obtain the same results as we would obtain by replacing the residence time equation, by using the *shadow CPU* technique. This technique gets its name from the fact that the single priority scheduled CPU in the actual system is represented in the model by C FCFS service centers, each visited by one class. Let CPU_c denote the c-th shadow CPU, which is visited only by class c. The service demand at CPU_c is set equal to $\dfrac{D_{c,CPU}}{1 - \sum_{j=c+1}^{C} U_{j,CPU}(\vec{I})}$. It should be apparent that the residence time of class c at its shadow CPU is given by equation (11.1): the service demand inflation caused by higher priority classes is captured in the redefinition of the service demand at the shadow CPU, and the queueing for customers of class c but not other classes is a consequence of the FCFS scheduling used at the shadow CPU, plus the fact that only class c visits there. Thus, we have created a queueing network amenable to the analysis techniques of Chapter 7 that represents the effects of priority scheduling.

Algorithm 11.1 describes the shadow CPU technique more precisely. Because the CPU utilizations of the various classes are not known beforehand, it is necessary to employ iteration. Initially, the throughput of each class is estimated to be zero. This corresponds to estimating that the CPU utilization of each class is zero. The model is evaluated, yielding an improved estimate for the throughput, and thus the CPU utilization, of each class. New model inputs are calculated based on these improved estimates. The iteration continues until successive estimates of the throughput of each class are sufficiently close. Extension of Algorithm

11.3. Priority Scheduling Disciplines

1. Given a K center model with a priority scheduled CPU, create a $K+C-1$ center model by replacing the original CPU center with C FCFS shadow CPU centers, each of which will be visited by only one class. Assume that the classes are ordered so that higher numbered classes have priority over lower numbered classes. Initially, assume that the throughput of each class c, X_c, is equal to zero.

2. Iterate as follows:

   2.1. Estimate the CPU utilization of each class c as:
   $$U_{c,CPU} = X_c D_{c,CPU}$$
   where $D_{c,CPU}$ is the "real" CPU demand of class c.

   2.2. Set the service demand of each class c at the j-th shadow CPU to:
   $$D_{c,CPU_j} = \begin{cases} \dfrac{D_{c,CPU}}{1 - \sum_{k=c+1}^{C} U_{k,CPU}} & c = j \\ 0 & c \neq j \end{cases}$$

   2.3. Evaluate the shadow CPU model using either the exact or the approximate algorithms given in Chapter 7.

   Repeat Step 2 until successive estimates of the X_c for each class c are sufficiently close.

3. The final performance measures for the system as a whole and for every center except the CPU are obtained directly from the last iteration. At the CPU, the residence time of each class, $R_{c,CPU}$, and the queue length of each class, $Q_{c,CPU}$, are obtained directly. The utilization of each class, though, is obtained as $U_{c,CPU} = X_c D_{c,CPU}$. (The utilizations reported for the C shadow CPUs are meaningless because of the way in which the service demands have been inflated.)

Algorithm 11.1 — Priority Scheduling at the CPU

11.1 to the case in which several centers are priority scheduled is straightforward.

Table 11.1 shows the results of applying Algorithm 11.1 to a particular example. We consider a system with four disks and a priority scheduled

Model Inputs:

$N_A = \text{<varying>} \quad Z_A = 10 \quad N_B = 6 \quad Z_B = 0$

	center				
	CPU	Disk 1	Disk 2	Disk 3	Disk 4
$D_{A,k}$	4	2	2	2	2
$D_{B,k}$	40	2	4	6	8

(all times are in seconds)

Class A Response Time:

solution technique	N_A				
	1	5	10	15	20
MVA	34.8	46.5	63.1	81.0	99.7
Algorithm 11.1	12.9	19.5	32.7	50.5	70.5
simulation	12.0	19.1	32.0	50.4	70.0

(all times are in seconds)

Table 11.1 — Priority Scheduling

CPU. There are two classes. Class A, which is of terminal type, has priority over class B, which is of batch type.

To assess the value of Algorithm 11.1 we would like to know whether its results are significantly better than those obtained by ignoring priority scheduling (i.e., by assuming that processor sharing is used). Unfortunately, we cannot determine exact performance measures for our example. Even though it has only five centers and two classes, it is too large to be analyzed using the global balance technique (described in Section 8.5.1). We have used simulation to obtain an estimate of the exact performance measures. As indicated in Section 8.5.2, simulation has two important drawbacks that make it less attractive than queueing network modelling for computer system analysis. First, the probabilistic nature of simulation causes the accuracy of its results to depend on the duration of the simulation. (For the duration used here, and in Sections 11.5 and 11.6, the error in the estimates obtained should be taken to be 5 to 10%.) Second, the computational expense of simulation is too great to allow it to be used regularly.

In the table we show the response time experienced by class A users for five different class A populations. The results obtained by ignoring the priority scheduling and applying mean value analysis directly are labelled "MVA" in the table, the results obtained by using Algorithm

11.1 are labelled "Algorithm 11.1", and the results obtained via simulation are labelled "simulation".

Comparing the results of MVA and Algorithm 11.1 illustrates the benefits of using Algorithm 11.1 rather than ignoring the priority scheduling. Comparing the results of Algorithm 11.1 and simulation illustrates the accuracy of Algorithm 11.1 for the specific example under consideration. Algorithm 11.1 will not always exhibit such close agreement to the results of simulation. Fortunately, though, the instances in which the algorithm may be unreliable are easy to identify. In most systems, priority scheduling is used to ensure that customers requiring short bursts of CPU service are not delayed excessively by customers requiring long bursts of CPU service. (Note that processor sharing is one step in this direction relative to FCFS scheduling, but that priority scheduling is one step further.) The technique presented in this section is designed to work well in this situation. It relies on the elongation of low priority service demands to reflect interruptions by high priority customers. This elongation is appropriate when service bursts of high priority customers are very short and very frequent relative to those of the low priority customers. However, whenever low priority service burst lengths are not significantly longer than high priority service burst lengths, the algorithm suggested in this section must be used with caution.

11.4. Variations on Priority Scheduling

While many operating systems permit specification of absolute priorities of the type discussed in the previous section, others support priorities of other natures. Two types of non-absolute priorities can be described as *biased processor sharing* and *goal-oriented scheduling*.

11.4.1. Biased Processor Sharing

Biased processor sharing describes a situation in which one class is favored over another by giving it longer bursts ("quanta") rather than by excluding the other class entirely when a customer of the higher priority class is present. Thus, a relative priority is associated with each class, and each customer receives service at a rate proportional to the relative priority of its class. For example, if the relative priorities of classes A and B are 2 and 1 respectively (a larger number indicating a higher priority), then with one customer of each class competing for service, the class A customer would progress at 2/3 the rate at which it would progress if alone at the center. With two class A customers and one of class B, each class A customer would progress at 2/5 of its full rate while the one class B customer would progress at 1/5 of its full rate.

An evaluation technique for this type of scheduling can be obtained by another modification of the residence time equation of the MVA algorithm:

$$R_{c,k}(\vec{I}) \approx D_{c,k} \left[\frac{\pi_c + \sum_{i=1}^{C} \pi_i Q_{i,k}(\overrightarrow{I-1_c})}{\pi_c} \right]$$

where π_i is the relative priority of class i. The quotient in parentheses is simply the inverse of the rate at which an individual class c customer receives service based on our expectation of the number of customers of each class at the center.

11.4.2. Goal-Oriented Scheduling

Goal-oriented scheduling differs from biased processor sharing in that dynamic scheduling priorities are used to ensure that each class attains specified performance objectives. For example, interactive users may be given general priority over a batch workload, subject to a constraint that batch throughput must have a certain minimum value. Such dynamic priorities are difficult to model in general, but creative use of transaction classes is helpful in some cases. For example, in the case described above, the model could initially give priority to the interactive class. If the solution indicates that the batch class attains its throughput goal, then no change to the model is needed. If the batch class fails to meet its throughput goal, however, we can assume that the goal-oriented scheduler would reduce the priority given to the interactive users enough to ensure the specified batch throughput. This can be reflected in the model by converting the batch workload to a transaction workload with its arrival rate set to the specified minimum throughput. For transaction classes, throughput is equal to arrival rate unless the system is saturated. Thus, the batch class is assured of the performance that it would attain under the goal-oriented scheduler, and the consequent degradation of service to the interactive class is represented.

11.5. FCFS Scheduling with Class-Dependent Average Service Times

If different classes have significantly different average service times per visit ($S_{c,k}$) at a FCFS center, our standard evaluation techniques from Chapter 7 may not provide acceptable accuracy. This situation is handled quite easily by another modification to the residence time equation of these techniques. The original form of the residence time equation is:

11.6. FCFS Scheduling with High Variability in Service Times

$$R_{c,k}(\vec{I}) = D_{c,k}\left[1 + Q_k(\overrightarrow{I-1_c})\right] = V_{c,k}\left[S_{c,k} + S_{c,k}Q_k(\overrightarrow{I-1_c})\right]$$

Since all classes must have the same service time per visit at a FCFS center (in a separable network), we can think of this equation as a shortened form of:

$$R_{c,k}(\vec{I}) = V_{c,k}\left[S_{c,k} + \sum_{i=1}^{C} S_{i,k}Q_{i,k}(\overrightarrow{I-1_c})\right]$$

Simply substituting non-identical $S_{i,k}$ into the above equation provides an intuitively appealing evaluation technique for FCFS centers at which different classes have different average service times per visit: each class i customer found ahead of an arriving class c customer is multiplied by a class i service time. With this small change to one equation of the standard MVA algorithm, substantially more accurate solutions are obtained for models involving FCFS centers at which average service times differ from class to class.

An example is shown in Table 11.2. We consider a system with four disks and a CPU scheduled FCFS. There are two classes. Class A is of terminal type and class B of batch type. In the table we show the response time experienced by class A users for five different values of class A service time per visit at the CPU. We obtain results in three different ways: by ignoring the class-dependent average service times and applying mean value analysis directly ("MVA" in the table), by using the algorithm suggested in this section ("Section 11.5" in the table), and by simulating the system ("simulation" in the table).

The results show that the effect of class-dependent average service times can be pronounced, and that the algorithm suggested here yields good results for the example under consideration.

11.6. FCFS Scheduling with High Variability in Service Times

In the previous section we presented a solution technique for FCFS centers where the average service times per visit differ among the customer classes. This technique was necessary because of the restrictions required for a model to be separable (see Sections 7.2 and 7.5), and thus amenable to analysis using the standard algorithms of Chapter 7. In this section we present a technique that overcomes another restriction of separable networks, that imposed by the service time homogeneity assumption (see Section 7.5). This assumption states that the rate of completion of customers from any service center does not depend on the state of the model as a whole (i.e., the locations of the other customers).

Model Inputs:

$$N_A = 10 \quad Z_A = 10 \quad\quad N_B = 6 \quad Z_B = 0$$

	center				
	CPU	Disk 1	Disk 2	Disk 3	Disk 4
$S_{A,k}$	<varying>	1	1	1	1
$V_{A,k}$	8	2	2	2	2
$S_{B,k}$	2	1	1	1	1
$V_{B,k}$	20	2	4	6	8

(all times are in seconds)

Class A Response Time:

solution	$S_{A,CPU}$				
technique	2	1/2	1/8	1/32	1/128
MVA	250.1	63.1	26.8	23.7	23.4
Section 11.5	250.1	133.1	104.4	97.2	95.4
simulation	250.1	131.1	98.0	97.9	92.0

(all times are in seconds)

Table 11.2 – FCFS with Class-Dependent Average Service Times

In modelling most computer systems, any violation of this assumption does not result in significant error. Therefore, it is only in unusual situations that the technique to be presented need be employed. (We discourage superfluous use of the technique because it requires more parameter values than the simpler separable models, and so the parameterization effort is increased.)

As a rule of thumb, we can expect separable models to perform satisfactorily when the variability in service times per visit at each FCFS center is moderate, that is, when the average and standard deviation of service times are comparable. Centers for which the use of the technique will yield a noticeable improvement in accuracy are characterized by having most service bursts (service acquired in a single visit) be of comparable duration, with occasional bursts of much longer duration. As an example, in a batch system the CPU service quantum might be set very long to reduce context switch overhead; this could result in many short service bursts during file access, followed by a single long period of computation once the data has been acquired. In such a situation a separable model would not capture the effect on performance of the occasional very long service bursts, even if the average service time in the model was set to the measured average of the system. The effect of these long bursts is

11.6. FCFS Scheduling with High Variability in Service Times

to increase the amount of queueing that occurs in the system. Thus, a separable model will tend to give optimistic results when used in these situations.

As in other cases, we suggest a solution technique based on modifying the MVA residence time equation, then using the modified equation in the basic MVA iteration. Residence time consists of service time plus queueing time. Consider a class c customer arriving at service center k. Service time per visit ($S_{c,k}$) is an input parameter, and so presents no problem. Since we are considering FCFS centers, queueing time is required for all jobs already present at the center. The arriving job must wait on average $S_{i,k}$ time units for each class i customer found in the queue but not yet in service. Finally, the arriving customer must wait for the customer currently in service to finish. We can summarize this as:

$$R_{c,k}(\vec{I}) \approx V_{c,k}\left[S_{c,k} + \sum_{i=1}^{C} S_{i,k}\left[Q_{i,k}(\overrightarrow{I-1_c}) - U_{i,k}(\overrightarrow{I-1_c})\right] + \sum_{j=1}^{C} r_{j,k} U_{j,k}(\overrightarrow{I-1_c})\right]$$

where $r_{j,k}$ is the average time until completion of a class j customer found to be in service by a class c arrival at center k. The first term in this equation represents the inherent service requirement of the class c job. The second term approximates the total time spent waiting for customers in the queue (the $Q_{i,k}(\overrightarrow{I-1_c})$ term) but not in service (thus the $-U_{i,k}(\overrightarrow{I-1_c})$ term). Interpreting $U_{j,k}(\overrightarrow{I-1_c})$ as the proportion of time that an arriving class c customer finds a class j customer in service, the final term approximates the time spent waiting for the customer in service to complete.

This equation is the basis for an MVA-like analysis technique for models containing FCFS centers with high service time variability. The remaining problem is to estimate $r_{j,k}$, which often is called the *residual service time* of class j at center k. To do so, we assume that a class c job is equally likely to arrive at any point during the class j service interval (that is, class c arrivals occur at random with respect to class j service intervals). Even with this simplification, a reasonable choice for $r_{j,k}$ is not immediately apparent. Intuitively, one might guess $r_{j,k} = S_{j,k}/2$. In fact, however, this is an extreme value (representing the smallest possible residual service time) occurring only when the class j service times of all visits to center k are exactly equal. Under our assumptions, the residual service time is given by:

$$r_{j,k} = \frac{S_{j,k}}{2} + \frac{\text{variance}}{2S_{j,k}}$$

where *variance* is the variance in the service times per visit of class j at center k. Thus, the actual residual can be any number at least as large as half the average service time (since it is possible for the variance to be any non-negative value). As an example, suppose class j experienced ten service bursts of length 1 for each burst of length 90. An arriving customer is then nine times as likely to arrive during the single long burst as during any of the short bursts. Thus, the residual service time is $(.1)(5) + (.9)(45) = 41$. In contrast, the average service time is $\frac{10}{11} 1 + \frac{1}{11} 90 = 9.09$. This surprising situation results from the fact that a customer is much more likely to arrive during a long burst than a short burst, even if many more bursts are short than long.

Table 11.3 presents an example of the use of this technique. We consider a system with four disks and a CPU. There is a single class of terminal type. In the table we show the response time experienced by users for five different degrees of variability in CPU service times. We obtain results in three different ways: by ignoring the high variability in CPU service times and applying mean value analysis directly ("MVA" in the table), by using the algorithm suggested in this section ("Section 11.6" in the table), and by simulating the system ("simulation" in the table).

The results show that the effect on performance of service time variability becomes more severe as this variability increases. The approach suggested in this section reflects the degradation in response time that occurs with increasing variability.

We note that this technique can be used whether the center we are considering has unusually high or low variance in service times per visit. While service time distributions with low variance also can be troublesome at FCFS service centers, their potential impact on model accuracy is more limited. Separable models tend to be slightly pessimistic for systems with low variance FCFS centers.

11.7. Summary

System configurations that include multiple processors or that use certain scheduling disciplines may require special techniques to obtain sufficiently accurate models. Tightly-coupled multiprocessors provide service at a total rate that depends on the number of jobs currently requiring CPU service. The set of processors is best represented as a single flow equivalent service center that provides service at a rate proportional to the number of busy processors, less a factor to account for interference among the processors. Loosely-coupled multiprocessors, on the other hand, require no such special treatment since each processor serves a separate job queue. Separate job classes can be used to distinguish jobs from different processors when they use shared I/O devices.

11.7. Summary

Model Inputs:

$$N = 10 \quad Z = 10$$

	center				
	CPU	Disk 1	Disk 2	Disk 3	Disk 4
S_k	0.5	1	1	1	1
V_k	8	2	2	2	2

(all times are in seconds)

Response Time:

solution	variance of S_{CPU}				
technique	.25	.5	1	2	4
MVA	32.6	32.6	32.6	32.6	32.6
Section 11.6	32.6	35.5	40.1	46.7	56.4
simulation	32.4	38.9	42.3	53.8	53.4

(all times are in seconds)

Table 11.3 − FCFS with High Variability in Service Times

Many operating systems use scheduling disciplines that are based on job class priorities, but priority scheduling is not compatible with separable models. Consequently, to obtain a model that can be validated, it may be necessary to employ a specialized technique for modelling priority scheduling. We have described a technique based on replacing the priority CPU by C "shadow" CPUs, each one visited by just one class. The service demand of each class at its shadow CPU is inflated to reflect the impact of higher priority classes. In some situations a different technique − based on hierarchical decomposition, a flow equivalent service center, and global balance − also may be applicable. Both of these techniques can be adapted to situations in which one, some, or all of the service centers are scheduled by priority. When priorities among classes are not absolute, it may be appropriate to model the discipline as biased processor sharing or goal-oriented scheduling. Techniques for treating these disciplines have been suggested.

Finally, FCFS scheduling also requires special treatment under some circumstances. If the average service requirement per visit to a center differs from class to class, then the model is not separable. Once again, a simple modification to the MVA algorithm produces good model solutions. Similarly, if there is high variability in the length of service times at each visit to a center, then FCFS scheduling cannot be accurately represented in a separable model. The high variability can be captured by adapting the MVA solution technique, and by making further assumptions that allow estimates for the residual service time of jobs found in service by an arriving customer.

The techniques described in this chapter are useful for the specific circumstances in which they have been described. An equally important reason for presenting them, however, is that they are indicative of the approaches that must be creatively applied to achieve efficient and accurate solutions to non-separable models.

11.8. References

Sauer and Chandy were the first to use flow equivalent service centers and a global balance solution of a two center model to evaluate non-separable models, including ones involving priority scheduling [Sauer & Chandy 1975]. They discuss other techniques for evaluating non-separable models elsewhere [Chandy & Sauer 1978; Sauer & Chandy 1980].

Bard first demonstrated the flexibility of the basic MVA algorithm in adapting to non-separable models, treating both priority models and models in which different classes have distinct average service requirements per visit to an FCFS center [Bard 1979]. Bard also has described a modelling approach capable of treating the dynamic priority scheduling used in IBM's VM/370 operating system [Bard 1981].

The shadow CPU technique described in Section 11.3 was developed by Sevcik [1977]. His approach involved identifying separable models that provide optimistic and pessimistic bounds on the performance of a (non-separable) model with a priority center.

The MVA-based approach to modelling high service time variability was proposed by Reiser and Lavenberg [1978]. An alternative approach is based on global balance and Cox's *method of stages* representation [Cox 1955]. Cox demonstrated that arbitrary service time distributions can be approximated as closely as desired by using a sufficient number of exponentially distributed stages with probabilistic selection. Sevcik, Levy, Tripathi, and Zahorjan describe three-parameter method of stages representations for both high variability and low variability distributions [Sevcik et al. 1977]. With these three-parameter representations, it is possible to match two characteristics (typically the mean and variance) of an arbitrary distribution. Lazowska has shown that more accurate models are obtained by matching the mean and some percentile (say the 90th) than by matching the mean and variance [Lazowska 1977]. Lazowska and Addison provide a technique for determining a method of stages

11.8. References

representation that matches the mean and an arbitrary number of percentiles of an arbitrary distribution [Lazowska & Addison 1979].

The simulation results reported in Tables 11.1, 11.2, and 11.3 were obtained from IBM's Research Queueing Package [Sauer et al. 1982].

[Bard 1979]
Yonathan Bard. Some Extensions to Multiclass Queueing Network Analysis. In M. Arato, A. Butrimenko, and E. Gelenbe (eds.), *Performance of Computer Systems*. North-Holland, 1979.

[Bard 1981]
Yonathan Bard. A Simple Approach to System Modelling. *Performance Evaluation 1*,3 (November 1981), 225-248.

[Chandy & Sauer 1978]
K. Mani Chandy and Charles H. Sauer. Approximate Methods for Analyzing Queueing Network Models of Computing Systems. *Computing Surveys 10*,3 (September 1978), 281-317.

[Cox 1955]
D.R. Cox. A Use of Complex Probabilities in the Theory of Stochastic Processes. *Proc. Cambridge Philosophical Society 51* (1955), 313-319.

[Lazowska 1977]
Edward D. Lazowska. The Use of Percentiles in Modeling CPU Service Time Distributions. In K.M. Chandy and M. Reiser (eds.), *Computer Performance*. North-Holland, 1977, 53-66.

[Lazowska & Addison 1979]
Edward D. Lazowska and Clifford A. Addison. Selecting Parameter Values for Servers of the Phase Type. In M. Arato, A. Butrimenko, and E. Gelenbe (eds.), *Performance of Computer Systems*. North-Holland, 1979, 407-420.

[Reiser & Lavenberg 1978]
Martin Reiser and Stephen S. Lavenberg. Mean Value Analysis of Closed Multichain Queueing Networks. Report RC-7023, IBM T.J. Watson Research Center, March 1978.

[Sauer & Chandy 1975]
Charles H. Sauer and K. Mani Chandy. Approximate Analysis of Central Server Models. *IBM Journal of Research and Development 19*,3 (May 1975), 301-313.

[Sauer & Chandy 1980]
C.H. Sauer and K. Mani Chandy. Approximate Solution of Queueing Models. *IEEE Computer 13*,4 (April 1980), 25-32.

[Sauer et al. 1982]
   Charles H. Sauer, Edward A. MacNair, and James F. Kurose. The Research Queueing Package, Version 2: Introduction and Examples. Report RA 138, IBM T.J. Watson Research Center, 1982.

[Sevcik 1977]
   Kenneth C. Sevcik. Priority Scheduling Disciplines in Queueing Network Models of Computer Systems. *Proc. IFIP Congress '77* (1977), 565-570.

[Sevcik et al. 1977]
   Kenneth C. Sevcik, Allan I. Levy, Satish K. Tripathi, and John Zahorjan. Improving Approximations of Aggregated Queueing Network Subsystems. In K.M. Chandy and M. Reiser (eds.), *Computer Performance.* North-Holland, 1977, 1-22.

11.9. Exercises

1. Consider a single class model of a dual processor system. The service demand at the CPU is 8 seconds (with each processor providing a portion of this service) and the service demands at each of the four disks are 2 seconds. The single customer class is of terminal type, with $Z = 20$ seconds.

   a. Compare the results obtained by modelling the dual processor as a single fast processor (with a service demand of 4 seconds) to the results obtained by using the FESC approach of Section 11.2 (with service rates of 0.125 with one customer in the queue, and 0.250 with more than one customer in the queue). Obtain solutions for populations of 5, 10, and 20 online users. (Use the MVA implementation of Chapter 18, extended to accommodate FESCs and terminal classes.)

   b. What do your solutions for the three population sizes indicate about the accuracy of the "single fast processor" approach in (a)? How well would you expect this approach to work if the configuration contained four processors rather than two?

2. Section 11.3 developed a technique for modelling preemptive priority CPU scheduling. Using this as a basis, develop a technique for modelling non-preemptive priority scheduling. Under non-preemptive priority, a job in service at the CPU receives a full service burst, even if a higher priority job arrives during that burst. When the service burst completes, the highest priority waiting job is selected for the next service burst.

11.9. Exercises

3. Consider a simple interactive computer system consisting of a CPU and four disks. Assume that the disks are scheduled FCFS, and that users can choose their I/O block size: the number of bytes transferred between a file and main storage on each access. Measurements of the system show that 75% of the users choose block sizes resulting in service times per disk visit of 32 milliseconds, and 25% choose sizes resulting in service times per disk visit of 44 milliseconds.

   a. Suppose that there are a total of 24 online users divided into two classes based on blocksize. Both classes have 20 second think times, and have interactions that require 4 seconds of CPU service and an average of 100 accesses to each of the four disks. Use the technique of Section 11.5 to estimate response times for each class.

   b. Using the throughput values obtained from (a), compute the average service time per I/O operation at each disk. Use this value to construct a model of the system with a single class of "average" users. This model can be evaluated using standard mean value analysis techniques.

   c. Compute the overall average response time in the two class model of (a). (Remember that the response times of the classes must be weighted by their throughputs.) Compare your result to the response time obtained in (b). What does this tell you about the effect on system performance of FCFS scheduling with class-dependent service times?

   d. Repeat (a) through (c) under that assumption that 75% of the users have disk service times of 12 milliseconds, and 25% have disk service times of 116 milliseconds. Compare your results to those obtained earlier. What does this tell you about the importance of reflecting service time variability in models of computer systems?

   e. Returning to the single class model, use the technique of Section 11.6 to model the high service time variability of an "average" job at each disk. To do so, you will need to estimate the variance of the service times at the disks. If proportion p of the total accesses require S_1 time units and proportion $1-p$ require S_2 time units, then the average service time S is equal to $pS_1 + (1-p)S_2$, and a reasonable estimate of the variance in service times is:

   $$\text{variance} = p(S_1 - S)^2 + (1-p)(S_2 - S)^2$$

   Calculate response times for the original set of disk service times and the modified set of (d), and compare these to the results obtained earlier. How do you account for the differences in the various estimates?

4. Discuss the treatment of scheduling disciplines in single class, separable queueing network models.
5. Discuss the treatment of scheduling disciplines in multiple class, separable queueing network models.
6. We have considered FCFS scheduling in four contexts: single class separable models, multiple class separable models, single class with high variability in service times, and multiple class with class-dependent average service times. Compare and contrast these.

Part IV

Parameterization

In Parts II and III we have discussed extensively the definition and evaluation of queueing network models. Here, in Part IV, we discuss the parameterization of these models. Parameterization is the heart of the modelling process, for the results of a study can be no more accurate than the parameter values provided to the queueing network evaluation algorithms.

Our presentation is divided into three parts. In Chapter 12 we discuss the construction of baseline models of existing systems. A validated baseline model is the starting point for any performance study of an existing system.

In Chapter 13 we discuss *modification analysis*: the process of adjusting parameter values to project performance for modified environments. The key to modification analysis is the ability to anticipate and represent primary effects. For this reason, modification analysis relies on the experience of the analyst to a significant extent.

In Chapter 14 we discuss the use of queueing network models to project the performance of proposed systems — systems for which baseline models cannot be constructed and validated. The process of designing a new system involves continuous tradeoffs between cost and performance. Queueing network models can help to quantify performance, and thus to guide the entire design process.

The divisions between these three chapters are artificial in many respects. The construction of a baseline model of an existing system must be guided by knowledge of the model's intended applications in projecting performance for the system as it evolves. The techniques for the successive refinement of workload characterizations that have been developed to model proposed systems can be extremely helpful in dealing with existing and evolving systems.

Chapter 12

Existing Systems

12.1. Introduction

In this chapter we discuss the construction of baseline models of existing systems. This activity relies on knowledge of the hardware, software, workload, and monitoring tools associated with the system under study. It also requires access to information recorded by accounting and software monitors during system operation. Here, we describe general approaches applicable to a variety of systems. In Chapter 17, we illustrate these approaches with an example based on a specific system (IBM's MVS) and a specific monitoring tool (RMF).

In Chapter 4 we divided the inputs of queueing network models into three groups: the *customer description*, the *center description*, and the *service demands*. The structure of the present chapter reflects this division.

Section 12.3 is devoted to the customer description: the correspondence of the workload components of the system to the customer classes of the model. In specifying the values of the customer description parameters, we are answering questions such as:

- How many customer classes are required?
- Of what type (transaction, batch, or terminal) should each class be?
- What should be the workload intensity value (λ, N, or N and Z) for each class?

Section 12.4 is devoted to the center description: the correspondence of the resources of the system to the service centers of the model. In specifying the values of the center description parameters, we are answering questions such as:

- What devices and subsystems should be included in the model?
- How should each of these entities be represented (e.g., as a queueing center, a delay center, or an FESC)?

Section 12.5 is devoted to the service demands: the description of the interactions between customers and centers. In specifying the values of the service demand parameters, we are answering the question:

12.2. *Types and Sources of Information*   275

- What proportion of the measured usage of each device should be attributed to the customers of each class?

We precede these three sections, in Section 12.2, with a survey of the information used to parameterize queueing network models: its types, its sources, and how it can be managed. We follow these sections, in Section 12.6, with a discussion of the validation of baseline models, indicating reasonable tolerances for various performance measures.

There is little reason to construct a model of an existing system unless this model is to be used for performance projection. Consequently, we cannot completely separate the task of constructing a baseline model of an existing system (the subject of this chapter) from the task of using the model to project performance for an evolving system (the subject of Chapter 13). Our (somewhat artificial) separation between the two tasks will be the following: problems that arise from limitations or shortcomings of current monitoring tools and techniques will be treated in this chapter, while problems that would persist even with ideal monitoring capabilities will be deferred to the next chapter.

12.2. Types and Sources of Information

The information required to specify parameter values for a queueing network model of an existing system includes *static* information about the system configuration and *dynamic* information extracted from records produced during system operation by various monitoring packages. Some information is recorded for purposes of accounting, while other information is recorded explicitly for performance evaluation purposes. Software packages of varying degrees of sophistication are available for storing, analyzing, and reporting the information recorded during system operation. In this section, we discuss briefly the information needed, how it can be obtained, and how it can be managed. Our intention is not to be comprehensive, but rather to highlight points of particular relevance to the construction and use of queueing network models.

One type of information required is a description of the hardware and software of the system. With respect to hardware, this information includes an enumeration of the components of the system (processors, channels, storage devices, communication devices, etc.) and an indication their interconnections (e.g., the paths over which data can be moved from a particular storage device to memory). With respect to software, this information includes the operating system in use, and the values of parameters that influence resource allocation. Examples of such parameters include CPU scheduling priorities for various workload components, placement of files on storage devices, etc.

This system description is relatively static, in that it changes only week to week or month to month. The information it provides about the hardware suggests what resources should be represented as centers in the model. The information it provides about the software and operating policies suggests appropriate modelling assumptions and helps in the interpretation of measurement data.

Another type of information that is required is recorded dynamically during system operation by various monitors. Accounting monitors write records at the termination of batch jobs or interactive sessions, indicating the system resources consumed by the job or session (CPU seconds, I/O operations, memory residence time, connect time, etc.). Software performance monitors write records describing resource usage and performance status from another point of view. At specified intervals, queue lengths or device status indicators may be *sampled* and the results written in a record. Also, certain *events* that are considered significant (such as swapping a customer out of main memory) may be documented in a record.

Because of their volume and their encoding, the records produced by accounting and software monitors are not usable directly. Rather, they must be processed by reporting routines that produce summary information for a specific purpose (e.g., accounting, workload forecasting, performance modelling). Most accounting and software monitors are packages that include both a *recording* component and a *reporting* component. For example, accounting records are written for each unit of work processed, and an accounting program periodically passes over the recent accounting records to determine charges for each account. Similarly, software monitors write records at certain events or sampling intervals, and a postprocessor later examines the records and produces reports organized to aid system tuning and performance evaluation.

The reports produced by accounting and software monitors usually are organized in one of two ways. Some reports are *class based*: they organize information by user or by workload component. Other reports are *resource based*: they organize information by system resource. Monitors that reliably break down resource usage by both workload component and resource are not used commonly in most systems. (Those that exist cause prohibitively high monitoring overhead.) Much of the effort in parameterization, as described in Sections 12.3 to 12.5, arises from the need to surmount the inadequacies of commonly available measurement information. As software monitors are improved, the parameterization task will become less burdensome, and some of the techniques described in this chapter will become unnecessary.

When using a reporting routine to obtain information, it is necessary to specify the interval of time over which information is to be gathered. Generally, it is appropriate to run the monitor during peak loads, as these

12.2. Types and Sources of Information

present the most significant performance problems. The duration of the observation interval should be long enough that *end effects* do not significantly affect the accuracy of the measurements. End effects are measurement errors caused by the fact that some customers are processed partly within and partly outside of the observation interval. In particular, it is typical to assume that the system operates in flow balance over the measurement interval, so that the job arrival and completion rates are equal. However, because some jobs arrive but do not complete during the interval, and other jobs arrive before but complete during the interval, flow balance may not hold. Clearly, measurements obtained from longer observation intervals are affected less by these end effects than are shorter intervals. Typically, observation intervals of thirty to ninety minutes are appropriate for obtaining software monitor data. If monitoring overhead is a concern, shorter intervals can be used, but the danger of anomalies is increased.

Other sources of useful information include hardware monitors and monitors specialized for particular application subsystems (such as database or telecommunications subsystems). Hardware monitors, because they are "external observers" of the system, obtain accurate measurements and do not perturb system operation. They are incapable, however, of associating resource usage with workload components. The specialized application subsystem monitors are helpful in assessing the performance of subsystems whose autonomy from the host operating system prevents standard monitors from being able to record their activity. (For example, special monitors are needed for IBM's IMS database system because RMF does not record information about individual IMS transactions.) While any information that is available from hardware and specialized application subsystem monitors should be exploited, our discussion in this chapter will be restricted to the kinds of information that are commonly reported in most medium or large computer installations.

Table 12.1 summarizes the information typically available from various sources. Information from different sources (accounting and software monitors, or even two different software monitors) may be based on different underlying assumptions. For this reason, and also because of end effect anomalies, information from different sources may appear to be contradictory. For example, consider a small interactive system in which monitors report that in a thirty minute observation interval:

— 7200 transactions were processed
— average response time was three seconds
— the sum of the queue lengths at the CPU and all disks was 18

We would conclude that throughput during the observation interval was:

$$\frac{7200 \text{ transactions}}{1800 \text{ seconds}} = 4 \text{ transactions/second}$$

type	information provided
system description	hardware configuration operating system (and version) resource allocation and scheduling strategies tuning parameter values
accounting monitor	CPU usage, by workload component logical I/O operation count, by workload component customer completions, by workload component
software monitor	measured busy time, by device physical I/O operation count, by device average queue length, by device throughput, by workload component average response time, by workload component
hardware monitor	observed busy time, by device

Table 12.1 — Sources of Information

Because the observation interval is long relative to the average response time, we could be confident that end-effects would not lead to significant errors in the estimates of throughput or response time. Considering Little's law, however, we would find the sum of the queue lengths (18) to be much higher than expected from the product of throughput (4 transactions/second) and response time (3 seconds). One possible explanation for such a situation is that the queue lengths include system tasks that are not counted in either the throughput or response time calculations. On the other hand, if the sum of the queue lengths had been reported as 8 (and other values remained the same), then Little's law would reveal a discrepancy in the other direction. A possible explanation for the second case would be that requests were queueing for admission to memory, thus spending a significant part of their response time where they were not included in the queue length of any device. The fundamental laws presented in Chapter 3 can be used to detect such apparent contradictions. System intuition and careful thought is required to resolve them.

Enhanced awareness of the problems of configuration management and capacity planning has led recently to some encouraging progress in the use and management of system measurement data. First, special reporting routines tailored to the requirements of queueing network modelling have been developed for some systems. These routines analyze records produced by existing accounting and software monitors. Some are capable of defining a queueing network in a format directly acceptable by particular queueing network modelling software packages.

While these routines are a great aid, intervention by an analyst still is necessary in most cases to obtain a validated model. This is true because of inadequacies in the measurement data, and the fact that the analyst's knowledge of the system is not available to the automated routine. (Further discussion of such routines appears in Chapter 16.)

Second, some of the newer reporting routines have been generalized to be capable of using and contributing to a *performance database*. The records written by various monitors constitute a rudimentary performance database. Merely organizing the records according to their types and source makes them easier to use. The utility of the database is further enhanced, however, if it is extended to include aggregated information produced by reporting routines. There are several advantages to maintaining such a performance database. For one, long-term trends can be examined if information aggregated on a month by month basis is included in the database. Also, information intended for management planning can be isolated from the more technically oriented information intended for system tuning. Finally, by having various aggregations of monitoring information available in a database, the need for regular printed reports is substantially reduced.

12.3. Customer Description

Most large computer systems have workloads consisting of several identifiable components. Performance studies often are intended to assess performance of each workload component, since system-wide average values for throughput and response time have little significance in systems that include such diverse workload components as background batch and foreground transaction processing. There are several goals to meet in deciding how to assign the workload components of the system to the customer classes of a queueing network model:

- Classes should consist of customers whose service demands are of comparable magnitude and similar balance across service centers, since input parameters to the model for all customers in the same class are identical. (For example, I/O bound customers should not ordinarily be in the same class as CPU bound customers.)

- Classes must distinguish workload components for which independent performance projections are desired as outputs of the model. (For example, if response time to database queries is of concern, then database queries should not be grouped in a single class with other workload components.)

- Classes may be made to correspond to accounting and performance groups. This facilitates the calculation of various parameter values, since accounting data is organized by accounting group.
- Classes may be used to distinguish work generated by various organizational units (e.g., divisions of a company). This permits unit-specific performance projections, and facilitates later modification analysis (since workload forecasts frequently are made on an organizational unit basis).

A first step in identifying customer classes is to group portions of the workload according to whether they are best represented as batch, terminal, or transaction types. Often, the nature of a workload component suggests an appropriate type: if requests arrive at a constant rate, then transaction; if requests are generated by a set of users that await the completion of service to one request before generating another, then terminal; if the number of active requests is constant, then batch. Variations are possible, though, especially in conducting a modification analysis. As one example, a workload component might in fact consist of users at terminals, but for planning purposes its intensity might be described in terms of a request arrival rate. In this case, the use of a transaction type might be appropriate. As another example, a system might have many workload components, only a few of which are of interest. The presence of the other components might be reflected in the model by a single "aggregate" class of transaction type (so that its throughput is guaranteed to equal the measured value).

Within each type of customer class, further separation of workload components may be desirable. Batch work of different priorities may be represented as distinct classes. Different interactive systems (e.g., APL and TSO in an IBM environment) may be treated as separate terminal classes. If trivial transactions (such as simple editing commands) can be distinguished from substantive transactions (such as complex database queries), then different classes can be used to distinguish the two groups.

The queueing network model input parameter C is simply the number of customer classes, determined according to the guidelines suggested above. Models of simple systems typically have just one or two classes, while models of complex multi-purpose systems may have eight or more. In some special situations it is useful to have a very large number of classes — say, twenty to forty.

One example of a situation in which a large number of classes was used is a model developed for projecting the performance of a hospital information system used in many hospitals. There were roughly thirty major transaction types (admit-patient, order-blood-test, set-dietary-restriction, etc.) each one of which was represented as a separate customer class. In this way, the arrival rate of each transaction type and the

12.3. Customer Description

priority assigned to the transaction type (reflecting its urgency in a particular hospital) could be represented directly in the model. The hospitals using the system differed substantially in size and in the hardware on which they ran the system. Also, they differed significantly in the particular mix of transactions that were processed. The model proved useful in configuration design. The response times for various transaction classes could be related to the arrival rates and priorities of the classes for various contemplated hardware configurations.

Having identified each workload component to be represented as a distinct customer class and determined the type of that class, the next step is to establish the workload intensity of each class. For a transaction class, the workload intensity is the transaction arrival rate. Over a reasonably long observation interval in a system that is not saturated, the arrival rate is essentially the same as the completion rate. Consequently, an estimate for the arrival rate of class c is:

$$\lambda_c = \frac{\text{measured completions of class } c}{\text{length of measurement interval}}$$

For a batch class, the workload intensity is given by the average number of batch customers active. An estimate for N_c, the number of class c customers, can be obtained in several ways:

- If jobs are processed in a fixed number of regions and memory queueing times are high (so that it is known that each region is busy throughout most of the observation interval), then N_c is the number of processing regions.
- If the software monitor provides an estimate of the average multiprogramming level of the class over the observation interval by sampling, then N_c can be taken to be that estimate.
- If accounting data provides the residence time of each job in the central subsystem, then N_c can be estimated by:

$$N_c = \frac{\sum_{\substack{\text{class } c \\ \text{jobs}}} \text{measured job residence time}}{\text{length of measurement interval}}$$

(This alternative is impractical without the use of a reduction package capable of automatically extracting this information from accounting records.)

For a terminal class, workload intensity is specified by the number of active terminals, N_c, along with the average think time, Z_c. Three possibilities for estimating N_c for terminal classes correspond directly to the three methods used for batch classes:

- If terminals connect to the system through a limited number of ports, and if all ports are busy throughout most of the observation interval, then N_c is the number of ports.
- If the software monitor provides the average number of active terminals over the observation interval, then N_c can be taken to be that number.
- If accounting data includes session lengths, then N_c can be estimated (over an observation interval that is long relative to average session length in order to restrict end effects) by:

$$N_c = \frac{\sum_{\substack{\text{class } c \\ \text{sessions}}} \text{measured session length}}{\text{length of measurement interval}}$$

The average think time of a terminal class often is one of the most difficult input parameters to estimate. There are several reasons. First, there are differing views of when think time starts and ends. We will adopt the one in which it starts with the arrival of the first character of a response from the system, and ends when the last character of the next request to the system is entered. Second, some systems allow a stream of commands to be entered without awaiting responses. Such systems can cause think times (as defined above) to be negative! Third, some think times become so long that they actually represent a loss of an active terminal. (This occurs when terminal users interrupt their work without logging off.) Fourth, average think time seldom is measured directly by performance monitors. Consequently, the best estimate of think time often is obtained by estimating Z_c from the response time law:

$$Z_c = \frac{N_c}{X_c} - R_c$$

where N_c is estimated as described above, and X_c and R_c are measured values. Because there often is less confidence in the estimate of think time than in the estimates of other parameters, it may be desirable to test the sensitivity of the model to this value.

When memory constraints are imposed on transaction or terminal classes, it is necessary to specify the capacity associated with each *domain* so that the modelling approach of Section 9.3 can be used. The capacity of each domain typically is known from the system description. Whether or not the domain was filled to capacity in a particular measurement interval is revealed by comparing the average number active among classes assigned to the domain (as reported by a monitor) to the domain capacity.

12.4. Center Description

The service centers of a queueing network model correspond to significant points of congestion or delay in the system. There are many ways of representing system resources by a set of service centers. Here we suggest only the most widely accepted methods, which have proven successful in a large number of modelling studies.

For systems with single CPUs and for tightly-coupled multiprocessors, a single service center is used to represent the CPU(s) in the queueing network model. Loosely-coupled multiprocessors are modelled by including one service center per processor. Front end communications processors and back end database machines also may be represented as separate service centers.

The representation of disk subsystems can be done in a variety of ways. (See the discussion in Chapter 10.) A number of components are involved in each disk I/O operation. The modelling approach that has proven most successful, however, is to use a single service center to represent each disk. Congestion due to other I/O subsystem components is represented by calculating an appropriate *effective service demand* for each center.

Other peripheral devices can be represented more simply than disks. Because tape drives are not capable of operation independent of the channel, a group of tape drives on a channel can be represented by a single service center. The service demands at the center can be established using channel utilization only, and ignoring the individual tape drives.

Unit record equipment typically is ignored in constructing queueing network models. This is justified in many systems because *spooling* makes the use of unit record devices asynchronous. Similarly, terminal controllers typically are not represented. If delays in the communications front end are thought to be important in a particular study, then a special approach must be used. This might involve a hierarchical model in which a conventional central subsystem model is evaluated, and then the delays due to communication are represented in a high-level model that includes an FESC representing the central subsystem.

12.5. Service Demands

The final set of values needed to parameterize a queueing network model are the service demands at each center of the customers belonging to each class. Obtaining these values can be a difficult and time consuming process. As a practical consideration, it is important to concentrate on obtaining accurate estimates for the most heavily utilized centers, because a small error in estimating the service demands at the bottleneck center will affect performance projections more than a much larger error at a lightly utilized center.

In estimating service demands, the three center types (delay, FESC, and queueing) are treated differently.

Delay centers have service demands that represent a delay that is not caused by congestion (e.g., a propagation delay in a communication network). It usually is not difficult to determine appropriate values for delay centers. In addition, errors in the service demands at delay centers are not "magnified" by queueing delay calculations when the model is evaluated.

For FESCs, the load dependent service rates can be determined in many ways, as described in Chapter 8. Two major approaches are evaluating low-level queueing network models (as illustrated in Chapter 9 for the case of memory constraints) and considering hardware characteristics (as illustrated in Chapter 11 for the case of tightly-coupled multiprocessors).

The remainder of this section is devoted to the case of queueing centers, by far the most common center type in queueing network models. Conceptually, estimating service demands for queueing centers is straightforward: at the conclusion of the measurement interval, the measured busy time for each class at each device is divided by the number of system completions for the class. In practice, however, two difficulties arise:

- In the multiple class case, the available data frequently is insufficient to apportion the measured busy time among the classes with certainty. The reasons and the remedies differ for various devices and various systems.

- A portion of the busy time attributed to each class is *intrinsic* to that class: its basic processing and I/O requirements. The remainder consists partly of *service demand inflation* and partly of *overhead*. Service demand inflation, introduced in Chapter 10, is the component of measured disk busy times due to contention in the I/O subsystem. (There is no service demand inflation for processors.) Overhead is work done by the operating system "on behalf of" the customers of the class. Part of the overhead component is *fixed*, in that it does not depend on system congestion (e.g., the CPU service required to initiate user I/O operations), and part of it is *variable* and typically increases with system load (e.g., paging I/O). In a baseline model these distinctions do not matter, but in conducting a modification analysis they can be crucial, for the service demand inflation and variable overhead components of the model usually change in a new environment.

12.5. Service Demands

This section is devoted to the first of these two difficulties: apportioning measured busy time among the various classes. We defer our discussion of the second difficulty to Chapter 13. The reader should understand, however, that while the techniques used to adjust the service demand inflation and variable overhead components of service demands are not required until projecting performance for an evolving system, they should be validated by examining several measurement intervals using the baseline model of the existing system.

Our discussion is organized into two subsections, the first devoted to processors and the second to I/O.

12.5.1. Estimating Processor Service Demands

Since the CPU typically is a heavily utilized resource, it is important to determine accurately the service demands of the various classes there. As noted in Table 12.1, monitor data often includes the CPU usage and the number of customer completions for each workload component. Unfortunately, the quotient of these quantities turns out in practice to yield a poor estimate of CPU service demand. The reason is that the CPU usage reported on a per class basis often fails to capture significant amounts of CPU activity. More specifically, the sum of the CPU busy times reported on a per class basis is likely to be considerably less than the total CPU busy time reported by a monitor that does not attempt to distinguish among classes. The ratio of attributed CPU usage for a class to the total CPU busy time due to activities initiated by that class is known as the *capture ratio*. Capture ratios typically range from .85 down to .40 for various systems and various workload components. For a particular system, the overall capture ratio can be estimated as suggested above: by dividing the sum of the CPU busy times reported on a per class basis (often by an accounting monitor) by the total CPU busy time reported by a monitor that does not attempt to distinguish among classes (often by a software monitor).

In the case of single class models, dividing the estimate of total CPU busy time from software monitor data by the estimate of total customer completions from either accounting or software monitor data will yield a good estimate for CPU service demand. In the case of multiple class models, though, techniques must be devised to apportion the unattributed CPU busy time among classes. This process has three steps:
- calculate the unattributed busy time during the interval
- decide how much to attribute to each class
- compute how much to attribute to each customer of each class

The second of these steps is the interesting one, and will be addressed in the paragraphs that follow.

Consider a system with a workload consisting of two components: batch jobs and interactive users. Assume that information comparable to that listed in Table 12.1 has been obtained. Let f_{BATCH} and f_{INTER} be (unknown) factors by which the attributed CPU busy time for each class must be multiplied so that all measured CPU busy time is attributed to some class. (Observe that f_c is the inverse of the capture ratio for class c.) This leads to the equation:

$$B_{CPU} = f_{BATCH} \times A_{BATCH,CPU} + f_{INTER} \times A_{INTER,CPU}$$

where $A_{c,CPU}$ is the CPU usage attributed to class c, and B_{CPU} is the total measured CPU busy time.

To determine unique values for f_{BATCH} and f_{INTER} we must establish a relationship between them in addition to this equation. Several possibilities exist:

- Assume that the ratio of total CPU time to attributed CPU time is the same for each class, yielding:

$$f_{BATCH} = f_{INTER} = \frac{B_{CPU}}{\left[A_{INTER,CPU} + A_{BATCH,CPU}\right]}$$

- Since the unattributed CPU busy time is likely to be overhead, use class based information on activities likely to cause CPU overhead (such as paging rate, swapping rate, spooling, user I/O, and job initiations) to determine a relative measure of total overhead for each class. For instance, assuming that overhead is due almost entirely to page fault handling, and letting OV_c (the relative overhead of class c) be the measured number of pages transferred because of class c faults, we have:

$$f_{INTER} = 1 + \frac{\frac{OV_{INTER}}{OV_{INTER} + OV_{BATCH}} \times \left[B_{CPU} - \left[A_{INTER,CPU} + A_{BATCH,CPU}\right]\right]}{A_{INTER,CPU}}$$

The second approach is the more reasonable. Unfortunately, more than one factor inevitably contributes to overhead. Thus, OV_c is better defined as the weighted sum of several factors:

$$OV_c = \sum_{all\ factors\ i} weight\ i \times factor\ i\ _c$$

When one attempts to apply this approach in practice, two common problems are apt to be encountered:

12.5. Service Demands

- Even for a single measurement interval, it may be difficult to determine which factors to consider, and what weights to assign to these factors. Iteration inevitably is required: estimate weights, calculate service demands, evaluate model, re-estimate weights, etc.
- If one truly is to have confidence in the weights selected, then data from a number of measurement intervals must be considered, and weights must be found that yield good model results when applied to each set of data. An ad hoc approach can be adopted, or linear regression techniques can be used.

Once f_{BATCH} and f_{INTER} have been determined, the service demands of the two classes can be estimated by the equation:

$$D_{c,CPU} = \frac{f_c \times A_{c,CPU}}{measured\ class\ c\ completions}$$

Note that the service demands determined in this way include intrinsic service, fixed overhead, and an amount of variable overhead that reflects the degree of system congestion in the interval covered by the measurement data.

12.5.2. Estimating I/O Service Demands

I/O activity in most current computer systems is dominated by operations on direct access storage devices (fixed head, movable head, and electronic disks). Tape I/O and I/O for staging data to and from mass storage devices plays a secondary role. Other types of peripheral devices typically are inconsequential with respect to performance. Our discussion in this section focuses on disk I/O, reflecting its importance.

In Section 10.7 we described how the lengths of certain portions of disk service requirements (seek, latency, rotation, and transfer) could be established from system knowledge (e.g., device characteristics) and measurement data. We assumed that both the visit counts and the service times per visit for each class at each disk were known. In this section, we suggest a method for determining these quantities. First we consider the visit counts, then the service times.

We distinguish two ways of viewing I/O operations. *Physical I/O* operations correspond to activations of I/O subsystem components to transfer data to or from peripherals. *Logical I/O* operations correspond to operating system calls by customers requesting access to blocks of information. For a number of reasons physical and logical I/O operations do not correspond directly to one another. Sometimes, a logical I/O operation may not result in a physical I/O operation; for example, a logical I/O operation may request access to a block of information that already is in memory. Sometimes, a logical I/O operation may result in several physical I/O operations; for example, errors detected in reading or writing a block may cause operations to be retried.

It is the physical I/O operations that correspond to the visit counts, but physical operations seldom are reported on a per class basis. Typically, logical I/O operations are broken down by class but not by device (often by an accounting monitor), while physical I/O operations are broken down by device but not by class (often by a software monitor).

The first step in confronting this situation is to estimate the ratio of physical to logical I/Os for each class. We now restrict consideration to a set of disk drives. Let P_k denote the physical I/Os at disk k, and let L_c denote the logical I/Os of class c over the set of disks. (Some monitors fail to distinguish logical disk I/Os from other logical I/Os. In such cases, we are forced to make some assumption such as that the fraction of all logical I/Os that are directed to the disks is the same as the fraction of all physical I/Os that are directed toward the disks, which is presumed to be available from measurements.) We define g_c to be the ratio of physical to logical I/Os for class c. (The assumption that the ratio depends on class but not device is realistic in most systems.) Estimating the g_c is a problem analogous to estimating the f_c in the case of the CPU. Possible approaches include:

- Assume that g_c is the same for each class, so that:

$$g_c = \frac{\sum_{\text{all disks } k} P_k}{\sum_{\text{all classes } j} L_j}$$

- Use generally accepted ratios for standard types of workloads for the architectural family of the system.

- For a number of observation intervals, determine the values for the g_c that best satisfy the set of equations:

$$\sum_{\text{all disks } k} P_k(i) = \sum_{\text{all classes } c} g_c \times L_c(i)$$

where (i) denotes values obtained during the i-th observation interval.

Once these g_c have been estimated, we proceed to determine the visit counts. In essence, we must satisfy the equations:

$$P_k = \sum_{\text{all classes } c} (\text{measured class } c \text{ completions}) \times V_{c,k}$$

$$L_c = (\text{measured class } c \text{ completions}) \times \sum_{\text{all disks } k} \frac{V_{c,k}}{g_c}$$

12.5. Service Demands

class	device			adjusted logical I/Os
	disk 1	disk 2	disk 3	
BATCH	?	?	?	$L_{BATCH} \times g_{BATCH}$
INTER	?	?	?	$L_{INTER} \times g_{INTER}$

physical I/Os	P_1	P_2	P_3

Table 12.2 — Physical Disk I/Os by Class and Device

Table 12.2 suggests a way of thinking about the problem of determining the number of physical I/Os by each class at each device, again for the case of two classes, batch (*BATCH*) and interactive (*INTER*). The central rows correspond to classes, while the central columns correspond to disks. The entry to be filled in at column k of row c is the number of physical I/Os by class c at device k ($V_{c,k} \times$ *measured class c completions*). The information available, however, is only that the columns must add to P_k while the rows must add to $L_c \times g_c$. This provides a number of equations equal to the sum of the number of classes and the number of disks, whereas the number of $V_{c,k}$ values that we must estimate is equal to the product of these quantities. (For instance, in Table 12.2 there are five constraints corresponding to the two row sums and three column sums, but there are six $V_{c,k}$ values to be determined.) Consequently, we must use additional information to specify the $V_{c,k}$ values uniquely. Alternatives include:

- The simplest assumption, which can be used in the absence of any other information, is that all classes use the various disks in the same proportions:

$$\frac{V_{c,k}}{V_{c,k'}} = \frac{V_{c',k}}{V_{c',k'}} \quad \text{for classes } c \text{ and } c', \text{ and disks } k \text{ and } k'$$

- The software configuration portion of the system description frequently indicates the location of various key data sets: paging files, swapping files, catalogs, files devoted to various applications, etc. If a particular class is known not to use a device, then its visit count there can be set to zero. If a particular class is known to be the exclusive user of a device, then its visit count there can be set to the measured physical I/O count of the device divided by the measured number of completions of the class. The remaining visit counts can be resolved in a series of stages. At each stage, the distribution of I/Os for the class for which the *least* flexibility remains is determined.

- In some systems there are software monitors capable of observing directly the number of physical I/Os broken down by both class and device. Although such monitors cause too much overhead to be used continuously, they can be used over short intervals (e.g., 10 minutes) to obtain an indication of the distribution of physical I/Os by class and device.
- Occasionally, the breakdown of logical I/Os by device as well as by class is known. This additional information makes it possible to proceed with greater confidence. In particular, if we can assume that the ratio of physical I/Os to logical I/Os is the same for each class, then the physical I/Os at a particular device can be attributed to classes in the same proportions as are the logical I/Os.

We turn now to the problem of determining the $S_{c,k}$. It is customary to assume that, at any particular disk, all classes have the same service time per visit. With this simplification, the service times are given by:

$$S_{c,k} = S_k = \frac{B_k}{P_k}$$

Situations in which one class has a substantially larger service time at a disk than another class typically arise when the former class uses a much larger block size. In such cases, disk characteristics (transfer rates, rotation times, and seek time functions) can be used to estimate the ratios $S_{c,k}/S_{c',k}$, for each pair of classes c and c' that use the disk. Those ratios, together with the equation:

$$B_k = \sum_{\text{all classes } c} V_{c,k} \times S_{c,k} \times (\textit{measured class c completions})$$

allow unique determination of the $S_{c,k}$. In both the cases of equal and unequal service times across classes, the service demands are given by:

$$D_{c,k} = V_{c,k} \, S_{c,k}$$

We now consider briefly the estimation of service demands for tape devices. As noted in an earlier section, it generally is appropriate to represent the tape channels rather than the individual drives. Further, it generally is appropriate to model all classes as using the various tape channels in the same proportions (although different classes will have different total amounts of tape I/O activity). Thus, the visit counts are given by:

$$V_{c,k} = \frac{L_c}{\sum_{\text{all classes } j} L_j} \times P_k \times \frac{1}{\textit{measured class c completions}}$$

where the P_k and L_c now are measured physical tape I/Os at center k and logical tape I/Os of class c, respectively. Assuming that all classes use essentially the same block size (so that they have the same service times), the service demands are given by:

$$D_{c,k} = V_{c,k} \frac{B_k}{P_k}$$

If block sizes differ significantly among classes, then service demands can be determined in a manner analogous to that suggested above for disks with class-dependent service times.

12.6. Validating the Model

Once values are established for all inputs, the model can be evaluated using the algorithms described in Part II, extended as described in Part III. This evaluation yields, for each class, estimates of system throughput and response time, and of device residence time, utilization, and queue length.

Model validation involves comparing these estimates with the measured values of the corresponding quantities. A model can be considered "validated" when it has been demonstrated that, in several (or many) measurement intervals, the differences between the estimates produced by the model and the measured quantities are sufficiently small.

In choosing observation intervals for use in validating the model, it is desirable to look ahead to the types of system changes to be investigated with the model. If the model is to be used to investigate the effect of an increased workload intensity, then the model should be validated on observation intervals representing a range of workload intensities. Similarly, if an increase in the size of main memory is to be considered, it is beneficial to validate the model on several different memory sizes. This could be done in a number of ways. Scheduling parameters could be adjusted to keep the number of active customers artificially low (thus underutilizing the memory). Alternatively, a portion of the memory could be disabled during an observation interval.

The correspondence between model estimates and measured quantities depends on several factors. Single class models can be validated with higher precision than multiple class models because their input parameter values can be determined from measurement data with greater accuracy. Some performance measures can be matched more easily than others. In validating multiple class models, it seldom is possible to reflect the behavior of every class at every device accurately. Clearly, it is desirable to have the model represent most accurately the behavior of the critical (mostly heavily used) resources. Similarly, if one class of customers is of particular interest in a modelling study, then validation of the model should place special emphasis on the performance measures of that class. Table 12.3 suggests rough guidelines for reasonable expectations of model accuracy during validation.

An important point to note is that queueing network models typically project percentage changes in performance with more accuracy than absolute levels of performance. For example, consider the projection of the effect on interactive response time of adding a batch workload to a system. Assume that the measured response time in the original system was six seconds, and the baseline model validated within 20%, giving a response time of five seconds. If the modified model then projected a ten second response time after the batch workload was added, we should anticipate a response time in the modified system of twelve seconds (rather than ten) since the model projected a doubling of the response time.

model type	system throughput	system response time	device utilizations	device queue lengths
single class	0 to 5%	5 to 20%	0 to 5%	5 to 20%
multiple class (per class)	5 to 10%	10 to 30%	5 to 10%	10 to 30%

Table 12.3 — Reasonable Tolerances in Validation

Often, even in well conceived and well executed modelling studies, an initial model will not satisfy the validation criterion. In such cases, reasonable modifications of the assumptions used in estimating input parameters (especially service demands) should be attempted. For example, by noting which classes have throughputs underestimated, the analyst may be guided in a reassessment of how overhead should be attributed to the various classes. This review is repeated until the model can be validated. It is not unusual for several iterations to be required at this stage. In some cases, however, no reasonable technique for estimating inputs yields acceptable results. This is a sign that some important aspect of the system's behavior has not been captured in the model. In many such cases, accuracy can be improved by adding more detail to the model.

It is important to realize the significance of validating a model successfully. If information from measurement data is used to establish values of model inputs, then the fact that the model outputs match the measurement data is, at first glance, not surprising. After a little thought, however, one realizes that success in validation carries the significant implication that the numerous assumptions made in establishing the model are acceptable in the context of the particular system under study. With a validated model, we are prepared to proceed to the modification analysis and performance projection, the subjects of the next chapter.

12.7. Summary

The inputs required by queueing network models can be divided into three groups: the customer description, the center description, and the service demands. The information required to determine the values of these inputs is obtained from a system description and data recorded and reported by various monitors. Many of the input values can be determined in a straightforward manner from this information. Other values, however, must be inferred. The bulk of this chapter has been devoted to techniques for doing so, for various inputs.

An appropriate modelling strategy is to start with the simplest model that might suffice, adding detail as necessary. The process of model validation may involve several iterations in which input values are revised and detail is added.

Thorough validation must be based on several measurement intervals. It also must be based on knowledge of the kinds of performance projection questions for which the model is to be used.

12.8. References

Several good books on computer system performance measurement techniques are available, such as [Ferrari 1978], [Ferrari et al. 1983], and [Svobodova 1976]. These, however, do not deal specifically with the needs of queueing network modelling.

Rose [1978] treats the queueing network parameterization problem in general, and also relates the techniques to various specific systems. Kienzle and Sevcik [1979] review the approaches to parameterization taken by a number of early queueing network modelling case studies.

Curtin [1979] describes a performance database which serves as a repository for measurement data, and which can be accessed by the SAS statistical analysis package to produce reports suitable for both managers and analysts. Lindsay [1980] reports on the accuracy of a software performance monitor by comparing its results to those of a hardware monitor.

Artis [1979] suggests a technique for identifying customer classes based on the similarity of their resource demand patterns. Cooper [1980] describes both the identification of customer classes and the use of capture ratios as part of his presentation of an overall capacity planning methodology. Anderson [1979] proposes a sophisticated method for apportioning unattributed device activity to classes using multiple linear regression.

The details of the parameterization process depend heavily on the system under consideration. Both the quantity and the quality of data varies widely among systems. Consequently, proceedings of "user group" conferences are good sources of papers describing techniques of relevance to a particular type of system.

[Anderson 1979]
Edwin Anderson. A Method for the Estimation of Resource Use for Queueing Models. *Proc. CMG X International Conference* (1979), 157-164.

[Artis 1979]
H. Pat Artis. A Technique for Establishing Resource Limited Job Class Structures. *Proc. CMG X International Conference* (1979), 249-253.

[Cooper 1980]
J.C. Cooper. A Capacity Planning Methodology. *IBM Systems Journal* 19,1 (1980), 28-35.

[Curtin 1979]
James P. Curtin. An MVS Performance Data Base and Reporting System Using SAS. *Proc. CMG X International Conference* (1979), 35-39.

[Ferrari 1978]
Domenico Ferrari. *Computer Systems Performance Evaluation.* Prentice-Hall, 1978.

[Ferrari et al. 1983]
Domenico Ferrari, Giuseppe Serrazi, and Alessandro Zeigner. *Measurement and Tuning of Computer Systems.* Prentice-Hall, 1983.

[Kienzle & Sevcik 1979]
Martin G. Kienzle and Kenneth C. Sevcik. Survey of Analytic Queueing Network Models of Computer Systems. *Proc. ACM SIGMETRICS Conference on Simulation, Measurement and Modeling of Computer Systems* (1979), 113-129.

[Lindsay 1980]
David S. Lindsay. RMF I/O Time Validation. *Proc. CMG XI International Conference* (1980), 112-119.

[Rose 1978]
Clifford A. Rose. A Measurement Procedure for Queueing Network Models of Computer Systems. *Computing Surveys 10,*3 (September 1978), 263-280.

[Svobodova 1976]
Liba Svobodova. *Computer Performance Measurement and Evaluation Methods: Analysis and Applications.* North-Holland, 1976.

12.9. Exercises

1. Section 2.2 describes two case studies in which queueing network models were used for performance projection in an IBM processing complex. In each case, the objectives and the results of the study were presented, but the details of the model were not. For each of these studies, use the available information to specify an appropriate structure for a model. Indicate the significant parameters of the model and suggest how their values might be established.

2. In a system with two workload components, batch and interactive, the following measurements were obtained in a 60 minute observation interval:

    observed CPU busy time: 50 minutes
    accounted batch CPU time: 20 minutes
    accounted interactive CPU time: 10 minutes

    a. Assuming that the "capture ratio" is the same for each workload component, what proportion of the observed CPU busy time should be attributed to each component?

    b. Assuming that the primary source of CPU overhead is page transfers and that 75% of all page transfers are for interactive customers, what proportion of the observed CPU busy time should be attributed to each workload component?

    c. In a second 60 minute observation interval, the observed CPU busy time was 45 minutes, while the accounted CPU times for batch and interactive were 15 and 10 minutes, respectively. Using the measurement data from both observation intervals simultaneously, what proportion of the observed CPU busy time should be attributed to each workload component?

3. In an observation interval, the number of logical I/Os (in thousands) for classes A, B, and C were 60, 50, and 30, respectively. In the same interval the number of physical I/Os (in thousands) at the two disk drives were 100 and 60, respectively. Determine an appropriate allocation to each class of the physical I/Os at each disk drive under each of the following assumptions:

    a. No further information is available.

    b. The ratios of physical to logical I/Os for classes A, B, and C are known to be approximately 13/12, 11/10, and 4/3, respectively.

Chapter 13

Evolving Systems

13.1. Introduction

We create and validate queueing network models of baseline systems, as described in Chapter 12, so that these models can be used to project the effects on performance of contemplated modifications to the workload, to the hardware, and to the operating policies and system software. In this chapter we will see how to represent such modifications by alterations to the inputs of the validated model. The accuracy and utility of the resulting performance projections depend on three factors:

- *how well the baseline model validates* — The construction and validation of baseline models was discussed in Chapter 12.
- *how accurately the modifications are forecast* — Anticipating the evolution of a system and its workload is a difficult task that is faced by organizational management. It lies beyond the scope of this book.
- *how well the anticipated modifications are represented as changes to the model inputs* — This is the subject of the present chapter.

In general, system modifications have both *primary* and *secondary* effects. For example, a CPU upgrade has the primary effect of reducing the CPU service requirement of each user (in seconds, rather than instructions), and may have one or more secondary effects, such as changing the number of times that each user is swapped, on the average. We will see that for many modifications it is relatively easy to anticipate and represent the primary effects, but harder to anticipate, and thus to quantify and represent, the secondary effects. For this reason, successful performance projection studies in which several alternatives are being considered often take the following form:

— Initially, each alternative is investigated by representing only its primary effects. This can be done quickly.
— The results may reveal that some of the alternatives are not worthy of further consideration. These alternatives are discarded.

13.2. *Changes to the Workload* 297

- The remaining alternatives are investigated in more detail, with attention paid to secondary as well as primary effects.

The organization of this chapter reflects this scenario. In Sections 13.2, 13.3, and 13.4, we discuss modelling the effects of modifications to the workload, to the hardware, and to the operating policies and system software, respectively. We concentrate in these sections on representing the primary effects of modifications, but also discuss certain secondary effects that are peculiar to a particular type of modification.

In practice, two or more modifications often will occur together. For example, if an increase in transaction processing volume is anticipated (a modification to the workload), one may wish to project performance under the assumption that the CPU is upgraded (a modification to the hardware). For clarity of presentation we will discuss such changes separately. To represent the effect of multiple modifications, the corresponding model input alterations can be applied serially.

In Section 13.5 we discuss some secondary effects that are common to most types of modifications. An example is the change in the level of variable overhead (CPU and I/O overhead due to swapping, for instance) that may accompany various modifications.

Finally, in Section 13.6, we describe three related case studies in which queueing network models were used to project the effects on performance of various modifications. In each case, the accuracy of the projection was assessed after actually implementing the modification. These three case studies are similar in spirit to the two studies of an IBM computing complex that we discussed in Section 2.2, where the modelling cycle was presented. A review of Section 2.2 would be worthwhile at this point.

13.2. Changes to the Workload

The workload presented to a computer system can change in several ways. First, the intensities of workload components can change. Second, the character of workload components (e.g., the service demands) can change. Third, the number of workload components can change. The following three subsections describe how the effects of each of these changes can be represented by adjustments to the inputs of a validated model. Both in this section and in the ones to follow, we will indicate modified input parameter values as primed quantities. For example, $D'_{c,k}$ will denote the modified service demand of class c at center k.

13.2.1. Changes in Workload Intensities

The most frequently studied workload changes are changes in intensity. Naturally, the primary effect of such a change is reflected by modifying the appropriate workload intensity input parameters.

For a transaction class, a typical workload forecast would be "a 30% increase in transaction volume". This can be represented in the model by $\lambda'_c \leftarrow 1.3\ \lambda_c$.

For a terminal class, a typical workload forecast would be "a 50% increase in the number of active users". This can be represented in the model by $N'_c \leftarrow 1.5\ N_c$. (In the absence of evidence to the contrary, it is reasonable to assume that average think time does not change.)

In the case of both transaction and terminal classes, increased competition for main memory will result from an increase in workload intensity. If the baseline model included a memory constraint ("at most twenty requests simultaneously active"), then we may assume that the same constraint still applies. If no such constraint were present in the baseline model, then the analyst must decide whether or not the increased central subsystem population that results from the parameter modification is realistic in light of the amount of memory available. If not, an appropriate memory constraint should be imposed. In either case, the variable component of overhead (e.g., paging and swapping service demands) may increase. This is discussed in Section 13.5.

For a batch class, it is unusual for a workload forecast to be phrased in terms of the multiprogramming level, N_c. (More likely, such phrasing would be used to describe the addition or re-allocation of memory.) Additional complexity arises from the fact that the value of this parameter in the baseline model can be due to several factors. At one extreme, there may be a persistent backlog of batch jobs, so that N_c reflects a memory constraint. In this case, an increase in the availability of batch jobs would only result in a larger backlog. At the other extreme, if sufficient memory is available to activate most batch jobs immediately when they arrive, the value of N_c is not related to a memory constraint. In this case, an increase in the availability of batch jobs would allow N_c to increase. Typically, a workload forecast for a batch class will be phrased in terms of throughput. The analyst must adjust N_c to achieve the forecast throughput, and then consider whether or not the increased central subsystem population is realistic with respect to the available memory.

13.2.2. Changes in the Character of Workload Components

Changes to application programs may lead to changes in the resource requirements of customers. Such changes would be represented in a model by adjusting service demands. Three examples are given in the following paragraphs.

It is proposed to modify an application program to do more checking of the validity and consistency of the input data it receives. The change is projected to increase the CPU path length of a transaction by 20%. The primary effect of this modification can be represented in the model by increasing the CPU service demand of transactions by 20%.

It is proposed to introduce data compression techniques to reduce the space occupied by a file that is processed sequentially by an application. The data transferred by the application will decrease, while its CPU requirements will increase (to translate data from compressed to uncompressed format and back again). To represent this modification in the model, the data transfer component of the service demand at the appropriate disk should be decreased, while the service demand at the CPU should be increased.

It is proposed to change the structure of a file used by an application. Initially, the file had three levels of indexing with the highest level kept in memory. The number of I/Os required to access any record was three: two index blocks plus the record itself. The new organization will be based on hashing, which is expected to decrease the average number of I/Os per record access to roughly 1.5. The primary effect of this modification can be represented in the model by halving the visit count at the appropriate disk (assuming that this is the only use of the disk by the class). A secondary effect of this modification might be an increase in the seek component of the service requirement at the disk, because the hashing technique would eliminate any locality of reference that might have existed under the indexed organization.

13.2.3. Changes in the Number of Workload Components

The primary effect of removing a workload component from a system is represented easily in the model by eliminating the corresponding customer class. The result will be a decrease in the activity at various devices, and a corresponding improvement in the performance of the remaining workload components.

Similarly, the primary effect of adding a workload component is represented by adding a new class. The result will be an increase in the activity at various devices, and a potential degradation in the performance of the original workload components. Of course, the workload intensity and service demands of the new class must be determined and specified. If a similar application runs at some installation with a similar hardware and software configuration, then measured service demands can be used. For a new application that cannot be measured, estimating service demands is much harder. This problem will be treated in Chapter 14.

Both the removal and the addition of workload components have a number of effects which, although of lesser importance than changes in device congestion, still can have considerable impact on performance. When a workload component is removed, memory becomes available for allocation to the remaining components. Knowledge of the operating policies of the system is required to determine how to represent this. When a component is added, it may be necessary to obtain memory at the expense of other components. Again, system knowledge is required.

As always, secondary effects arise in the realm of variable overhead. These will be considered in Section 13.5.

Modelling changes in the number of workload components is of particular benefit in multiple mainframe installations composed of several machines of the same architecture using the same operating system. In such environments, a large part of capacity planning involves projecting the performance resulting from various ways of assigning workload components to machines. The service demands measured for a class on one system can be translated for other systems, using known speed ratios. An example of capacity planning in a multiple mainframe environment was considered in Section 2.2.

13.3. Changes to the Hardware

New hardware products based on recent technological developments are announced with great frequency. This makes capacity planning and configuration management a continuing challenge. Fortunately, queueing network models are well suited to quickly evaluating configuration modifications.

In the subsections to follow we describe how CPU upgrades, memory expansions, and I/O subsystem modifications can be represented as modifications to model input parameter values.

13.3.1. CPU Upgrades

Perhaps the most common configuration change is the upgrade of a CPU within a family of processors of the same architecture. Fortunately, this also is one of the easiest changes to evaluate using queueing network models. The relative instruction execution rates among processors within a family generally are known and publicized by vendors and user groups. Consequently, the primary parameter change is to multiply the CPU service demand by the ratio of old CPU's processing rate (r_{OLD}) to that of the new (r_{NEW}):

$$D'_{c,CPU} \leftarrow \frac{r_{OLD}}{r_{NEW}} \times D_{c,CPU} \quad \text{for each class } c$$

A common secondary effect of a CPU upgrade is a change in variable overhead (considered in Section 13.5). Additional memory or I/O equipment often accompanies such an upgrade (later subsections suggest ways to reflect these changes).

Rather than acquiring a faster CPU, it sometimes is possible to acquire a second processor to form a tightly coupled multiprocessor system. As we discussed in Chapter 11, the primary corresponding change to model parameters would be to represent the processor complex as an FESC with service rate approximately twice as great with two or more customers present as with only one customer present. An important secondary effect is the interference between the processors in accessing memory or shared data structures. This interference causes the capacity of a dual processor to be considerably less than twice the capacity of a single processor. If appropriate measurement data is available, the service rates of the FESC can be set to reflect the degree of interference. An example in Section 13.6 treats the change from a uniprocessor to a dual processor.

13.3.2. Memory Expansions

Since additional memory can be allocated in a number of ways, representing the effect of a memory expansion requires knowledge of the operating policies of the system.

The most common way to employ additional memory is to permit an increase in the central subsystem population of various classes. For batch classes, the parameter N_c would be changed. For transaction or terminal classes, the memory constraint would be adjusted upwards. The key, of course, is to estimate the extent to which each class will be affected. To some extent, this is under the control of installation-dependent tuning parameters. A few, well chosen experiments with smaller memory sizes can help to determine the effect of operating policies. Changes in swapping and paging activities can result; these secondary effects are discussed in Section 13.5.

Additional memory also can be used to permit workload components to run more efficiently at existing central subsystem populations. In this case, the entire effect of the memory upgrade would be felt as a decrease in variable overhead (see Section 13.5).

A third use of additional memory is to make frequently accessed files permanently resident in memory. Examples include system routines or indices. If measurement data indicates frequency of use for these files, then disk service demands can be decreased by an appropriate amount to represent fixing them in memory.

As a final example, additional memory can be used to increase the size of the disk cache employed by many operating systems. Experimentation with a few different cache sizes would indicate the relationship between disk cache size and disk cache hits (and thus I/O activity).

13.3.3. I/O Subsystem Modifications

Each generation of disks can be characterized by basic quantities such as capacity, seek time, latency time, and transfer rate. From these characteristics it is possible to estimate the changes in disk service demands that will result from replacing one type of disk with another. For example, due to faster seeks and higher transfer rates, service demands are reduced by 25% to 30% when converting from IBM 3350 to IBM 3380 disks. The exact speed ratio depends on block size, seek pattern, and I/O subsystem contention.

A secondary effect to consider in this case is the fact that, because the capacity of a 3380 is nearly double that of a 3350, there is a temptation to reduce the number of drives as part of a conversion effort. The resulting change in seek patterns may cause the average seek distance to increase, making it more difficult to forecast service demands.

Recently, solid state drums have provided a new alternative in I/O subsystems. These devices have limited capacity, but provide much faster access times than conventional disks or drums (factors of 4:1 currently). In modelling the addition of a solid state drum to a system, several steps are required:
  − Identify the files to be placed on the drum. (Typically, these will be small, highly active files.)

- Reduce the service demands on the disks from which these files will be removed.
- Add a new center to the model and set the service demand there to be a fraction of the service demands removed from the disks, determined by the relative speeds of the devices.

An I/O subsystem can be upgraded by increasing the numbers of channels and controllers or by changing the interconnections among existing components, as well as by adding storage devices. Changes of this sort would be expected to reduce contention in the I/O subsystem by creating alternate paths between the CPU and the disks. Consequently, the contention component of effective disk service demands would be reduced. The techniques suggested in Chapter 10 are oriented towards assessing the effect of this sort of modification.

13.4. Changes to the Operating Policies and System Software

Operating systems typically leave a great deal of flexibility to installations with respect to certain operating policies that can have a significant influence on performance: placement of files on devices, assignment of workload components to memory domains, setting of scheduling priorities, etc. The first three subsections that follow discuss the representation of modifications to such operating policies in queueing network models. The fourth subsection discusses the representation of the effect of operating system upgrades.

13.4.1. File Placement

Performance often can be improved by altering the assignment of files to devices, with the objective of balancing the load across disks and other I/O subsystem components. The parameter changes to represent such modifications in the model are straightforward. If the disks involved are identical, then the primary effect can be represented (in the case of three disks) by:

$$D'_{each\ disk} \leftarrow \frac{D_{Disk\ 1} + D_{Disk\ 2} + D_{Disk\ 3}}{3}$$

If a decrease in the contention component of the effective disk service demands is expected, then the techniques of Chapter 10 should be used, with the analyst balancing the seek, latency, and transfer components as above.

If the devices involved differ in speed, more effort is required. The service demands in the baseline model must be viewed as the product of visit counts and service times per visit. The service demand at each of k disks after balancing is given by the equations:

$$D'_{each\ disk} = V'_{Disk\ 1}\ S_{Disk\ 1} = \ldots = V'_{Disk\ k}\ S_{Disk\ k}$$

and:

$$\sum_{j=1}^{k} V'_{Disk\ j} = \sum_{j=1}^{k} V_{Disk\ j}$$

Thus, we assume that balancing the load does not change the service times at the devices substantially (e.g., by changing seek patterns), and that the total number of physical I/O operations does not change. The service demand for each disk will be:

$$D'_{each\ disk} = \frac{\sum_{j=1}^{k} V_{Disk\ j}}{\sum_{j=1}^{k} (1\ /\ S_{Disk\ j})}$$

Thus, we would seek an assignment of files to disks such that capacity constraints are not exceeded and the visit count to files assigned to each disk approximately satisfy:

$$V'_{Disk\ j} = \frac{D'_{each\ disk}}{S_{Disk\ j}}$$

The approach described above generally will succeed only in approximately balancing the I/O load. The service times at the various disks in fact will change due to altered seek patterns and other secondary effects. Also, carefully balancing the I/O load according to access patterns observed during one period of the day will not lead to a balanced load throughout the day. Consequently, in doing I/O balancing, peak load periods should be given most consideration, but implications for other periods should be considered.

When representing the addition of disks to a configuration, it is appropriate to attempt I/O load balancing at the same time. An example in Section 13.6 illustrates the evaluation of the effect of I/O load balancing through altering the placement of user files.

13.4.2. Memory Allocation

The allocation of memory is critical to performance. An operating system typically requires substantial memory for its own use, devoted to resident code and data structures, transient routines, and I/O buffers.

13.4. Changes to the Operating Policies and System Software

The remaining memory is allocated to user programs. As noted in Chapter 9, it is typical to define domains with limited capacities and to assign workload components to these domains. This approach regulates competition for memory so that thrashing does not occur.

The primary effect of altering the allocation of memory can be represented by changing the domain capacities in the model (the multiprogramming level, in the case of batch classes). The problems that arise are similar to those that arise in modelling the addition of memory, which were discussed in an earlier section. Especially in a virtual memory system, it can be difficult to determine the number of jobs that can be accommodated in a specific amount of memory. Limited benchmarking can be of assistance in determining how the rate of paging depends on the amount of main memory available for each active customer.

13.4.3. Tuning Parameters

In most operating systems, many of the scheduling and resource allocation activities are controlled by tuning parameters. Among other things, such parameters control the dispatching and initiation priorities of various workload components, and the amount of service guaranteed to customers before they are eligible to be swapped out. Queueing network models can be used to gain an understanding of the effects of changing certain tuning parameters. The major benefit of such studies is to estimate the extent to which performance might be affected by a particular parameter.

Representing the effect of changes in the relative priorities of workload components is straightforward, using the techniques described in Chapter 11. Chapter 16 includes an example of such a study.

The swapping quantum (the amount of service guaranteed a customer before becoming eligible for swapping) is another example of an important tuning parameter. The case study of Section 9.6.2 illustrates the incorporation of this parameter in a queueing network model.

13.4.4. Operating System Upgrades

Operating systems provide certain services to the programs that execute under them. The variety of services available and the efficiency with which they are delivered differs from one system to another. The operating systems for most major computer systems evolve continually. Each version (or "release") typically provides some new functions, and possibly improves the efficiency with which earlier functions are delivered.

To model the effect of an operating system upgrade, the analyst must determine the relative efficiency of various functions by relying either on statements by the vendor or on experience of early users ("beta test" sites). Given this information, modification of the model is straightforward. For example, if it is claimed that CPU path lengths for user I/O processing will be decreased by a factor of two, the analyst first must determine this overhead component of CPU service demand for the workload on the existing system, then divide it by two to represent the effect of the new release.

Operating system efficiency also is of importance when comparing various systems under consideration for the support of a new workload. In this case, it is necessary to translate a workload description in system-independent terms into service demands for each candidate system. In the case of CPU service demands, for example, the relative CPU execution rates of the various systems tell only part of the story: the efficiency of operating software can have a dramatic effect on performance. As we showed in an example in Section 2.4, simple, single-thread benchmarking experiments are appropriate and useful in quantifying software efficiencies for incorporation in queueing network models.

13.5. Secondary Effects of Changes

Previous sections have concentrated on the representation of the primary effects of system changes. In the present section we consider the representation of certain secondary effects that are common to a number of the modifications we have discussed.

To a certain extent, these issues already have been addressed in Part III of the book. In Chapter 9, we showed one approach to estimating the change in swapping activity that would accompany various system modifications. We also showed how variability in paging activity could be incorporated in a model. In Chapter 10, we developed algorithms to estimate path contention in complex I/O subsystems as a function of other system characteristics. In Chapter 11, we mentioned the representation of the CPU overhead that accompanies all other activities.

In this section, we will talk in more general terms about techniques to forecast the level of CPU and I/O overhead present in a system. Our approach will be one that was suggested in earlier chapters: to extrapolate from the results of a few measurement intervals.

13.5.1. Changes in Variable Overhead

Almost every contemplated change to a system will, as a secondary effect, change the variable overhead incurred in system operation. The most significant examples of this in many systems are changes in paging

13.5. Secondary Effects of Changes

and swapping rates, which involve both CPU and I/O activity. CPU upgrades, memory expansions, increases in workload intensities, even changes in the priority structure among classes, all have the secondary effect of changing paging and swapping rates.

As we have noted in earlier chapters, when a model is used to project performance for relatively minor modifications (a 10% increase in workload intensity, a 25% increase in CPU capacity), changes in variable overhead need not be considered. The more significant the modification under consideration, the more important it is to attempt to quantify these changes. This is a difficult task; in some cases it will be necessary to employ a sensitivity analysis to indicate the range of anticipated performance.

---

1. Obtain measurements from several observation intervals, preferably including a range of degrees of system congestion.
2. For each interval, determine the service demand at each center.
3. For the measure of system congestion of greatest concern (e.g., workload intensity), for each center, fit a simple curve to the observed service demands as a function of the measure of concern.
4. Use the simple curve for each center to extrapolate service demand for unobserved situations.

Algorithm 13.1 − **Variable Overhead in Single Class Models**

An approach to characterizing variable overhead for single class models based on measurements from several observation intervals is given as Algorithm 13.1. The simplest curve to use in Algorithm 13.1 is a straight line. This suffices for representing variable overhead as long as the range of congestion being investigated is not extreme. Assume that measurements are available for two observation intervals in which the workload intensities are $I^{(1)}$ and $I^{(2)}$, respectively, and in which the observed service demands at device k are $D_k^{(1)}$ and $D_k^{(2)}$, respectively. Assuming that variable overhead increases linearly with workload intensity, an appropriate estimate for the service demand at device k for a new workload intensity I' is given by:

$$D'_k = D_k^{(1)} + (I' - I^{(1)}) \times \left[\frac{D_k^{(2)} - D_k^{(1)}}{I^{(2)} - I^{(1)}} \right]$$

Approximating the dependence of variable overhead on workload intensity by more complex curves typically yields slightly greater accuracy, particularly if workload intensity changes are large, but this gain may not justify the added complexity.

Careful treatment of variable overhead is more difficult in multiple class models. There are several issues involved:

- In the multiple class case, more observation intervals are necessary, because the workload intensity now is a vector. For example, if the workload consists of two major components, interactive and batch, we might consider four observation intervals: heavy batch and heavy interactive, heavy batch and light interactive, light batch and heavy interactive, and light batch and light interactive.
- Within each observation interval, it is difficult to attribute variable overhead to the classes accurately, because of the inadequacy of measurement tools. Techniques such as those described in Section 12.5 can be used.
- Where the single class case involved fitting a curve through some points, the analogous procedure for the multiple class case with C classes involves fitting a C-dimensional surface. Such multidimensional surface fitting, however, is too complex to be justified considering other limitations on the accuracy of this technique. In almost all cases, a sequence of one-dimensional extrapolations based on changes to one workload component at a time will suffice.

From the preceding discussion, it should be apparent that estimating changes in variable overhead is difficult, and cannot be done with high confidence. Consequently, it often is appropriate to evaluate the model under both optimistic and pessimistic assumptions in order to assess the importance of accurately estimating overhead in projecting performance. For example, when memory size is increased, paging and swapping activity typically are reduced. Because it is difficult to determine the extent of this reduction, we might evaluate the model once assuming no change in paging and swapping activity, and again assuming that all paging and swapping activity is eliminated.

13.5.2. Changes in I/O Service Times

Many modifications have the secondary effect of changing the seek, transfer, and contention components of effective disk service time. The contention component was considered in Chapter 10. Here we discuss the others.

Relocating files from one disk to another can cause the seek patterns to change on each disk. Typically, the average seek time will increase on the disk to which the file is moved and will decrease on the other. If all files do not have the same block size, then the average transfer times at both disks also will be altered.

Similar considerations arise in such system modifications as increasing the block size of a file or increasing the workload intensity of a class (which can alter the seek pattern and change the average transfer time if the class accesses some files particularly heavily).

13.6. Case Studies

In this section we describe three case studies conducted over a period of several years on an evolving UNIVAC 1100 system running the Exec 8 operating system. Initially the system was configured as an 1100/41 (a uniprocessor) with the following I/O subsystem structure:

channel 0	1 FH-1782 drum
channel 1	1 FH-1782 drum
channel 2	4 tape drives
channel 3	8 8424 disk drives
channel 4	4 8433 disk drives

In each of the three case studies, synthetic benchmarks designed to reflect actual workloads were used, and the same experimental procedure was followed:

- The benchmark was run on the existing configuration and measurements were taken with UNIVAC's SIP (the Software Instrumentation Package).
- A baseline queueing network model was developed and validated.
- The model was modified to project the effect on performance of a specific proposed change to the system.
- This change was implemented, and the benchmark was run again.
- The performance projected by the model was compared to the performance measured on the modified system.

Note that this experimental procedure follows closely the modelling cycle described in Section 2.2. It is this aspect that makes these three case studies particularly interesting in the context of the present chapter. On the one hand, the parameter adjustments used to project performance occasionally were somewhat simplistic, in that obvious secondary effects were ignored. On the other hand, retrospective attempts were made to attribute discrepancies between projections and measurements to specific secondary effects. The sequence of case studies thus is a good example of how lessons learned in one study can be used to improve the accuracy of subsequent studies. In a production environment where decisions are made after the performance projection step, there is a tendency to omit the final two steps of the procedure outlined above. These steps are important, however.

13.6.1. Moving to a Dual Processor

In the first study, a baseline model of a uniprocessor system (an 1100/41) was modified to project the performance of a dual processor system (an 1100/42). The model contained a single class of batch type and six service centers: one representing the CPU (or pair of CPUs) and five representing the five I/O channels of the system. The use of centers to represent channels rather than disks differs from the approach suggested in Chapter 10. This case study pre-dates that approach. Further, the channels had considerably higher utilizations than the disks in this system, and thus were thought to be the principal constraints on performance. Also, reliable measurements of busy times were available for the channels but not for the disks.

In this study six different benchmarks were used. After each benchmark was run on the uniprocessor system, measurement data was used to parameterize the baseline model, as follows:

— The service demand at the CPU center was set to the CPU busy time divided by the number of job completions.
— At the five centers representing the channels, the service demands were set to the corresponding channel busy times divided by the number of job completions. (Note that the seek component of disk service times was not represented in this model.)
— SIP provided an estimate of multiprogramming level that was known to be unreliable. Consequently, the value of N was adjusted until the throughput of the model exactly matched that of the system.

The technique of establishing the value of some parameter according to what yields the best results is called *calibration*. It should be avoided unless legitimate uncertainty exists concerning the value of a single parameter.

This baseline model then was modified to reflect the addition of the second CPU. This was done by replacing the CPU center with an FESC. With one customer present, the FESC service rate was the same as that of the uniprocessor. With two or more customers present, it was double this value. That is:

$$\mu(n) = \begin{cases} \dfrac{1}{D_{CPU}} & n = 1 \\ \dfrac{2}{D_{CPU}} & n > 1 \end{cases}$$

where D_{CPU} was the processor service demand in the baseline model.

13.6. Case Studies

benchmark	throughput			
	original	projected	actual	error
1	48.2	53.8	48.0	+ 12%
2	47.8	67.3	48.9	+ 38%
3	48.9	55.0	50.9	+ 8%
4	39.9	56.8	50.1	+ 14%
5	33.7	47.0	45.9	+ 2%
6	40.4	57.1	59.9	− 5%

Table 13.1 − Moving to a Dual Processor

Table 13.1 compares the projections of the model to the measured performance after the second processor was added for each of the six benchmarks. The error in projected throughput was 15% or less in five of the six cases, but 5% or less in only two. This cannot be viewed as successful, especially in light of the 38% discrepancy in the sixth case.

A retrospective analysis revealed that the CPU upgrade caused the average multiprogramming level to drop substantially − to one half its former value for four of the six benchmarks. This likely was the reason for the counter-intuitive fact that the addition of the second processor made essentially no difference in measured performance for the first three benchmarks. Even with the benefit of hindsight, it was difficult to understand why this drop in multiprogramming level occurred. (Conceivably it was indicative of a shortcoming in the system's job scheduler.) The assumption that the multiprogramming level would not change with the addition of the second processor played a substantial role in the optimistic throughputs projected for five of the six benchmarks.

A second factor that contributed somewhat to the optimistic projections was that interference between the two processors was not taken into account in determining the rates of the FESC. As was noted in Chapter 11, the full power of the second CPU is not realized in dual processor systems; $\mu(n)$ for n greater than one should have been set to a value less than $2/D_{CPU}$.

Finally, no change in the number of swaps per job was anticipated or represented in modifying the model parameter values. In fact, the number of swaps per job decreased, possibly due to the reduced multiprogramming level. This meant that the average number of visits made by a job to channel 1 (the location of the swapping drum) decreased, and also that the average service time per visit was reduced (because swapping operations had much higher average service times than did user I/O operations at this device). This effect was not large, and was more than offset by the other, optimistic discrepancies.

13.6.2. Altering File Placement

The second case study was an investigation of the effect of balancing the load across channels by altering the placement of user files. By the time of this study, the configuration had evolved somewhat. Specifically, the disk channels had been converted to "dual channels": two disks on the same dual channel could be active (in any phase, even data transfer) simultaneously. Thus, performance measures of the two studies are not directly comparable.

The model employed was similar to that used in the first study. Once again, there was a single class of batch type. Again, an FESC was used to model the dual processor of the UNIVAC 1100/42 configuration. The other centers in the model corresponded to channels. Each of the dual channels was modelled as an FESC that behaved similarly to the FESC used to represent the dual processor CPU.

A single benchmark was run on the system with the original assignment of user files to devices. Data from both SIP and UNIVAC IOTRACE was used to parameterize the baseline model. (IOTRACE reported the channel busy time due to accesses of each individual file.) Most of the model parameters were established in conventional ways. The centers representing the dual channels required special attention, however. One channel in each pair was the *primary* and was used whenever available. The other was the *secondary* and was used only when necessary. The service rates of each FESC were calculated as:

$$\mu(n) = \begin{cases} \dfrac{C_{prim}}{B_{prim}} & n=1 \\ \dfrac{C_{prim}}{B_{prim}} + \dfrac{C_{sec}}{B_{sec}} & n>1 \end{cases}$$

where *prim* and *sec* denote the primary and secondary channels of the pair, C_k is the measured number of operations on channel k, and B_k is the measured busy time of channel k. (Note that this calculation ignores the fact that the secondary channel is blocked if the request it is serving happens to access the same disk as the request being served by the primary channel.) Once again it was necessary to determine the multiprogramming level N by calibrating on throughput.

User files accounted for only 30% of the measured I/O accesses. The other 70% of the accesses were to system files whose placement was considered fixed in this experiment. Two alterations in the existing placement of user files were considered:

13.6. Case Studies

- Place all user files on the 8433 disks associated with channel 4. A careful analysis indicated that this would result in the greatest performance improvement — a "best case" scenario.
- Place all user files on devices attached to the most heavily utilized channel. It was believed that this would result in the greatest performance degradation — a "worst case" included for comparison.

The parameters of the baseline model were adjusted to represent each of these file placements, using techniques similar to those suggested earlier in this chapter. After model projections were obtained for each case, the files actually were moved, and the benchmark was run again for each case.

case	quantity	original	projected	actual	error in projection
best	U_{CPU}	.843	.888	.881	+0.8%
	X	79.5	86.7	84.5	+2.6%
worst	U_{CPU}	.843	.762	.640	+19.1%
	X	79.5	68.0	59.6	+14.1%

Table 13.2 — Altering File Placement

The results for both cases are shown in Table 13.2. The last column indicates the error in the projection relative to the observed value. The results show that throughput for the best placement of user files, which account for only 30% of I/O accesses, is roughly 5% greater than for the existing placement, and roughly 35% greater than for the worst placement. The accuracy of the model for the best placement is quite good, while for the worst placement it is acceptable but not good.

Retrospectively, it was observed that the major source of error was the fact that the model ignored changes in swapping behavior that accompanied the alterations in file placement. The worst case scenario caused many user files to be located on the drum containing the swap data set. Swap operations took longer, the CPU was left idle more often (because jobs were not available for service while being swapped), the scheduler activated more jobs to try to keep the CPU busy, and swapping (and the associated channel congestion) increased. The model, which assumed that swapping would be unaffected, underestimated the deterioration in performance. (The best placement caused some files to be removed from the swapping drum, leading to some reduction in swapping, but the effect was not significant.)

13.6.3. Moving Swapping Activity from Drum to Disk

A third study of the same system considered the effect of moving swapping activity from drum to disk. The disks were under-utilized relative to the drums, and their newer technology and dual channel capability made them competitive in terms of performance. By moving swapping activity to disk, the drums could be used for temporary data sets, accessed frequently during their short lifetimes.

In constructing the baseline model, additional detail in the representation of the I/O subsystem was incorporated. Centers were included to represent each disk, in addition to the FESCs representing the two dual channels. So that no component of I/O service demand would be duplicated at the disk and channel centers, the disk centers represented only seek times, while the channel centers represented latency and transfer times. Because this approach tends to yield optimistic results (in the model, one customer's seek activity at a disk can be overlapped with another customer's latency and transfer activity at the same disk), the disk centers were represented as FESCs whose service rates *decreased* when more than one customer was present.

Remembering the lessons from the first two studies, thought was given to examining both primary and secondary effects of the proposed modification. The procedure used to adjust the parameters of the baseline model to reflect the movement of swapping from drum to disk was iterative in nature:

- Assume initially that the level of swapping activity will remain unchanged after the modification.
- Knowing that the operating system tends to place temporary files on faster devices, estimate the visit counts at drums and disks that would result from moving all swapping activity to disk.
- Knowing the files placed on each device, the relative access frequencies to files, and the average transfer size for each file, adjust the service demands at the centers representing the drums, and the service rates at the FESCs representing the disks and dual channels.
- Evaluate the model.
- Use an empirically derived relationship between throughput, multiprogramming level, and swapping activity to estimate the change in the level of swapping activity resulting from the modification.
- Return to the second step, iterating until convergence is achieved.

As in the two earlier case studies, the change to the system was implemented and the benchmark was run once again. Table 13.3 displays the results. This experiment was successful in producing usefully accurate performance projections.

quantity	original	projected	actual	error in projection
CPU utilization	.609	.665	.679	− 2.1%
CPU queue length	1.08	1.32	1.39	− 5.0%
throughput	101	115	121	− 5.0%

Table 13.3 − Moving Swapping Activity from Drum to Disk

13.7. Summary

The principal value of a validated queueing network model of a baseline system is its utility as a basis for performance projection. In this chapter we have indicated, through discussion and example, how to modify the parameters of a baseline model to represent various common changes to the workload, to the hardware, and to the system software and operating policies.

A key point to keep in mind in conducting a modification analysis, especially as part of a study in which a large number of alternatives must be considered, is the need to identify those effects of the modification that are primary, and those that are secondary.

Primary effects typically are easy to anticipate and to represent. In the early stages of a study, alternatives can be compared on the basis of their primary effects alone.

Secondary effects typically are less easy to anticipate, and even once anticipated, less easy to quantify and represent. Several approaches can be adopted:

- Extreme assumptions can be evaluated; for example, the addition of memory at worst leaves swapping unaffected, and at best eliminates it.
- A more careful estimate of secondary effects can be made, based on measurements from several observation intervals.
- A sensitivity analysis can be used to assess the extent to which the projections of a model depend upon the assumptions that have been made.

We have tried to indicate the importance of the "verification phase" of the modelling cycle, described in Chapter 2. Expertise and confidence in conducting modification analyses is best acquired by learning from prior modelling experiences.

13.8. References

There have been a number of case studies in which a baseline model was constructed, performance projections were obtained, and the accuracy of these projections was checked after the system had been modified. The three studies described in Section 13.6 all were carried out at the University of Maryland, using facilities available at the computer center there. The study of moving to a dual processor was conducted by Dowdy, Agrawala, Gordon, and Tripathi [1979]. The study of altering file placement was conducted by Dowdy and Budd [1982]. The study of moving swapping activity from drum to disk was conducted by Dowdy and Breitenlohner [1981].

Several similar studies from production environments were reviewed in Chapter 2: by Lo [1980] on the effect of reallocating workloads among systems in a multiple mainframe environment, and by Lazowska [1980] and Sevcik, Graham, and Zahorjan [1980] on evaluating various candidate systems for specified applications.

There are several related papers that we have not discussed specifically. Tibbs and Kelly use quadratic fits obtained by non-linear regression to forecast the change in overhead in doing performance projections for a UNIVAC 1100 [Tibbs & Kelly 1982]. Bard [1978] describes a performance projection tool for systems running IBM's VM/370 operating system. Buzen presents a queueing network model of systems running IBM's MVS operating system, which includes the effects of shared memory domains and performance periods [Buzen 1978]. Models of DECsystem-10 systems running TOPS-10 have been described by Saxton and Lamont [1978] and by Sanguinetti and Billington [1980]. Dowdy, Stephens, and Perez-Davila [Dowdy et al. 1982] have done a study of performance projection in a UNIX environment, in which the treatment of memory management was the principal issue.

In the realm of workload forecasting, Artis proposes a way of estimating what the workload of a system would be if sufficient capacity were provided to handle it [Artis 1981]. Cooper [1980] describes an approach to capacity planning in an organization, which integrates business planning forecasts with the use of models of computer system performance.

[Artis 1981]
    H. Pat Artis. Estimating Latent Demand for Random Arrival Batch Workloads. *Computer Performance* 2,1 (March 1981), 26-29.

13.8. References

[Bard 1978]
Y. Bard. The VM/370 Performance Predictor. *Computing Surveys 10*,3 (September 1978), 333-342.

[Buzen 1978]
Jeffrey P. Buzen. A Queueing Network Model of MVS. *Computing Surveys 10*,3 (September 1978), 319-331.

[Cooper 1980]
J.C. Cooper. A Capacity Planning Methodology. *IBM Systems Journal 19*,1 (1980), 28-45.

[Dowdy & Breitenlohner 1981]
Lawrence W. Dowdy and Hans J. Breitenlohner. A Model of Univac 1100/42 Swapping. *Proc. ACM SIGMETRICS Conference on Measurement and Modeling of Computer Systems* (1981), 36-47. Copyright © 1981 by the Association for Computing Machinery.

[Dowdy & Budd 1982]
Lawrence W. Dowdy and Rosemary M. Budd. File Placement Using Predictive Queueing Models. In R.L. Disney and T.J. Ott (eds.), *Applied Probability - Computer Science: The Interface, Vol. II.* Birkhauser, 1982, 459-476.

[Dowdy et al. 1979]
Lawrence W. Dowdy, Ashok K. Agrawala, Karen D. Gordon, and Satish K. Tripathi. Computer Performance Prediction via Analytical Modeling — An Experiment. *Proc. ACM SIGMETRICS Conference on Simulation, Measurement and Modeling of Computer Systems* (1979), 13-18. Copyright © 1979 by the Association for Computing Machinery.

[Dowdy et al. 1982]
Lawrence W. Dowdy, Lindsey E. Stephens, and Alfredo Perez-Davila. Performance Prediction in a UNIX Environment. *Proc. 18th CPEUG Meeting* (1982), 205-211.

[Lazowska 1980]
Edward D. Lazowska. The Use of Analytic Modelling in System Selection. *Proc. CMG XI International Conference* (1980), 63-69.

[Lo 1980]
T.L. Lo. Computer Capacity Planning Using Queueing Network Models. *Proc. IFIP W.G.7.3 International Symposium on Computer Performance Modelling, Measurement, and Evaluation* (1980), 145-152.

[Sanguinetti & Billington 1980]
John Sanguinetti and Richard Billington. A Multi-Class Queueing Network Model Of An Interactive System. *Proc. CMG XI International Conference* (1980), 50-55.

[Saxton & Lamont 1978]
Harold E. Saxton and Gary B. Lamont. Validation of a DEC-10 Closed Queueing Network Model. *Proc. CMG IX International Conference* (1978), 143-151.

[Sevcik et al. 1980]
K.C. Sevcik, G.S. Graham, and J. Zahorjan. Configuration and Capacity Planning in a Distributed Processing System. *Proc. 16th CPEUG Meeting* (1980), 165-171.

[Tibbs & Kelly 1982]
Richard W. Tibbs and John C. Kelly. The Application of Analytic and Simulation Models to Size a Large Computer System. *Proc. 18th CPEUG Meeting* (1982), 231-257.

13.9. Exercises

1. Expand on Exercise 1 of Chapter 12. For each of the case studies, indicate how the model you proposed could be modified to represent the primary effects of the system change being investigated. In addition, consider what secondary effects should be represented.
2. A group of files are stored on a disk and a drum with service times of 30 and 10 milliseconds per access, respectively. Currently, the service demands at the disk and drum are 6 and 3 seconds, respectively. Consider each of the following scenarios for changing the system:
   a. Knowing the relative access counts for the files, indicate how you would relocate files in order to balance the demand on the two devices.
   b. If the disk were replaced by a second drum, and the demand were balanced across the two devices, what would the service demand at each be?
   c. If all files on the disk were moved to a solid state drum with a service time of 2 milliseconds per access, what would be the resulting service demand?
3. Consider a system with a single batch class in which each customer has a CPU service demand of 30 seconds and does 1000 I/O operations involving a total of four files: 400 accesses to file W, 300 to file X, 200 to file Y, and 100 to file Z. The files can be placed on three I/O devices with service times per access of 10 milliseconds at device

13.9. Exercises

1, 30 milliseconds at device 2, and 50 milliseconds at device 3. Using a single class queueing network solution package such as the one provided in Chapter 18, determine how to assign the files to the storage devices to maximize throughput, for each of the following situations:

a. Multiprogramming level is 1.

b. Multiprogramming level is 4.

c. Multiprogramming level is 12.

(Assume that each device has sufficient capacity to accommodate whatever files you choose to assign there.)

4. Three observation intervals yield the following information:

quantity	measurement interval		
	1	2	3
jobs completed	600	800	500
CPU busy time	14400	20800	11500
multiprogramming level	4	6	3

In projecting performance for a multiprogramming level of 10, what service demand should be used to reflect a simple linear model of variable overhead?

5. Suppose that you had two single class models, one open and one closed. Suppose that these models were "equivalent" in the sense that they had identical service centers, identical service demands, and identical throughputs and utilizations.

a. Would you expect these models to have identical queue lengths and residence times? Why or why not?

b. If you were to modify the open model by doubling the arrival rate and the closed model by doubling the population, how would you expect the changes in performance measures to differ between the two models?

c. Doubling the arrival rate of an open model and doubling the population of a closed model correspond to two very different "scenarios" about the future of a system. State the system change that is addressed by each of these modifications.

Chapter 14

Proposed Systems

14.1. Introduction

The preceding two chapters have discussed the parameterization of queueing network models of existing systems and evolving systems. In this chapter we consider models of proposed systems: major new systems and subsystems that are undergoing design and implementation.

The process of design and implementation involves continual tradeoffs between cost and performance. Quantifying the performance implications of various alternatives is central to this process. It also is extremely challenging. In the case of existing systems, measurement data is available. In the case of evolving systems, contemplated modifications often are straightforward (e.g., a new CPU within a product line), and limited experimentation may be possible in validating a baseline model. In the case of proposed systems, these advantages do not exist. For this reason, it is tempting to rely on seat-of-the-pants performance projections, which all too often prove to be significantly in error. The consequences can be serious, for performance, like reliability, is best designed in, rather than added on.

Recently, progress has been made in evolving a general framework for projecting the performance of proposed systems. There has been a confluence of ideas from software engineering and performance evaluation, with queueing network models playing a central role. The purpose of this chapter is to present the elements of this framework. In Section 14.2 we review some early efforts. In Section 14.3 we discuss, in a general setting, some of the components necessary to achieve a good understanding of the performance of a proposed system. In Section 14.4 we describe two specific approaches.

14.2. Background

User satisfaction with a new application system depends to a significant extent on the system's ability to deliver performance that is acceptable and consistent. In this section we describe several early attempts at assessing the performance of large systems during the design stage. Some common themes will be evident; these will be discussed in the next section.

In the mid 1960s, GECOS III was being designed by General Electric as an integrated batch and timesharing system. After the initial design was complete, two activities began in parallel: one team began the implementation, while another developed a simulation model to project the effects of subsequent design and implementation decisions.

The simulation modelling team came out second best. The model was not debugged until several months after a skeletal version of the actual system was operational. Thus, many of the design questions that might have been answered by the model were answered instead by the system. The model could not be kept current. The projections of the model were not trusted, because the system designers lacked confidence in the simulation methodology.

This attempt to understand the interactions among design decisions throughout the project lifetime failed. Other attempts have been more successful.

In the late 1960s, TSO was being developed as a timesharing subsystem for IBM's batch-oriented MVT operating system. During final design and initial implementation of the final system, an earlier prototype was measured in a test environment, and a queueing network model was parameterized from these measurements and from detailed specifications of the final design.

The average response time projected by the model was significantly lower than that measured for prototype. However, the design team had confidence in the model because a similar one had been used successfully for MIT's CTSS system (see Section 6.3.1). The team checked the prototype for conformance with specifications and detected a discrepancy: the scheduler had been implemented with an unnecessary locking mechanism that created a software bottleneck. When this was corrected, the projections of the model and the behavior of the prototype were compatible.

In the early 1970s, MVS was being designed and developed as a batch-oriented operating system for IBM's new family of virtual memory machines. A simulation model was developed for an early version of this system, OS/VS2 Release 2. The model's purpose was to provide performance information for system designers.

Model validation was a problem. In the design stage, key model parameters were represented only as ranges of values. Performance projections were checked for reasonableness, to ensure that the model represented the functional flow of work through the system. This type of sensitivity analysis compensated for the lack of precise parameter values. The system was changing constantly during design and implementation. To reduce this problem, the model builders maintained a close working relationship with the system designers and implementors.

This modelling effort was considered to be a success, because several of its recommendations had direct, beneficial effects on the design of the system.

In the mid 1970s, the Advanced Logistics System (ALS) was under development for the U.S. Air Force. After the design was completed, during initial implementation, a modelling study was undertaken to determine the bottlenecks in the design and to recommend alternate designs yielding better performance. Hierarchical modelling, as described in Chapter 8, was applied. Four major subsystems were identified in ALS: CPU and memory, system disks, database disks, and tapes. A hierarchical model was structured along these lines, dividing the modelling task into manageable components. Parameter values came from a combination of measurements and detailed specifications.

Both analytic and simulation solutions of the model were obtained. Most ALS features could be captured in the analytic solution. Simulation was used to validate the analytic results and to explore certain system characteristics in more detail.

The modelling study predicted that as the workload increased, the first bottleneck would be encountered in the system disk subsystem, and the next in the CPU and memory subsystem. Both predictions were verified in early production operation. Thus, the study was judged a success.

Each successful project that we have described used a different underlying approach: an analytic model for TSO, a simulation model for MVS, and hierarchical analytic and simulation models for ALS. However, these projects shared a number of underlying principles. In the next section, we include these and other principles in a general framework for studying the performance of proposed systems.

14.3. A General Framework

Unfortunately, it is not common to attempt to quantify the performance of proposed systems. There are two major reasons for this:
- Manpower devoted to performance projection is viewed as manpower that otherwise could be devoted to writing code and delivering the system on time.
- There is no widely accepted approach to integrating performance projections with a system design project.

The first of these points is rendered invalid by the false sense of economy on which it is based: the implications of misguided design decisions for the ultimate cost of a system can be enormous. The second of these points is becoming less significant as aspects of a general framework begin to emerge. These are the subject of the present section.

14.3.1. The Approach

Performance is not the domain of a single group. Thus, performance projection is best done in a team environment, with representation from groups such as intended users, software designers, software implementors, configuration planners, and performance analysts. By analogy to software engineering, the team would begin its task by conducting a *performance walkthrough* of a proposed design. A typical walkthrough would consist of the following steps:

- The intended users would describe anticipated patterns of use of the system. In queueing network modelling terms, they would identify the workload components, and the workload intensities of the various components.
- The software designers would identify, for a selected subset of the workload components, the path through the software modules of the system that would be followed in processing each component: which modules would be invoked, and how frequently.
- The software implementors would specify the resource requirements for each module in system-independent terms: software path lengths, I/O volume, etc.
- The configuration planners would translate these system-independent resource requirements into configuration-dependent terms.
- The performance analysts would synthesize the results of this process, constructing a queueing network model of the system.

Various parts of this process would be repeated as the performance analysts seek additional information, as the design evolves, and as the results of the analysis indicate specific areas of concern. An important aspect of any tool embodying this approach is the support that it provides for this sort of iteration and successive refinement.

It should be clear that what has been outlined is a methodical approach to obtaining queueing network model inputs, an approach that could be of value in any modelling study, not just an evaluation of a proposed system. (For example, see the case study in Section 2.4.)

It also should be clear that this approach, since it forces meaningful communication between various "interested parties", can be a valuable aid in software project management.

14.3.2. An Example

Here is a simple example that illustrates the application of this general approach. A store-and-forward computer communication network is being designed. Our objective is to project the performance of this network, given information about the planned usage, the software design, and the supporting hardware.

The topology (star) and the protocol (polling) of the network are known. The system is to support three kinds of messages: STORE, FORWARD, and FLASH. From the functional specifications, the arrival rate, priority, and response time requirement of each message type can be obtained. Each message type has different characteristics and represents a non-trivial portion of the workload, so it is natural to view each as a separate workload component and to assign each to a different class. Given knowledge of the intended protocol, a fourth class is formulated, representing polling overhead. Further refinements of this class structure are possible during project evolution.

The software specifications for each class are imprecise in the initial stages. Only high-level information about software functionality, flow of control, and processing requirements are available. A gross estimate of CPU and I/O resource requirements for each class is obtained. The CPU requirement specifies an estimated number of instructions for each message of the type, and an estimated number of logical I/O operations. For STORE messages, as an example, the I/O consists of a read to an index to locate the message storage area, a write to store the message, and a write to update the index. No indication is given here about file placements or device characteristics. Instead, the logical properties of the software are emphasized, to serve as a basis for further refinement when the software design becomes more mature.

14.3. A General Framework

Physical device characteristics are identified: speed, capacity, file placement, etc. A CPU is characterized by its MIPS rate and its number of processors. A disk is characterized by its capacity, average seek time, rotation time, transfer rate, and the assignment of files to it. From consideration of the software specifications and the device characteristics, service demands can be estimated. As a simple example, a software designer may estimate 60,000 CPU instructions for a STORE message, and a hardware configuration analyst may estimate a CPU MIPS rate of .40. This leads to a STORE service demand for the CPU of .15 seconds. This admittedly is a crude estimate, but it serves as a basis, and more detail can be incorporated subsequently.

At this point, a queueing network model of the design, incorporating classes, devices, and service demands, can be constructed and evaluated to give an initial assessment of performance. Alternatives can be evaluated to determine their effect on performance. Sensitivity analyses can be used to identify potential trouble spots, even at this early stage of the project.

One of the strengths of this approach is the ability to handle easily changes in the workload, software, and hardware. In the example, no internal module flow of control was specified and processing requirements were gross approximations. As the design progresses, the individual modules begin to acquire a finer structure, as reflected in Figure 14.1. This can be reflected by modifying the software specifications. This structure acquires multiple levels of detail as the design matures. The submodules at the leaves of the tree represent detailed information about a particular operation; the software designer has more confidence in the resource estimates specified for these types of modules. The total resource requirements for a workload are found by appropriately summing the resource requirements at the various levels in the detailed module structure. Software specifications thus can be updated as more information becomes available.

The important features we have illustrated in this example include the identification of workload, software, and hardware at the appropriate level of detail, the transformation of these high-level components into queueing network model parameters, and the ability to represent changes in the basic components.

14.3.3. Other Considerations

The design stage of a proposed system has received most of our attention. This is where the greatest leverage exists to change plans. However, it is important to continue the performance projection effort during the life of the project. Implementation, testing, and

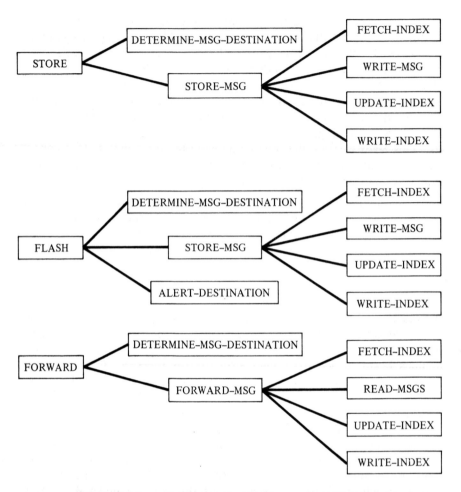

Figure 14.1 — Refinement of Software Specifications

maintenance/evolution follow design. Estimates indicate that the largest proportion of the cost of software comes from the maintenance/evolution stage.

Given the desirability of tracking performance over the software lifetime, it is useful to maintain a repository of current information about important aspects of the project (e.g., procedure structure within software modules). If the repository is automated in database form, software designers and implementors are more likely to keep it current.

A prerequisite for the success of the approach we have outlined is that management be prepared to listen to the recommendations rather than adopting an expedient approach. Budgeting time and manpower for performance projection may lengthen the development schedule somewhat, but the benefits can be significant.

A final important factor is the ability to turn this general framework into specific working strategies. In the next section, we describe two recent tools that are examples of attempts to do so.

14.4. Tools and Techniques

14.4.1. CRYSTAL

CRYSTAL is a software package developed in the late 1970s to facilitate the performance modelling of proposed and evolving application systems.

A CRYSTAL user describes a system in three components: the *module specification*, the *workload specification*, and the *configuration specification*. These specifications are inter-related, and are developed in parallel. They are stated in a high-level *system description language*.

- The *module specification* describes the CPU and I/O requirements of each software module of the system in machine-independent terms: path lengths for the CPU, and operation counts to various files for I/O.
- The *workload specification* identifies the various components of the workload, and, for each component, gives its type (i.e., transaction, batch, or terminal), its workload intensity, and the modules that it uses.
- The *configuration specification* states the characteristics of hardware devices and of files.

From these specifications, CRYSTAL calculates queueing network model inputs. These are supplied to an internal queueing network evaluation algorithm, which calculates performance measures.

We illustrate some of the important aspects of CRYSTAL by describing its use in modelling a proposed application software system. An insurance company is replacing its claims processing system. CRYSTAL is used to determine the most cost-effective equipment configuration to support the application.

As a first step, the workload components are identified in the workload specification. Many functions are planned for the proposed system, but the analyst determines that five will account for more than 80% of the transactions. These include, for example, Claims Registration. (This information comes from administrative records.)

Since the planning of this system is in its preliminary stages, it is not possible to say with certainty how the system will be structured into modules. The analyst decides initially to define one module corresponding to each of the five workload components. This information is represented in both the workload and the module specification. (Naturally, this is an area where the appropriate level of detail will vary as knowledge of the system evolves.)

For each module, resource requirements are stated in the module specification. The units of CPU usage are instructions executed. There are two components: application path length and support system path length. In the case of the example, benchmarks of similar modules currently in use provide information for application path length. Where no benchmark exists, the logical flow of the software is used to provide estimates. For support system path length, major system routines are examined in detail to provide estimates; some benchmarks also are done. The units of I/O usage are number of physical I/O operations. The analyst determines these, beginning from a logical view of each module, and taking into account the file structure to be used.

The major application files and their sizes are part of the configuration specification. (These files correspond to those referred to in the I/O component of the module specification.) Initially, a simple file structure is proposed, but eventually file indices and database software will be introduced. In addition, a series of entries describe the devices of the system, e.g., for a disk, its transfer rate, seek time, rotation time, and a list of its files.

When the system description is complete, CRYSTAL can calculate queueing network model inputs and obtain performance measures. For example, response times can be projected for the baseline transaction volume and hardware configuration. If the results are satisfactory when compared to the response time requirement stipulated for the application, projections can be obtained for increased transaction volume by adjusting the arrival rates of the relevant workloads in the workload specification. Hardware alternatives can be investigated in a similar manner.

This concludes our description of CRYSTAL. The major activities in using this tool are completing the module specification, the workload specification, and the configuration specification. The study described here occurred during the initial stages of a project. As noted before, additional benefits would arise if the study were extended through the lifetime of the project. Better resource estimates would be available from module implementation, and the ability of the configuration to meet the response time requirement could be re-evaluated periodically.

14.4. Tools and Techniques

14.4.2. ADEPT

The second technique to be discussed is ADEPT (A Design-based Evaluation and Prediction Technique), developed in the late 1970s.

Using ADEPT, resource requirements are specified both as average values and as maximum (upper bound) values. The project design is likely to be suitable if the performance specifications are satisfied for the upper bounds. Sensitivity analyses can show the system components for which more accurate resource requirements must be specified. These components should be implemented first, to provide early feedback and allow more accurate forecasts.

The software structure of the proposed application is determined through performance walkthroughs and is described using a graph representation, with software components represented as nodes, and links between these components represented as arcs. Because the software design usually results from a top-down successive refinement process, these graphs are tree-structured, with greater detail towards the leaves. An example is found in Figure 14.2, where three design levels are shown. Each component that is not further decomposed has a CPU time estimate and a number of I/O accesses associated with it.

The graphs are analyzed to determine elapsed time and resource requirements for the entire design by a bottom-up procedure. The time and resource requirements of the leaf nodes are used to calculate the requirements of the nodes one level up, and so on up to the root node. A static analysis, assuming no interference between modules, is performed to derive best case, average case, and worst case behavior. The visual nature of the execution graphs can help to point out design optimizations, such as moving invariant components out of loops.

Additional techniques handle other software and hardware characteristics introduced as the design matures. These characteristics include data dependencies (for which counting parameters are introduced), competition for resources (for which queueing network analysis software is used), and concurrent processing (in which locking and synchronization are important).

ADEPT was used to project the performance of a database component of a proposed CAD/CAM system. Only preliminary design specifications were available, including a high-level description of the major functional modules. A small example from that study will be discussed. A transaction builds a list of record occurrences that satisfy given qualifications, and returns the first qualified occurrences to the user at a terminal. It

330 **Parameterization:** Proposed Systems

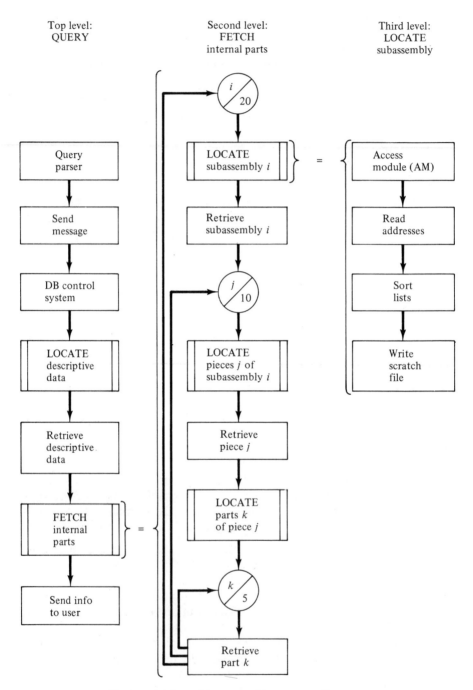

Figure 14.2 — Example Execution Graphs

14.4. Tools and Techniques

issues FIND FIRST commands to qualify record occurrences and FIND NEXT commands to return the occurrences. The execution graphs for the FIND commands have the structure shown in Figure 14.3.

The performance goal for processing this transaction was an average response time of under 5 seconds, when the computing environment was a Cyber 170 computer running the NOS operating system. A performance walkthrough produced a typical usage scenario from an engineering user and descriptions of the processing steps for the FIND commands from a software designer. Resource estimates for the transaction components were based on the walkthrough information. Many optimistic assumptions were made, but the best case response time was predicted to be 6.1 seconds, not meeting the goal (see Figure 14.3). About 43% of this elapsed time (2.6 seconds) was actual CPU requirement. Thus, it was clear at the design stage that response times would be unacceptably long because of excessive CPU requirements.

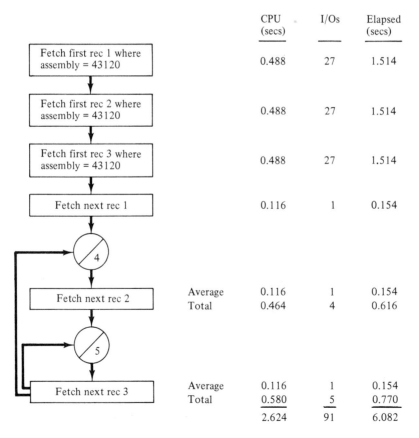

Figure 14.3 − Transaction Steps and Projections

The application system has been implemented. Although actual parameter values were different in the running system than in the design, CPU bottlenecking still was present, more than a year after it was predicted. This demonstrates the success of the ADEPT approach. (The specific corrective advice provided by the performance analysts using ADEPT was not acted on, because it would have caused slippage in the delivery schedule for the system. However, the performance problems that arose resulted in delay and dissatisfaction anyway.)

This study shows that it is possible to predict with reasonable accuracy resource usage patterns and system performance of a large software system in the early design stage, before code is written. It also is possible to achieve these benefits without incurring significant personnel costs. This example was part of a project staffed by a half-time performance analyst and took approximately one person-month of work.

14.5. Summary

The process of design and implementation involves continual tradeoffs between cost and performance. Quantifying the performance implications of various alternatives is central to this process. Because doing so is challenging and requires resources, it is tempting to rely on seat-of-the-pants performance projections. The consequences of doing so can be serious, for user satisfaction with a system depends to a significant extent on the system's ability to deliver acceptable performance.

We began this chapter with a description of several early experiences in projecting the performance of proposed systems. We then discussed various aspects of a general approach to the problem. Finally, we studied two recent attempts to devise and support this general approach. We noted that projecting the performance of proposed systems requires a methodical approach to obtaining queueing network model inputs, an approach that could be of value in any modelling study. We also noted that the process of performance projection can be a valuable project management aid, because it serves to structure and focus communication among various members of the project team.

14.6. References

The early attempts at projecting the performance of proposed systems, discussed in Section 14.2, were directed not towards devising general approaches, but rather towards addressing the particular problems of specific systems. The study of GECOS III was described by Campbell and

14.6. References

Heffner [1968]. The study of TSO was described by Lassettre and Scherr [1972]. The study of OS/VS2 Release 2 was described by Beretvas [1974]. The study of ALS was described by Browne et al. [1975]. A good summary of these attempts appears in [Weleschuk 1981].

We have described two recent attempts at devising and supporting general approaches. CRYSTAL was developed by BGS Systems, Inc. [BGS 1982a, 1982b, 1983]. The examples in Sections 14.3.2 and 14.4.1 come from internal BGS Systems reports, as does Figure 14.1. ADEPT was developed by Connie U. Smith and J.C. Browne. The case study in 14.4.2 was conducted by Smith and Browne [1982]; Figure 14.3 comes from this paper. Other good sources on ADEPT include [Smith 1981] (the source of Figure 14.2), and [Smith & Browne 1983].

[Beretvas 1974]
T. Beretvas. A Simulation Model Representing the OS/VS2 Release 2 Control Program. *Lecture Notes in Computer Science 16.* Springer-Verlag, 1974, 15-29.

[BGS 1982a]
CRYSTAL/IMS Modeling Support Library User's Guide. BGS Systems, Inc., Waltham, MA, 1982.

[BGS 1982b]
CRYSTAL/CICS Modeling Support Library User's Guide. BGS Systems, Inc., Waltham, MA, 1982.

[BGS 1983]
CRYSTAL Release 2.0 User's Guide. BGS Systems, Inc., Waltham, MA, 1983.

[Browne et al. 1975]
J.C. Browne, K.M. Chandy, R.M. Brown, T.W. Keller, D.F. Towsley, and C.W. Dissley. Hierarchical Techniques for Development of Realistic Models of Complex Computer Systems. *Proc. IEEE 63,*4 (June 1975), 966-975.

[Campbell & Heffner 1968]
D.J. Campbell and W.J. Heffner. Measurement and Analysis of Large Operating Systems During System Development. *1968 Fall Joint Computer Conference Proceedings, AFIPS Volume 37* (1968), AFIPS Press, 903-914.

[Lassettre & Scherr 1972]
Edwin R. Lassettre and Allan L. Scherr. Modeling the Performance of the OS/360 Time-Sharing Option (TSO). In Walter Freiberger (ed.), *Statistical Computer Performance Evaluation.* Academic Press, 1972, 57-72.

[Smith 1981]
   Connie Smith. Increasing Information Systems Productivity by Software Performance Engineering. *Proc. CMG XII International Conference* (1981).

[Smith & Browne 1982]
   Connie Smith and J.C. Browne. Performance Engineering of Software Systems: A Case Study. *1982 National Computer Conference Proceedings, AFIPS Volume 51* (1982), AFIPS Press, 217-244.

[Smith & Browne 1983]
   Connie Smith and J.C. Browne. *Performance Engineering of Software Systems: A Design-Based Approach.* To be published, 1983.

[Weleschuk 1981]
   B.M. Weleschuk. Designing Operating Systems with Performance in Mind. M.Sc. Thesis, Department of Computer Science, University of Toronto, 1981.

Part V

Perspective

We have provided a general overview of computer system analysis using queueing network models (Part I), a discussion of the algorithms used for evaluating separable networks (Part II), a look at specialized techniques for the detailed modelling of particular subsystems (Part III), and a guide to parameterizing queueing network models (Part IV). Here, in Part V, we attempt to "fit the pieces together".

Chapter 15 shows, through example, how the techniques that we have presented can be used in non-traditional contexts: modelling computer communication networks, local area networks, software resources, database concurrency control, and operating system algorithms. As computer systems continue to evolve, it is important to recognize that the applicability of queueing network technology extends well beyond the confines of centralized systems with simple characteristics.

Chapter 16 discusses the structure and use of queueing network modelling software. This is a fitting conclusion to the book, for such software can embody many of the techniques covered in Parts I - IV.

Two natural additional components of Part V would be a review of the important differences between queueing network modelling and other approaches to computer system analysis, and a discussion of various considerations that arise in the course of conducting a modelling study. These topics were addressed, by way of introduction, in Chapters 1 and 2. We suggest reviewing those chapters at the conclusion of Part V, since you then will be in a position to appreciate them fully.

Chapter 15

Extended Applications

15.1. Introduction

In this chapter we will illustrate how the techniques developed in Parts II and III can be used to model systems and subsystems whose characteristics are significantly different from those of the centralized systems previously used as examples. Our objective is twofold: to convey the range of applicability of these techniques, and to indicate the sorts of "creative approaches" that have proven successful.

Our presentation will consist of five example application areas: computer communication networks (IBM's SNA), local area networks (Ethernet), software resources, database concurrency control, and operating system algorithms (the SRM in IBM's MVS system). Each application is discussed in a separate section. The sections are brief; further details can be obtained from the papers cited in Section 15.8.

15.2. Computer Communication Networks

Computer communication networks use a variety of *flow control policies* to achieve high throughput, low delay, and stability. Here, we model the flow control policy of IBM's System Network Architecture (SNA).

SNA routes messages from *sources* to *destinations* by way of *intermediate nodes* which temporarily buffer the messages. Message buffers are a scarce resource. The *flow control policy* regulates the flow of messages between source/destination pairs in an effort to avoid problems such as *deadlock* and *starvation*, which could result from poor buffer management.

SNA has a *window* flow control policy. The key control parameter is the *window size*, W. When a source starts sending messages to a particular destination, a *pacing count* at the source is initialized to the value of W. This pacing count is decremented every time a message is sent. If the pacing count reaches zero, the transmission of messages is halted. When the first message of a window reaches the destination, a *pacing*

15.2. Computer Communication Networks

response is returned to the source. Upon receipt, the source increments the current value of the pacing count by W. Another pacing response is sent to the source by the destination each time an additional W messages have been received. Thus, the maximum number of messages that can be *en route* from source to destination at any time is $2W-1$.

Our objective is to model the "response time" of messages between a single source/destination pair — the average time required for messages to flow from source to destination. The most convenient model, in terms of simplicity and ease of evaluation, is an open queueing network. There are M centers, representing the source node, the destination node, and $M-2$ intermediate nodes. (Obviously, M is determined by adding two to the number of intermediate nodes.) Customers, which represent messages, arrive at the source node at rate λ. They flow from node to node, requiring D units of service at each node. This model is shown in Figure 15.1.

Figure 15.1 — **Open Model of SNA Flow Control** (© 1982 IEEE)

Response times can be calculated easily for this model. Unfortunately, the model makes a significant simplifying assumption which impacts the applicability of the results: there is no representation of the flow control policy! The source continues to transmit, regardless of the number of outstanding messages.

A more realistic approach, therefore, is to use a closed model, in which it will be possible to represent the limit on the number of outstanding messages. Figure 15.2 shows this model. There are $2W-1$ customers, representing the possible outstanding messages. As in the open model, there are M centers corresponding to the source node, the destination node, and $M-2$ intermediate nodes. Customers have service demand D at each of these centers. In addition, there is a "message generation" center and a "pacing box". Together, the message generation center and the pacing box mimic the flow control policy, in the following way.

The pacing box "stores" up to $W-1$ messages. When the W-th message arrives, it triggers the discharge of all W messages into the queue of the message generation center. The message generation center has service rate λ; as long as its queue is non-empty, it will generate message traffic at this rate. A bit of thought will reveal that the arrival of the W-th message to the pacing box corresponds to the source's receipt of a pacing response from the destination; such receipt carries with it the right to initiate W additional messages.

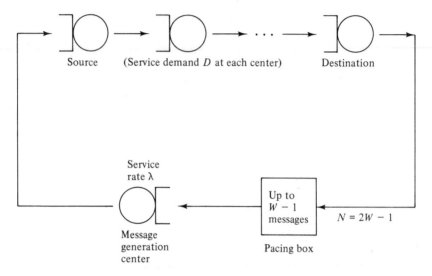

Figure 15.2 — **Closed Model of SNA Flow Control** (© 1982 IEEE)

The model of Figure 15.2, while realistic, is not separable, because of the unusual characteristics of the pacing box. The model could be evaluated directly using the global balance approach, described in Chapter 8. However, the potentially large size of the model makes this approach infeasible in general. A viable alternative, also described in Chapter 8, is to replace the M centers representing the source, destination, and intermediate nodes with an FESC. The resulting three node model of Figure 15.3 still is not separable, but it is small enough for global balance to be practical.

The load dependent service rates of the FESC are estimated in the usual way. A closed, separable model consisting of the M centers representing the nodes, each with service demand D, is evaluated for each feasible message population, from 1 to $2W-1$. Throughputs are determined, and used to define the FESC. Once this has been accomplished, writing the global balance equations and numerically evaluating them to obtain the equilibrium state probabilities is tedious but straightforward. These probabilities yield system throughput and average queue length at the FESC. Little's law then can be applied to determine average response time.

15.3. Local Area Networks

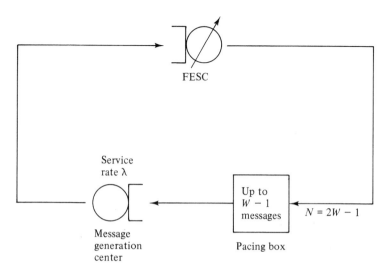

Figure 15.3 − **FESC Representing the Message Path** (© 1982 IEEE)

One assumption made by this modelling approach is that the only traffic passing through a node is due to the source/destination pair of interest. This unrealistic assumption can be eliminated by modifying the separable model used to estimate the load dependent service rates of the FESC. As one approach, if the traffic at each node due to other source/destination pairs is known, it can be represented as an open class whose presence will impede the progress of messages associated with the source/destination pair of interest, with a resulting decrease in FESC rates.

Comparisons with detailed simulations indicate that this simple modelling approach yields good accuracy.

15.3. Local Area Networks

Computer communication networks such as SNA are designed to perform well over long distances at moderate bandwidths. Local area networks, on the other hand, are optimized for use over moderate distances (say, 1 km.) at high bandwidths (10 MHz. or greater). Ethernet is perhaps the most widely known and used local area network. In this section we will describe how to incorporate a representation of Ethernet into a queueing network model of a locally distributed system.

Ethernet uses a single coaxial cable to interconnect *stations* (computers). A station wishing to communicate with another station *broadcasts* a *packet* on this *channel.* (Long messages are decomposed into multiple packets prior to transmission.) The packet contains the address of the destination station, plus the desired data. All stations will "see" the packet, but only the station to which the packet is addressed will copy the packet into its local memory.

Since the channel is shared by all stations, the key to Ethernet is the way in which access to the channel is controlled. Ethernet uses *carrier sense multiple access with collision detection* (*CSMA-CD*). *Multiple access* means that all stations share the same channel. *Carrier sense* means that no station will begin to transmit a packet if it hears data from another station on the channel. Of course, a *collision* still can occur, because two stations can begin transmitting simultaneously (or, in fact, at times that differ by as much as the propagation delay of the channel). *Collision detection* means that stations "listen" while they are transmitting, stop if they detect such a collision, and *retry* at some point in the future. In Ethernet, the average amount of time that a station delays before such a retry increases with the load on the channel, with the result that stability is achieved.

The implementation of Ethernet is complex, and an attempt to incorporate a detailed representation in a queueing network model would be ill-advised. However, Ethernet is based on a simple underlying policy. It is possible to represent the behavior of this policy in a queueing network model. Further, simulation results and measurements indicate that such a model yields accurate results. The approach that we will use is a two-level hierarchical model. At the low level we will determine the *efficiency* of Ethernet (the proportion of its bandwidth devoted to useful work) as a function of the *instantaneous load* (the number of stations simultaneously desiring to transmit packets). The results of this analysis will be used to define an FESC, which will be used to represent the channel in a system-level model.

Imagine time to be divided into *slots* whose duration, S, is equal to the round-trip propagation time of the channel. (This is the time required for a collision to be detected by all stations.) Consider a slot during which some number of stations $n > 0$ desire to transmit packets. If no station transmits, the slot is wasted. If exactly one station transmits, that station *acquires* the channel and continues transmitting until it has finished sending its packet. If more than one station transmits, a collision occurs and the slot is wasted. The Ethernet control policy attempts to maximize the probability that exactly one station transmits during a slot by allowing each station to transmit with probability $1/n$ when n stations desire to use the channel. (The actual implementation differs from this policy because the value of n is not known, and must be estimated by each station.)

15.3. Local Area Networks

If n stations desire to use the channel and each transmits independently with probability $1/n$, then the probability that any of the stations successfully acquires the channel during a particular slot is equal to the probability that exactly one station transmits, or:

$$A = \left[1 - \frac{1}{n}\right]^{n-1}$$

The average number of slots devoted to contention (collisions) before a successful acquisition by some station is:

$$C = \sum_{i=1}^{\infty} i\, A\, (1-A)^i = \frac{1-A}{A}$$

For $n > 0$ the channel has, by definition, no idle periods; time consists of contention intervals interleaved with transmission intervals. The efficiency of the channel at instantaneous load n can be expressed as:

$$E(n) = \frac{\text{length of a transmission interval}}{\text{length of a transmission interval} + \text{length of a contention interval}}$$

The length of a transmission interval equals the average packet length in bits, P, divided by the network bandwidth in bits per second, B. The length of a contention interval equals the expected number of slots devoted to contention, C, multiplied by the slot duration, S (a parameter of the configuration, related to the length of the network). In other words:

$$E(n) = \frac{P/B}{P/B + C \times S}$$

Given P, B, and S, efficiencies are calculated algebraically for each feasible value of n. An FESC then is defined as follows:

$$\mu(n) = B/P \times E(n)$$

In other words, the rate at which the Ethernet delivers packets is equal to its maximum theoretical capacity in packets per second (B/P) multiplied by the proportion of that capacity that is devoted to useful work when there are n stations desiring to transmit packets ($E(n)$). This FESC is used to represent the Ethernet in a system-level model.

As noted earlier, comparisons with simulation results and with measurements indicate that this simple modelling approach yields good accuracy. The analysis can be extended to represent the (non-negligible) effect of packet size variability on performance. The same two-level hierarchical approach can be used to represent other local area networks. For example, a corresponding analysis has been done for the Cambridge ring.

15.4. Software Resources

The usual viewpoint in constructing queueing network models is that service centers correspond to hardware resources. It also is the case, though, that queueing delays in computer systems can arise from contention for *software resources*: operating system critical sections, non-reentrant software modules, etc. In this section we consider the use of queueing network models to evaluate software system structures.

Our approach will be to define a *software-level* queueing network model in which customers, as usual, correspond to users, but in which service centers correspond to software modules. The service demand at each center will be equal to the time the customer spends executing the corresponding software module. The queueing delay at each center, calculated when the model is evaluated, will be an estimate of the time the customer is blocked awaiting access to the corresponding software module. A reentrant software module will be represented as a delay center, since a customer is never blocked awaiting access. A non-reentrant module will be represented as a queueing center, since only one customer can be executing it at a time.

Obviously, the service demand at each center in the software-level model includes various service requirements and queueing delays incurred in executing the corresponding software module on the underlying computer system. This service demand can be thought of as the "response time" of the user once access to the software module has been granted. This service demand will be estimated using a more conventional *hardware-level* queueing network model, in which customers correspond to users executing software modules, and centers correspond to hardware resources. The service demands are easily obtained for this hardware-level model, but the customer population is not known, because the degree of concurrency at the hardware level depends upon the extent to which users are blocked awaiting access to modules at the software level. Thus, an iterative solution is required, in which the hardware-level model provides service demand estimates for the software-level model, which in turn provides customer population estimates for the hardware-level model.

A simple example of a software-level model is shown in Figure 15.4. There are centers corresponding to various software activities: editing, compilation, linking, loading, and execution. There are various possible "execution sequences": edit and compile; compile, link, and execute; load and execute; etc. Each execution sequence is represented as a separate customer class. The number of customers in each class is the number of users performing the corresponding execution sequence.

15.5. Database Concurrency Control

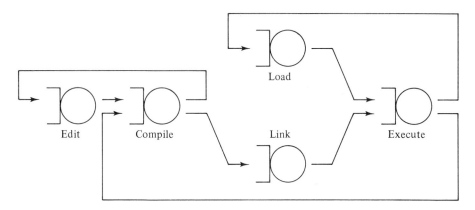

Figure 15.4 - A Software-Level Queueing Network Model

Once the service demands for the centers in the software-level model are known, the model can be evaluated. From the results, the average number of users concurrently executing each software module can be estimated. If a module is reentrant it will be represented as a delay center, and the average population at that delay center will be the average number of users concurrently executing the module. If a module is non-reentrant it will be represented as a queueing center, and the utilization of that center will be the proportion of time that a user is executing the module.

To estimate the service demands for the centers in the software-level model, we use the hardware-level model. As noted earlier, customers in this model correspond to users executing software modules. One class represents each module. The service demands of the various classes at the various centers are determined by the resource requirements of the corresponding software modules. The response time of a class in this hardware-level model determines the service demand at the center corresponding to the same software module in the software-level model. The iteration proceeds in the obvious manner. (The think time of a terminal workload can be represented at either level in this approach, although the software level is a more natural place.)

This approach, and several related ones, have proven quite successful in practice.

15.5. Database Concurrency Control

In any database system, many users will wish to access and update the database concurrently. Problems may arise if this concurrency is undisciplined:

- The database may become *inconsistent* because of an unfortunate interleaving of reads and writes by various users.
- Even if the database remains consistent, individual users may "see" inconsistent views, again because of an unfortunate interleaving of activity.

As an example, Table 15.1 illustrates an inconsistency that might arise if two users concurrently attempted to transfer $50 from their individual bank accounts (u1 and u2, respectively), each initially containing $75, to a shared bank account (sh), initially containing $50: their original total assets of $200 are decreased to $150!

time	user 1		database			user 2	
	action	local copy	u1	sh	u2	local copy	action
0			$75	$50	$75		
1	read u1	$75					
2						$75	read u2
3	− $50	$25					
4						$25	− $50
5	write u1		$25				
6					$25		write u2
7	read sh	$50					
8						$50	read sh
9	+ $50	$100					
10						$100	+ $50
11	write sh			$100			
12				$100			write sh

Table 15.1 − **Effect of Undisciplined Concurrency**

To free the user from concern for problems such as these, the concept of a *transaction* has been devised. The key property of a transaction is *atomicity*:

- The user executing a transaction is guaranteed a single, consistent view of the database, regardless of the activities of other users.
- Other users perceive a transaction as a single action, rather than as a series of separate reads and writes of data items.

The job of a *concurrency control mechanism* is to allow transactions to be executed concurrently while guaranteeing that the consistency of the database is preserved. A crude concurrency control mechanism would grant exclusive access to the entire database to one transaction for its duration. (Concurrency is restricted unnecessarily by this simple solution: two transactions that reference entirely different sets of data items would be

15.5. Database Concurrency Control

unable to proceed concurrently.) A more reasonable mechanism would grant exclusive access to various data items to one transaction for its duration. Other possibilities exist. Clearly, the presence of a concurrency control mechanism can have a significant effect on system performance — an effect somewhat analogous to that of a memory constraint. Equally clearly, a queueing network model that represents the concurrency control mechanism directly will be non-separable: customers may be blocked when data items they require are held by other customers. Techniques comparable to those developed in Part III and in the other sections of this chapter are required.

In this section we consider the evaluation of a database system employing a particular, simple concurrency control mechanism. The processing of a transaction under this concurrency control mechanism is described in Table 15.2. Consider the banking example in Table 15.1. Under the concurrency control mechanism, the activities of user 1 and user 2 would constitute separate transactions. User 1 would obtain locks on data items u1 and sh, and would proceed without concern for interference from others. User 2 would be granted a lock on u2 but denied a lock on sh, and would abort, releasing the lock on u2. Subsequently, user 2's transaction would be re-submitted. (We assume that aborted transactions are re-submitted after some delay.) User 1 would be finished, so the lock on sh would be granted to user 2, who would find the value of sh equal to $100.

The effect of the concurrency control mechanism on performance is evident from this example and from Table 15.2. Some transactions abort because they are unable to obtain a lock on a required data item. From the point of view of the system, a transaction that aborts consumes resources (although not to the extent of a successful transaction). From the point of view of a user, several attempts may be required to complete a transaction successfully.

Estimating the proportion of transactions that abort and the service demands of these transactions are the keys to modelling the system. Initially, though, let us assume that conflicts never occur, so all transactions complete successfully. In this case, a traditional separable queueing network model is suitable. Users at terminals submit transactions. The service demands of transactions can be calculated by considering their complexity: number of items read, number of items written, processing requirements, overhead of lock manipulation required for concurrency control, etc. Evaluating this model yields the average transaction response time and other performance measures of interest.

How can this model be extended to represent the effect of conflicts between transactions? As noted, we must estimate the proportion of transactions that abort, $P[abort]$, and the service demands of these

> *locking phase*
> - Request a *read lock* on each data item whose value is required. A read lock will be granted if no other transaction currently holds a *write lock* on the item.
> - Request a *write lock* on each data item that is to be written. A write lock will be granted if no other transaction currently holds either a read lock or a write lock on the item.
> - If any lock is refused, *abort*, releasing all locks previously granted to the transaction.
>
> *processing phase*
> - Read the values of the required data items.
> - Based on these values, compute the values of the data items to be written.
> - Update the values of the data items to be written.
>
> *termination phase*
> - Release all read and write locks held by the transaction.

Table 15.2 − Steps in Processing a Transaction

transactions. Given this information, we could adjust the service demands of transactions in the model to be:

$$(1 - P[abort]) \times \text{(service demands of a successful transaction)} +$$
$$P[abort] \times \text{(service demands of an aborted transaction)}$$

The model could be evaluated using this parameterization to yield response times for each submission of a transaction. To compute the effective response time to successfully complete a transaction we would multiply the response time of each submission by the average number of submissions required. (Obviously, a homogeneity assumption is employed here.) The average number of submissions required is:

$$1 \times (1 - P[abort]) +$$
$$2 \times (1 - P[abort]) \times P[abort] +$$
$$3 \times (1 - P[abort]) \times P[abort]^2 +$$
$$\vdots$$
$$= \frac{1}{1 - P[abort]}$$

The proportion of transactions that abort depends on many factors, including the average number of active transactions (if few transactions are active simultaneously, then the probability of conflict is low) and the average number of data items read and written by each transaction, relative to the total number of items in the database (if each transaction locks a very small proportion of the items in the database, then the probability of conflict is low). As an example, one particularly simple approach is to assume that each transaction requires read locks on r of the I items in the database, chosen at random, and write locks on w of these I items, also chosen independently and at random. A probabilistic analysis then yields $P[abort]$. This analysis is based on reasoning such as the following: If N transactions are active, they hold $N \times w$ write locks. An arriving transaction will be able to acquire all of its r required read locks with probability:

$$\frac{\begin{bmatrix} I - N \times w \\ r \end{bmatrix}}{\begin{bmatrix} I \\ r \end{bmatrix}}$$

(More accurate estimates of $P[abort]$ can be obtained from more detailed submodels, evaluated either probabilistically or using simulation.)

The service demands of an aborted transaction can be estimated roughly as one half of the lock manipulation overhead of a successful transaction. (We expect half the required locks to be obtained before one is denied; these must be released when the transaction aborts.) In addition, by assumption aborted transactions are re-submitted after some delay. This can be represented by adding a delay center to the model, or by adjusting the "think time" downwards in a manner analogous to that used for service demands.

The average number of active transactions, which is a key parameter required to estimate $P[abort]$, is an output of the model. This suggests the iterative evaluation scheme outlined in Algorithm 15.1. We have left many details unspecified, and have made a number of simplifying assumptions concerning the nature of the system and of the concurrency control mechanism. The basic iterative approach of Algorithm 15.1 is relatively general, however.

15.6. Operating System Algorithms

During the design of an operating system, extremely subtle performance questions may arise that require certain subsystems to be modelled at a level of detail greater than we have considered thus far. Examples of such questions include the design of complex resource allocation

> 1. Construct a traditional separable queueing network model with basic transaction service demands calculated as suggested in the text. Initially, assume that the average number of active transactions is zero. (This will cause $P[abort]$ to be estimated as zero in the first iteration, so the model will be evaluated without adjustment.)
> 2. Iterate as follows:
>   2.1. Based on various input parameters plus the average number of active transactions, use a submodel to determine the proportion of transactions that abort. This submodel may involve probabilistic or simulation analyses, as described in the text.
>   2.2. Calculate revised transaction service demands, as described in the text.
>   2.3. Evaluate the queueing network model. Obtain the average number of active transactions.
>
>   Repeat Step 2 until successive estimates of the average number of active transactions are sufficiently close.
> 3. Obtain performance measures from the final iteration, as described in the text.

Algorithm 15.1 − Concurrency Control in the Rough

algorithms that coordinate the control of paging, swapping, and processor scheduling.

On the one hand, queueing network models are not ideally suited to answering these extremely detailed questions. (Fortunately, such questions arise very infrequently!) On the other hand, queueing network models offer such tremendous advantages over alternative techniques (such as simulation or experimentation) that there is a strong motivation to use them to the greatest possible extent. Often, the solution is to employ *hybrid modelling*, as described in Chapter 8.

In this section we describe a hybrid model of IBM's MVS operating system. This model was designed to study the internal details of the MVS System Resources Manager (SRM). Under MVS, each installation classifies its workload components into *performance groups*. Within each performance group, customers pass through a sequence of *performance periods* as service is acquired. For each performance period, *service objectives* are established. Customers are served at various resources at a rate that depends on the service objectives specified for their current performance period. (For example, a customer's susceptibility to swapping will depend on that customer's current performance period.) In addition, goals are established for the utilizations of various resources. These goals impose additional constraints on scheduling decisions. It is the job of the SRM to reconcile these many objectives by making appropriate long-term and short-term resource allocation decisions.

Figure 15.5 illustrates the structure of the two-level hierarchical hybrid model that allowed the internal algorithms of the SRM to be represented. There are two workload components: TSO (timesharing) and batch. In the high-level model, customer arrivals and the operation of the SRM are represented. Two principal SRM modules are represented explicitly. Swap Analysis keeps track of the attained service of each customer and determines if a swap is to be performed. Resource Monitor calculates target multiprogramming levels, invokes Swap Analysis if necessary, and collects various statistics. These statistics are used in an overhead submodel to determine the overhead service demands of the operating system. The high-level model is evaluated using simulation.

In the low-level model, the central subsystem is represented. Paging activity is determined by a submodel that has knowledge of the particular paging policy of interest. The low-level model is evaluated using techniques from Parts II and III.

The hybrid solution of this model proceeds iteratively. The high-level model determines the multiprogramming mix and the overhead service demands, and supplies these to the low-level model. The low-level model determines throughputs and utilizations, which allow the high-level model to calculate the time of the next completion and to make resource allocation decisions.

Of course, representing the internal algorithms of the SRM is a level of detail far beyond that which is appropriate for capacity planning and performance projection applications. Still, this hybrid model was successful at answering detailed questions concerning SRM behavior. Evaluation of the model was estimated to be 30 to 100 times faster than would have been possible using a pure simulation approach. The modelling approach led to greater flexibility than would have been possible in direct experimentation on an MVS system.

15.7. Summary

This chapter has used five examples to illustrate that the applicability of queueing network models extends well beyond the confines of centralized systems with simple characteristics. We have studied models of computer communication networks, local area networks, software

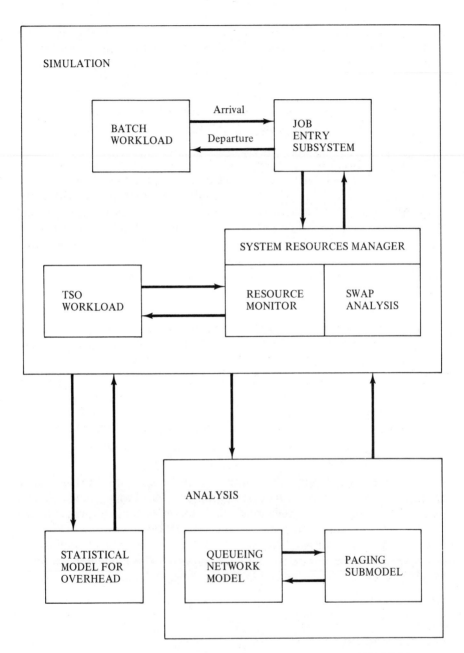

Figure 15.5 — A Detailed Hybrid Model of the MVS SRM

resources, database concurrency control, and operating system algorithms. These models have employed techniques such as FESCs with global balance, FESCs whose service rates are determined through probabilistic analysis, two-level hierarchical iteration, and hybrid modelling. These techniques, combined with good knowledge of the system being modelled and a modicum of inventiveness, can solve a wide variety of computer system analysis problems.

15.8. References

Queueing theory has been used widely in the detailed analysis of computer communication network protocols. The use of queueing network models to evaluate networks and to represent them in system-level models is more recent. A useful general discussion of this area is the book by Schwartz [1977]. The SNA flow control model in Section 15.2 was constructed by Schwartz [1982]; this paper is the source of Figures 15.1, 15.2, and 15.3.

Local area networks have received widespread attention recently. Ethernet was described originally by Metcalfe and Boggs [1976]. The Ethernet model in Section 15.3 was developed by Almes and Lazowska [1979]. King and Mitrani [1982] discuss a similar model of the Cambridge ring.

The technique for modelling software resources described in Section 15.4 is similar to one described by Agre and Tripathi [1982]. Other approaches are described by Smith and Browne [1980], Agrawal and Buzen [1983], and Jacobson and Lazowska [1983].

The modelling of database concurrency control mechanisms is the subject of much recent research activity. Sevcik [1983] provides a survey of various approaches. An excellent discussion of the issues involved, including a framework for classifying mechanisms, is provided by Bernstein and Goodman [1981].

An overview of an early version of the MVS SRM is given by Lynch and Page [1974]. The hybrid hierarchical model in Section 15.6 was developed by Chiu and Chow [1978]; their paper is the source of Figure 15.5. Buzen [1978] describes a queueing network model of MVS that is better suited to capacity planning applications.

[Agrawal & Buzen 1983]
    Subhash C. Agrawal and Jeffrey P. Buzen. The Aggregate Server Method for Analyzing Serialization Delays in Computer Systems. *Transactions on Computer Systems 1*,2 (March 1983), 116-143.

[Agre & Tripathi 1982]
Jon R. Agre and Satish K. Tripathi. Modelling Reentrant and Non-Reentrant Software. *Proc. ACM SIGMETRICS Conference on Measurement and Modeling of Computer Systems* (1982), 163-178.

[Almes & Lazowska 1979]
Guy T. Almes and Edward D. Lazowska. The Behavior of Ethernet-Like Computer Communication Networks. *Proc. 7th Symposium on Operating Systems Principles* (1979), 66-81. Copyright © 1979 by the Association for Computing Machinery.

[Bernstein & Goodman 1981]
Philip A. Bernstein and N. Goodman. Concurrency Control in Distributed Database Systems. *Computing Surveys 13*,2 (June 1981), 185-221.

[Buzen 1978]
Jeffrey P. Buzen. A Queueing Network Model of MVS. *Computing Surveys 10*,3 (September 1978), 319-331.

[Chiu and Chow 1978]
Willy W. Chiu and We-Min Chow. A Performance Model of MVS. *IBM Systems Journal 17*,4 (1978), 444-462.

[Jacobson & Lazowska 1983]
Patricia A. Jacobson and Edward D. Lazowska. A Reduction Technique for Evaluating Queueing Networks with Serialization Delays. *Proc. IFIP W.G.7.3 International Symposium on Computer Performance Modeling, Measurement, and Evaluation* (1983), 45-59.

[King & Mitrani 1982]
Peter J.B. King and Israel Mitrani. Modelling the Cambridge Ring. *Proc. ACM SIGMETRICS Conference on Measurement and Modeling of Computer Systems* (1982), 250-258.

[Lynch & Page 1974]
H.W. Lynch and J.B. Page. The OS/VS2 Release 2 System Resources Manager. *IBM Systems Journal 13*,4 (1974), 274-291.

[Metcalfe & Boggs 1976]
Robert M. Metcalfe and David R. Boggs. Ethernet: Distributed Packet Switching for Local Computer Networks. *CACM 19*,7 (July 1976), 395-404.

[Schwartz 1977]
Mischa Schwartz. *Computer Communication Network Design and Analysis*. Prentice-Hall, 1977.

15.8. References

[Schwartz 1982]
    Mischa Schwartz. Performance Analysis of the SNA Virtual Route Pacing Control. *IEEE Transactions on Communications COM-30*,1 (January 1982), 172-184. Copyright © 1982 IEEE.

[Sevcik 1983]
    Kenneth C. Sevcik. Comparison of Concurrency Control Algorithms Using Analytic Models. *Proc. IFIP Congress '83* (1983).

[Smith and Browne 1980]
    Connie Smith and J.C. Browne. Aspects of Software Design Analysis: Concurrency and Blocking. *Proc. IFIP W.G.7.3 International Symposium on Computer Performance Modeling, Measurement, and Evaluation* (1980), 245-253.

Chapter 16

Using Queueing Network Modelling Software

16.1. Introduction

A variety of techniques for evaluating queueing network models have been described. The general techniques of bounding analysis, single and multiple class analysis, and hierarchical modelling were presented in Part II. Specific techniques for memory, disk I/O, and processor subsystems were discussed in Part III. This collection of techniques comes together for the computer system performance analyst in the form of queueing network modelling software. Such software frees the analyst from the algorithmic portion of the modelling process, allowing concentration on important issues such as model construction and validation, performance projection, and capacity planning.

Most queueing network modelling software can be understood in terms of the structure illustrated in Figure 16.1. There is a sequence of software layers, each transforming input received from the layer above into output suitable for the layer below. In Section 16.2 we will refer to this structure in describing the major components of queueing network modelling software. In Section 16.3 we give an example of conducting a performance study using such software.

16.2. Components of Queueing Network Modelling Software

16.2.1. The Core Computational Routine

The job of the *core computational routine*, situated at the lowest level in Figure 16.1, is simply stated. Given a separable queueing network model defined by its customer description, center description, and service demands, this routine produces performance measures. In other words, the core computational routine is an embodiment of the techniques described in Part II of the book. The core routine may be based on either exact or approximate algorithms.

16.2. Components of Queueing Network Modelling Software

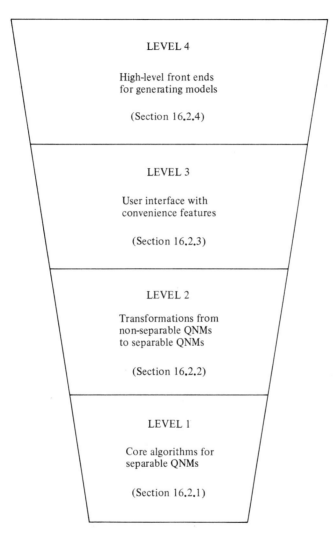

Figure 16.1 − The Structure of Queueing Network Modelling Software

16.2.2. The Approximation Transformations

When viewed at a detailed level, many subsystems have characteristics that lead to non-separable queueing network models. The software layers immediately above the core computational routine, the *approximation transformations* (level 2 in Figure 16.1), translate these non-separable models into a form that is suitable for the core routine. In other words, the approximation transformations correspond to the techniques described in Part III. Many of these techniques require an iterative relationship with the core routine: the transformations provide input suitable to the core routine, and the core routine's output is used as additional input to the transformations.

Consider the treatment of C classes with independent memory constraints in Section 9.3.2. In a sense, there are C different separable low-level models, and C different separable high-level models. The low-level model corresponding to class c has each class other than c represented by its average central subsystem population, while class c's population is varied from 1 to M_c, its memory constraint. The high-level model corresponding to class c is a single class model using the FESC obtained from the low-level model. Each model is evaluated by the core routine, with the transformation layer using the outputs of some models to define the inputs of others.

Another example is the treatment of RPS disks in Section 10.3. The service demands at the centers representing the disks must reflect more than the seek, latency, and data transfer requirements. Approximation transformations estimate the component due to path contention, as in Algorithm 10.2, and calculate *effective* service demands, which are passed to the core routine. Again, this process is iterative.

Additional examples of transformations include those used to represent priority scheduling (Section 11.3) and distributions of multiprogramming level (Section 9.2).

16.2.3. The User Interface

An important attribute of queueing network modelling software is the convenient expression of performance models. The *user interface*, level 3 in Figure 16.1, must bridge the gap between the world of queueing network models and the world of computer systems, so that performance studies can be conducted efficiently by analysts whose primary training is in computer systems.

In many cases, queueing network concepts correspond directly to computer system concepts: e.g., classes to workload components, centers to devices, customers to users or jobs. These queueing network concepts

16.2. Components of Queueing Network Modelling Software

are made visible to the analyst with little intervention by the user interface layer.

In other cases, the correspondence is not so direct. In particular, the user interface layer smoothes the analyst's interaction with the transformation layer. As an example, when multipathing I/O is modelled, as described in Section 10.5, the user interface layer allows the analyst to describe the system in terms of channels, controllers, heads of string, disks, and logical channels (in the case of an IBM configuration) and translates this information into a form acceptable to the transformation layer. The transformation layer then uses this information to estimate effective service demands for the various disks, interacting iteratively with the core routine, which evaluates a sequence of separable models provided by the transformation layer. Finally, the transformation and user interface layers provide performance measures in a meaningful form.

A related facility allows the analyst to associate a "type" with a device, and to specify the relevant characteristics of this type to the software. This is useful during modification analysis. As a simple example, an existing system may have an IBM 3081-D CPU, and one of the contemplated modifications may be an upgrade to a 3081-K. If the relative speeds of these two processors are known to the software (in fact, they are roughly 4.0 and 5.5), then the following sequence of interactions would be possible:

- analyst provides all inputs for baseline model
- analyst identifies CPU as 3081-D
- baseline model is validated
- analyst specifies that CPU type should be changed to 3081-K
- software adjusts CPU service demands based on internal information

A similar approach can be applied to other devices.

The user interface layer typically provides the ability to save, recall, and edit model definitions during an interactive session, since model modification is the major activity in interacting with queueing network modelling software. Output reports also need to be stored for postprocessing.

A final example of a facility provided by the user interface layer is a means for the analyst to "program" the package using a simple language, similar to the "exec file" or "command file" facility provided by most contemporary timesharing systems. A simple application of this facility would be to perform automatically a parametric study (for example, on the effect of increasing the number of active terminals). Of course, an analyst could conduct such a study by interacting with the software directly, issuing separate commands to evaluate the model with each population of interest. A better approach, though, would be to write a simple program that, when interpreted by the package, accomplishes this task. A more sophisticated application would be to implement new specialized evaluation techniques for subsystems peculiar to a particular environment, similar to the established techniques described in Part III.

16.2.4. High-Level Front Ends

Given measurement data for a system of realistic size, it takes a significant amount of time to calculate and enter the inputs of a queueing network model. The process is error-prone, because of the large volume of data. Many of the actions are repetitive. All of these factors argue strongly in favor of a computer program that partially automates the construction of baseline models of existing systems. This is one major example of a *high-level front end*, level 4 in Figure 16.1. Other examples of application areas in which high-level front ends are of great utility include performance projection for proposed systems, for database systems, and for communication networks. We touch upon each of these in turn.

Modelling Existing Systems

In Chapter 12 we discussed the parameterization of baseline models of existing systems. To be sure, parts of this process require subtle judgements on the part of the analyst. For any particular system, though, it is possible to automate a large proportion of the labor involved in translating measurement data (stored in a specified format in a performance database) into queueing network model inputs.

The feasibility of constructing such a high-level front end relies on the fact that the general structure of queueing network models, while system dependent, is not highly installation dependent. For example, in constructing queueing network models of IBM MVS systems it is reasonable to equate "performance groups" with customer classes. In general, such front ends construct fairly simple models, which subsequently are adjusted by the analyst. Overall, the savings in time can be significant.

Modelling Proposed Systems

Performance projection for proposed systems was discussed in some detail in Chapter 14, where we illustrated the syntax of two high-level front ends. The interface for this application is based on the system designer's point of view. It deals with the natural units of the application (e.g., estimated number of disk reads on the software side, transfer rate on the hardware side) and translates them into a form acceptable to the core routine of the queueing network modelling software.

16.2. Components of Queueing Network Modelling Software 359

Modelling Database Systems

In systems where database processing has a major influence on performance, the use of a specialized high-level front end can expand the scope of performance questions that can be addressed conveniently.

The presence of a significant database workload component does not detract from the applicability of the techniques described in Parts II and III to the problem of projecting performance. In fact, several of the case studies described earlier treat systems in which database activity is significant. There are performance questions that arise in the context of database systems, however, that cannot be addressed conveniently using the interface provided by general queueing network modelling software.

The efficient operation of database systems depends on many design decisions, including:

- representation of the logical data model as a set of files
- specification of links that relate records to one another
- selection of indices to facilitate access to records having various values for a certain attribute in a file
- choice of query processing strategy
- placement of indices and files in main memory and on various storage devices
- allocation of buffer space for various purposes

All of these decisions have a substantial influence on the service demands of database transactions. The purpose of a high-level front end for database systems is to support investigation of decisions such as these.

In most database systems, a few transaction types are dominant. If we represent each such transaction type as a customer class in a model, the class can be characterized by its workload intensity and its typical pattern of accesses to items in the database. The front end can calculate the number of physical block accesses per transaction from the pattern of accesses to logical data items, by taking into account the representation of the logical database as files and links between them. In processing a query, the order in which operations are done (the query processing strategy) and the presence of indices on selected attributes also influence the number of block accesses. Finally, data placement decisions determine the fraction of block accesses directed to each device, and buffer strategies determine the fraction of block accesses that require a physical data transfer. Thus, by considering the characteristics of a database environment and the design decisions made, the front end can transform the high-level specification of transaction types into a specification suitable for input to a standard queueing network modelling package.

Specific database management systems impose specific restrictions. For example, IMS and System 2000 support particular ways of linking and indexing files, and provide particular tuning parameters (such as priority specifications, and the allocation of buffer space to various uses). A high-level front end tailored to a specific database system is most convenient for an analyst because all the decisions resolved within the system can be built into the front end.

Modelling Computer Communication Networks

Many issues need to be considered in the design of computer communication networks: network bandwidth, multiplexing and concentration strategies, network protocol layers, flow control policies, routing strategies, and buffering strategies. A network designer is interested in knowing if a proposed network can handle a projected workload intensity while providing an acceptable level of performance. A front end for this application accepts system descriptions in terms of entities such as cluster controllers, line speeds, host/satellite topology, and protocols. These are transformed for the queueing network model into service demands at various centers (e.g., the communications controller). Hierarchical modelling, as described in Chapter 8, is useful here.

16.3. An Example

This section gives an example of the way in which advanced queueing network modelling software can be used by an analyst to develop and modify a model of a large contemporary computer system. We have three objectives:

- We wish to illustrate the levels of detail that are appropriate for building a model and using that model for performance projection.
- We wish to illustrate the relationship between modelling concepts, evaluation algorithms, and modelling software.
- We hope to indicate how such software can increase the productivity of the analyst.

To achieve these objectives, it is necessary to include example commands for specific queueing network modelling software. We have chosen the package MAP for our example. Other packages have similar features.

16.3.1. Description of the Example System

The system we treat is an Amdahl 470 V/7 with 16 million bytes of main memory. It runs IBM's MVS operating system with two important

16.3. *An Example* 361

workload components, batch and TSO (interactive). Other workload components are present, but the performance questions of interest concern only batch throughput and TSO response time.

Approximately 200 disk drives, IBM 3350s, are accessible through eight physical channels. Other devices are attached to the system (e.g., unit record devices and tape devices), but they have little influence over performance in the current system or any contemplated future system.

16.3.2. Building a Baseline Model

The first step in our modelling process is to construct a baseline model that validates well on batch throughput and TSO response time. To present the concepts clearly, we will be showing how an analyst might interact directly with the package in building the baseline model, rather than using a high-level front end to partially automate the process, as described in Section 16.2.4. As a practical consideration, though, some software assistance is a necessity in building large models. We assume that all necessary parameter values have been obtained from measurement data in accordance with the procedures suggested in Chapters 12 and 17.

Our discussion of model building treats classes, centers, domains, and service demands in turn.

Classes

Because of the performance questions of interest, the batch and TSO workload components must be represented as separate classes. Multiple class algorithms similar to those of Chapter 7 will provide the necessary class based performance measures. In the actual system, other workload components interfere with these two important components. We must include this effect in the model. We do so with an aggregated artificial third class. (Neglecting these other components would yield optimistic results for the classes of interest.)

The definition of a class includes its name, its type, and its workload intensity. TSO is a class of TERMINAL type with an associated number of terminals and average think time. PRODUCTION is a class of BATCH type with an associated average multiprogramming level. The artificial class OTHER is specified as a class of TRANSACTION type with an associated arrival rate, set equal to the rate at which jobs of the other workload components complete. (This approach guarantees that the modelled throughput of the OTHER class will equal the aggregated throughputs of the other workload components.)

362   Perspective: Using Queueing Network Modelling Software

MAP commands to define the classes are:

CLASS TSO	(create class TSO)
TYPE TERMINAL	(state its type)
ONLINE_USERS 68	(specify the number of active terminal users)
THINK 12.53	(think time of 12.53 seconds)
CLASS PRODUCTION	(create class PRODUCTION)
TYPE BATCH	(state its type)
AVG_MPL 8	(set its average multiprogramming level)
CLASS OTHER	(create class OTHER)
TYPE TRANSACTION	(state its type)
ARRIVAL_RATE .11	(arrival rate is .11 jobs/sec.)

Additional commands that further specify class attributes will be given shortly, when other necessary components of the model have been defined.

Centers

The model includes a center representing the CPU and a center representing each disk drive. The definition of a center includes its name and optional attributes such as its scheduling discipline and its manufacturer's model designation. The MAP commands to define the CPU and I/O devices are:

CENTER CPU	(create a center named CPU)
SCHEDULE PRIORITY	(specify priority scheduling)
MODEL V/7	(inform MAP that it is an Amdahl V/7 CPU)
CENTER SYS001	(create center SYS001)
MODEL 3350	(inform MAP that is an IBM 3350 disk)
CENTER PAG001	(create center PAG001)
MODEL 3350	(it is an IBM 3350)
⋮	

If no scheduling discipline is specified, processor sharing is assumed at a CPU center, and FCFS is assumed at other centers. Here, the CPU has been made a priority center. The relative scheduling priorities of classes at this center must be given. In MAP, this is done using the PLEVEL command:

16.3. *An Example*

CLASS TSO PLEVEL 3 (TSO has highest priority)

CLASS PRODUCTION PLEVEL 2 (PRODUCTION has middle)
                                                                                                                          (priority)

CLASS OTHER PLEVEL 1 (OTHER has lowest priority)

Higher PLEVEL values indicate higher scheduling priorities. The inclusion of priority scheduling will cause MAP to evaluate the model using a technique similar to the one described in Section 11.3.

Domains

In the actual system, the TSO workload component is subject to a memory constraint of nine processing regions, meaning that at most nine TSO users can be competing for CPU and I/O service at once (other users must queue for memory). In MAP, this is modelled by the DOMAIN feature. The definition of a domain consists of its name and its capacity. It also is necessary to indicate which classes are constrained by each domain. The MAP commands to do this for our example are:

DOMAIN DOM_TSO CAPACITY 9 (create domain DOM_TSO;)
                                                                                                               (set its capacity to 9 jobs)

CLASS TSO INDOMAIN DOM_TSO (associate TSO with domain)
                                                                                                               (DOM_TSO)

The other classes are not associated with domains. The use of the DOMAIN feature will cause MAP to evaluate the model using a technique similar to the one described in Section 9.3.

A final memory related command is MEMSIZE, which informs MAP of the amount of main storage in the base configuration:

MEMSIZE 16   (16MB of main storage in the base system)

This information is used by MAP during modification analysis, as we will see.

Service Demands

The final components we define in the baseline model are the service demands of all classes at all centers. A convenient way of entering the service demand values is to have the package prompt for them in a systematic manner. In MAP, this is accomplished by specifying class and center values of ALL. Then, in response to the DEMAND command, MAP will print class and center names and accept the corresponding service demands, as in:

CLASS ALL   (ALL will cause MAP to prompt with all class names)
CENTER ALL   (ALL will cause MAP to prompt with all center)
                 (names)

DEMAND   (indicate to MAP that we want to specify service)
                 (demands)

TSO: (MAP prompt for TSO, to which the following)
                 (prompts apply)
  CPU: (MAP prompt for CPU service demand)
  .09   (user-specified)
  SYS001:   (MAP prompt for SYS001 service demand)
  .04   (user-specified)
  ⋮

(Prompts printed by MAP are shown in italics.)

Performance Reports

Having defined the model, a number of performance reports can be obtained for the baseline system. Examples are shown in Table 16.1.

	System Performance Measures			
Class	Response Time	Time in System	Memory Wait	Throughput
TSO	5.6423	2.3868	3.2555	3.7420
PRODUCTION	12.5391	12.5391	0.0000	0.6380
OTHER	7.2880	7.2880	0.0000	0.1100

	Device Utilizations			
Center	TSO	PRODUCTION	OTHER	Total
CPU	0.3368	0.5104	0.0220	0.8692
SYS001	0.1497	0.1276	0.0033	0.2806
PAG001	0.1123	0.2233	0.0330	0.3686
⋮				

Table 16.1 − **Example Performance Report**

16.3. *An Example* 365

The model can be validated by checking its calculated performance measures against those from several measurement periods, as described in Chapters 2 and 12. This often is an iterative process in which the model and its parameters are refined as a better understanding of the system is gained.

At this point, we assume that the model has been validated successfully. Once this has been accomplished, the model definition can be saved:

SAVE BASELINE   (save model definition in permanent file)

The model could be retrieved subsequently, either in this MAP session or in other sessions, using:

READ BASELINE   (read the named model definition file)

16.3.3. Reflecting Anticipated Changes

In Chapter 13 we discussed the parameterization of models of evolving systems. Many model modifications are possible. In this subsection we will discuss modifications to the workload, hardware, and software components of the baseline model. In doing so, we will illustrate many of the convenience features of contemporary queueing network modelling software.

Workload

One standard workload change is an increase in workload intensities. To specify a new value for the PRODUCTION multiprogramming level, the AVG_MPL command would be issued with a new value as its operand:

CLASS PRODUCTION AVG_MPL 12   (set the new value)

A more usual specification involves a relative change, e.g., a 20% increase in the TSO workload intensity (number of terminals), as in:

CLASS TSO ONLINE 1.2*   (multiply the previous value by 1.2)

(Any unique prefix is an acceptable abbreviation in MAP; ONLINE is a shortened form of ONLINE_USERS.)

Other workload changes might involve the service demands. New application software might reduce the CPU path lengths for the TSO class. This might be reflected in the model by:

366 Perspective: Using Queueing Network Modelling Software

CLASS TSO CENTER CPU DEMAND .90* (CPU demand is)
(reduced by 10%)

The service demands at the disks might change as well. New blocking strategies for files might reduce the number of I/Os per transaction, resulting in reduced access time per byte transferred because of smaller total seek and latency requirements. Disk service demands can be modified in a manner similar to above.

Having modified the parameters of the model as appropriate, performance estimates for the proposed system can be obtained using the PERFORMANCE command, as was done during the validation phase of the study.

Hardware

A typical hardware change is the upgrading of a device to a more powerful one. In our example, the V/7 CPU might be upgraded to a 5860. Because a MODEL has been specified for the CPU center, specifying a new MODEL has the effect of changing the service demands of all classes at the CPU:

CENTER CPU MODEL 5860   (represent upgrade to Amdahl 5860)

This change is done automatically within the package based on built-in knowledge of the relative speeds of these two CPUs.

Just as a CPU can be upgraded, so can disks. If some or all of the 3350 disks are changed to 3380 disks, the MODEL command can be used to alter the service demands of all classes at the devices upgraded, as in:

CENTER SYS001 MODEL 3380   (upgrade from IBM 3350 to)
(IBM 3380)

Another typical upgrade involves a memory expansion. Going from 16 megabytes to 24 megabytes allows additional space to be allocated to the user workloads. How this space is allocated is dependent on the operating system memory policies. MAP uses built-in knowledge to estimate how a memory expansion will affect various classes. The command:

MEMSIZE 24   (increase main memory size to 24MB)

causes MAP to alter automatically the average multiprogramming level of the PRODUCTION class and the domain capacity of the TSO class to reflect the use of increased main memory. (The values computed by MAP are estimates and account only for first-order effects; the analyst might want to modify them on the basis of more detailed knowledge.)

16.3. An Example

As a final hardware change, consider the introduction of an additional controller to the I/O subsystem to reduce path contention. (See Chapter 10 for a more complete description of the I/O subsystem architecture being considered.) Our basic model contains no explicit representation of I/O paths (the effect of contention for I/O paths is reflected in the disk service demands, which contain a path contention component), so to model this change a more detailed representation of the I/O subsystem is required. We illustrate the process of creating a detailed model of this sort with an example. For purposes of exposition, we will keep the example small.

Suppose two strings of disks can connect to memory through two controllers, and these two controllers can connect through two channels. This connection scheme comprises what we call a *logical channel*, and must be represented in our detailed model. Before defining the logical channel in MAP, the basic components should be created, as in:

CHANNEL CH1	(define a channel)
CHANNEL CH2	(define another channel)
CONTROLLER CTLA	(define a controller)
CONTROLLER CTLB	(define another controller)

The logical channel now can be defined:

LCHANNEL LCH	(define the logical channel; MAP)
	(now will prompt for path descriptions)
CHANNEL:	(MAP prompt for channel at head of this path)
CH1	(user-specified)
CONTROLLER:	(prompt for controller on this path)
CTLA	(user)
CONTROLLER:	(prompt for another controller on this path)
CTLB	(user)
CONTROLLER:	(prompt)
	(user null line indicates end controller list)
CHANNEL:	(prompt for another channel to head)
CH2	(another set of paths)
CONTROLLER:	
CTLA	
CONTROLLER:	
CTLB	
CONTROLLER:	
	(user null line, to end controller list)
CHANNEL:	
	(user null line, to end channel list)

Disks are grouped into sets called strings, with all disks on the same string accessible over the same set of I/O paths (i.e., the same logical channel). At this point, strings and associated information must be defined, as in:

STRING STR1	(create a string)
ONLCHANNEL LCH	(inform MAP of the paths by)
	(which disks on the string)
	(are accessible)
STRING STR2	(create another string)
ONLCHANNEL LCH	(place STR2 on paths LCH)
CENTER SYS001 ONSTRING STR1	(put SYS001 on string STR1)

⋮

To model the addition of a controller to the system, the logical channels affected by the new controller could be redefined to include it, and the model re-evaluated. MAP automatically estimates the new level of path contention, and uses this to alter the disk service demands.

Other changes, such as the addition of channels and strings or the movement of disks among strings, can be modelled similarly.

Software

By software changes we mean changes in the operating policies or parameters of the system, not changes in the intrinsic workload demand. As an example, consider attempting to balance resource usage by moving a set of TSO files from one disk, where these files are responsible for one third of the accesses, to another. This can be represented in MAP using:

CLASS TSO MOVE .33 SYS001 SYSTMP

We can change the priority scheduling structure at the CPU simply by specifying new PLEVEL values, as in:

CLASS TSO PLEVEL 2	(reduce TSO's priority level)
CLASS PRODUCTION PLEVEL 3	(give PRODUCTION priority)
	(over TSO)

PRODUCTION now has priority over TSO at the CPU.

Unlimited variations are possible, but the essence of constructing and modifying a model should be evident. The interactions between user and package that have been illustrated are typical of those involved in performance projection studies. The usual goal of such studies is to estimate the performance of an existing system subjected to new workloads and configurations. To support this activity, the software allows the analyst to modify classes, centers, domains, and service demands in a simple manner. The ability of the software to represent device-specific and system-specific information is especially advantageous. A wide range of alternatives can be investigated rapidly and interactively.

16.4. Summary

Queueing network modelling software can be viewed as consisting of four levels. From lowest to highest, they are:

1. The *core computational routine*, which evaluates separable queueing network models as described in Part II.
2. The *approximation transformations*, which interact with the core routines to evaluate detailed, non-separable models of subsystems such as memory, disk I/O, and processors, as described in Part III.
3. The *user interface*, which allows the analyst to use the terminology of computer systems, rather than the terminology of queueing networks, and which also supports facilities such as the filing and retrieval of model definitions and output reports, and the programmability of the software.
4. *High-level front ends*, which partially automate specific tasks described in Part IV: producing queueing network model inputs from system measurement data, iteratively evolving the system specifications needed for projecting the performance of proposed systems, etc.

All of these levels need not be present; indeed, simple queueing network modelling software often consists only of the first level. However, the higher levels are important to the professional computer system performance analyst. The four levels need not be packaged together in a single piece of software; it is typical for the fourth level to be separate from the other three.

An obvious question that arises is whether queueing network modelling software should be developed in-house, using information from sources such as this book, or obtained from a vendor. Most of the arguments support the latter choice. Many of these arguments are managerial, but one is technical, and we will consider it briefly.

Queueing network modelling technology has advanced rapidly in the recent past, and can be expected to continue to do so in the near future. That portion of a computer system performance analyst's time not devoted to computer system analysis is better spent staying abreast of advances in computer systems than staying abreast of advances in queueing network technology.

A quick historical review in support of this point may be of interest. Table 16.2 shows that even at the relatively well understood level of the core computational routine, advances have been recent, rapid, and significant. At the level of the approximation transformations, progress has been even more recent. For example, the techniques for evaluating multiple class memory constrained queueing networks (Section 9.3) and for evaluating multipathing I/O subsystems (Section 10.5) both were developed during the two year gestation period of this book. In other words, extensive changes have taken place recently even in the algorithms at the lower levels of queueing network modelling software.

rough date	development
1965	First application of queueing network models to computer systems: a two center model with a population of a few customers, evaluated using the global balance technique.
1970	First efficient evaluation algorithm, an exact technique (the "convolution method") for single class separable models.
1975	Extension to multiple class separable models.
	Concept of flow equivalent service centers using load dependent servers.
1980	Mean value analysis and, subsequently, the highly efficient MVA-based iterative approximate evaluation techniques for separable models.

Table 16.2 − Advances in the Core Computational Routine

16.5. Epilogue

Given that much of what has been discussed in this book can be − and has been − packaged in queueing network modelling software, why have you and we together labored so long over this material? The reason is simple: the effectiveness with which such software can be applied is multiplied many times by an understanding of the principles and techniques upon which it is based. Briefly:

- In the case of Part I, you learned that Little's law and its relatives, which provide the technical foundation of queueing network modelling, are reasonable, and are of extremely broad applicability. This knowledge, along with an awareness of the widespread success of per-

16.5. Epilogue

formance studies using queueing network models, provides confidence in the approach.

- In the case of Part II, you learned the techniques used to evaluate separable queueing network models, and the assumptions upon which these techniques rely. The robustness of these techniques with respect to the assumptions was indicated. This knowledge again builds confidence in the approach and, more importantly, it indicates the range of applicability of queueing network models, and provides insight into the ways in which detailed models of specific subsystems can be constructed using separable queueing network models as a basis.

- In the case of Part III, you learned a collection of techniques for augmenting the algorithms of Part II to evaluate detailed models of specific subsystems, where it often is necessary to represent the effects of system characteristics that violate the separability assumptions. This knowledge will help you understand the homogeneity assumptions made by queueing network modelling packages in specific cases (for example, in representing multipathing I/O subsystems), so that you will know if these assumptions should be a source of concern in a particular performance study. The techniques presented in Part III also can serve as a model for similar techniques that you might devise yourself when confronted with a unique modelling problem.

- In the case of Part IV, you learned how to parameterize queueing network models to conduct studies of existing, evolving, and proposed systems. Often a data reduction tool will not be available, and you will be forced to work from raw measurement data. Even if such a tool is available, it often will be desirable to augment it to accommodate the requirements of a particular installation. You now have the knowledge to do so.

- In the case of Part V, you learned through example how queueing network models can be applied in "non-traditional" contexts. As computer systems continue to evolve, it is important to recognize that the applicability of queueing network technology, and of existing queueing network modelling software, extends well beyond the confines of centralized systems with simple characteristics.

To repeat some comments made in the Preface, queueing network models, while not a panacea, are the appropriate tool in a wide variety of applications. Computer system analysis using queueing network models is a blend of art and science, requiring both education and experience. We hope to have contributed, and wish you success in applying queueing network models in your work.

16.6. References

Our discussion of the structure of queueing network modelling software is based on the treatment by Graham, Lazowska, and Sevcik [1982]; this paper is the source of Figure 16.1.

BEST/1, available since the late 1970s and considerably updated since then, is one of the earliest commercial queueing network modelling packages [BGS 1982a]. More recent packages include CMF/Model [Boole & Babbage 1983], RESQ [Sauer et al. 1982], and MAP [QSP 1982a, 1982b].

An example of a high-level front end for modelling existing systems is CAPTURE/MVS [BGS 1982b], which prepares BEST/1 input from MVS performance monitor data.

A view of how to structure a hierarchical tool for database performance projection, including a sequence of database workload descriptions, is provided by Sevcik [1981]. A front end interface for projecting performance of System 2000 databases is described by Casas Raposo [1981].

[BGS 1982a]
BEST/1 User's Guide. BGS Systems, Inc., Waltham, MA, 1982.

[BGS 1982b]
CAPTURE/MVS User's Guide. BGS Systems, Inc., Waltham, MA, 1982.

[Boole & Babbage 1983]
CMF/Model. Boole & Babbage, Inc., Sunnyvale, CA, 1983.

[Casas Raposo 1981]
I. Casas Raposo. Analytic Modeling of Data Base Systems: The Design of a System 2000 Performance Predictor. Technical Note 25, Computer Systems Research Group, University of Toronto, July 1981.

[Graham et al. 1982]
G.S. Graham, E.D. Lazowska, and K.C. Sevcik. Components of Software Packages for the Solution of Queueing Network Models. *Proc. CPEUG '82* (1982), 183-187.

[QSP 1982a]
MAP User Guide. Quantitative System Performance, Inc., Seattle, WA, 1982.

[QSP 1982b]
MAP Reference Guide. Quantitative System Performance, Inc., Seattle, WA, 1982.

16.6. References

[Sauer et al. 1982]
Charles H. Sauer, Edward A. MacNair, and James F. Kurose. The Research Queueing Package, Version 2: Introduction and Examples. Report RA 138, IBM T.J. Watson Research Center, 1982.

[Sevcik 1981]
K.C. Sevcik. Data Base System Performance Prediction Using an Analytical Model. *Proc. 7th VLDB Conference* (1981), 182-189.

Part VI

Appendices

The four Appendices contain detailed information omitted from the body of the book.

Chapter 17 uses an example based on a specific system (IBM's MVS) and a specific monitoring tool (RMF) to illustrate the parameterization of queueing network models, which was described in general terms in Chapter 12.

Chapter 18 contains a Fortran program implementing the exact mean value analysis algorithm for evaluating single class queueing network models. Chapter 19 extends this program to multiple classes. It is our intention that these programs be used for educational experimentation with simple models; we advocate the use of commercial queueing network modelling software for "serious" computer system analysis.

Chapter 20 discusses the evaluation of queueing network models containing load dependent service centers, and describes how the programs in Chapters 18 and 19 can be modified to accommodate such centers.

Chapter 17

Constructing a Model from RMF Data

17.1. Introduction

In Chapter 12 we described in general terms how to determine the input parameter values of a queueing network model from knowledge of a system, measurement data, and accounting data. In this chapter we are more specific: we consider the determination of parameter values for models of computer systems running IBM's MVS operating system, using information obtained from the MVS Resource Measurement Facility (RMF). We choose MVS for special treatment for several reasons:

- More large installations run MVS than any other single operating system.
- Many performance analysts who work with other systems have MVS experience in their backgrounds.
- The measurement and monitoring facilities associated with MVS have greater variety and sophistication than those of most other systems.

While this chapter will be of greatest utility to those involved with MVS and RMF, the techniques and difficulties that we illustrate are similar to those that arise in the context of many systems. To facilitate understanding by persons not familiar with MVS and RMF, Sections 17.2 and 17.3 provide introductions to the concepts and terminology associated with them. Of necessity, our discussions are relatively superficial, and pertain to specific releases of MVS and RMF.

Following the structure of Chapter 12, Sections 17.4, 17.5, and 17.6 treat customer description, center description, and service demands, respectively. Section 17.7 indicates how performance measures can be derived from RMF reports for the purpose of model validation. In each of these sections we first describe the techniques used to determine the corresponding parameter values from RMF reports, and then, in a "double-boxed" paragraph, illustrate these techniques in the context of a specific example. This example is based on standard RMF reports from an installation running MVS on an Amdahl 470 V/8 with 16 megabytes of main memory, 12 physical channels, and roughly 150 IBM 3350 disk drives. The workload consists of two components: interactive (TSO) and batch. The RMF reporting interval was one hour during an afternoon peak load period. This system is much simpler than many large MVS installations (for example, it has fewer workload components). It suffices, however, to illustrate the basic parameter determination techniques.

As we have emphasized throughout the book, the goals of a particular modelling study must be taken into account in the construction of a model. In our example we assume that the system changes to be investigated are moderate changes to workload intensities. There are two significant implications of this assumption:

- We need not include a sophisticated representation of the disk I/O subsystem, as described in Chapter 10, since we would not expect the various components of disk service demands to change significantly under the modifications being investigated.
- We do not require a careful breakdown of paging activity by workload component (difficult to obtain from RMF alone), since we would not expect the level of paging activity to change significantly.

17.2. Overview of MVS

In this section we introduce some aspects of IBM's MVS operating system, which runs in many major computer installations. MVS has several components and features that relate to performance, and thus are important to the modelling of MVS systems.

Workload components in MVS are defined in the *Installation Performance Specification (IPS)*. A set of *performance groups* is established, each of which optionally is divided into a set of *performance periods*. An incoming transaction, based on its identity, enters the first performance period of some particular performance group.

A *service objective* is associated with each performance period, which states the desired rate at which *service units* are acquired by transactions belonging to that period. Service units are computed as the weighted sum of logical I/O operations, main storage occupancy (in units of 50 kilobyte-seconds), and CPU service (measured in 100ths of a second and adjusted by a factor reflecting the speed of the processor). The weights, called *service definition coefficients*, are set by the installation manager.

The *System Resources Manager (SRM)* controls the allocation of resources. The SRM's decisions are based on the progress of transactions relative to their associated service objectives. As transactions reach specified thresholds of attained service, they move from one performance period to the next. The service objectives of successive performance periods call for lower and lower priority for resource allocation. The thresholds are chosen so that most of the transactions that enter a period complete within that period. This has the effect of providing good response times to short transactions by discriminating against longer transactions.

Competition for memory is handled by defining a set of *domains*, establishing a limit on the number of transactions that can be active simultaneously in each domain (the *domain capacity*), and associating each performance period with some domain. Transactions in periods associated with the same domain compete with one another for memory.

Typical performance groups defined in MVS installations include batch (possibly split into components such as production and test) and TSO. Other performance groups correspond to *started tasks* (jobs that never terminate). Started tasks may include such major subsystems as the IMS Control Region and associated Message Processing Regions, CICS, TCAM, and JES, as well as lesser tasks such as RMF and other performance monitors.

17.3. Overview of RMF Reports

In Chapter 12 we described software monitors in general. RMF is a software monitor that records information during system operation with modest overhead (typically, 2% to 3%). RMF uses a combination of *sampling* and *event recording*. For example, queue length distributions at various resources are determined by sampling, while the number of physical I/O operation (SIOs) is accumulated by event recording.

RMF generates a number of standard reports. These reports provide far more data than is needed for our purposes. Table 17.1 (which is divided into four parts) presents some relevant data items excerpted from various RMF reports for our example system. The following paragraphs briefly describe the content of the various reports.

The *CPU Activity Report*, shown in abbreviated form in Table 17.1a, provides information on the CPU and its usage. It includes the length of the observation interval, the CPU model number, the percentage of time that the CPU was idle (WAIT TIME PERCENTAGE), the average number of TSO users (TSO AVG ASIDS), and the CPU queue length distribution (not shown in the table) broken down in various ways.

Note that RMF always expresses percentages out of 100 (e.g., 32.09 rather than .3209 in Table 17.1a); we would calculate CPU busy time as:

$$B_{CPU} = \text{INTERVAL} \times (1 - (\text{WAIT\_TIME\_PERCENTAGE} / 100))$$

17.3. *Overview of RMF Reports* 379

CPU Activity	
INTERVAL	60 min.
CPU MODEL	0470
WAIT TIME PERCENTAGE	32.09%
TSO AVG ASIDS	78.5

Table 17.1a − RMF CPU Activity Report

The *Channel Activity Report*, shown in abbreviated form in Table 17.1b, has two parts. The first reports on each physical channel, indicating its type, volume of activity (CHANNEL ACTIVITY COUNT, the total number of operations during the observation interval), and utilization, among other quantities. The second reports on each logical channel, indicating the physical channels to which it corresponds, its activity rate (measured in operations per second), and its queue length distribution (only the average queue length is shown in the table). There also is information on the percentage of requests that are delayed by congestion in the I/O subsystem, and the components responsible for such delays.

The *Direct Access Device Activity Report*, shown in abbreviated form in Table 17.1c, describes individual devices. It provides information on logical channel attachment, physical I/O count, utilization, queue length distribution (only the average queue length is shown in the table), and average service time (not shown in the table). There also is information (not shown) concerning the causes of request delays and the proportions of time that devices are used in various ways. The *Tape Device Activity Report* (not shown) provides similar information for tape devices. Note that, in contrast to the convention used throughout this book, the queue length information reported in the Device Activity Reports does *not* include the customer in service.

The *Workload Activity Report*, shown in abbreviated form in Table 17.1d, differs from the reports described so far in that it presents resource usage information broken down by performance group and performance period. In the table, group 0 (with one period) is overhead tasks, group 1 (with two periods) is batch work, and group 2 (with three periods) is TSO. Three parts of this report are shown in the table. The first part indicates the installation's service definition coefficients. As described in Section 17.2, resource consumption is reported by RMF in service units that are determined by these coefficients. The second part indicates resource consumption. The service units acquired during the observation interval by each performance group (GRP NUM) and performance period (GRP PER) are given for I/O (IOC), main storage occupancy (MSO), and CPU (CPU), as well as in total (TOT) and on a per-second basis (PER SEC). (RMF sometimes reports a fourth form of service

Physical Channel Activity

CHANNEL NUMBER	CHANNEL TYPE	CHANNEL ACTIVITY COUNT	PERCENT CHANNEL BUSY
0	BYTE MPX	392	0.00
1	BLOCK MPX	130696	29.46
2	BLOCK MPX	70544	12.41
3	BLOCK MPX	77220	15.82
4	SELECTOR	64	0.00
5	BLOCK MPX	65876	12.58
6	BLOCK MPX	904	12.27
7	BLOCK MPX	130120	27.76
8	BLOCK MPX	87212	15.38
9	BLOCK MPX	70648	15.82
A	SELECTOR	23296	5.25
B	BLOCK MPX	98156	1.83

Logical Channel Activity

LOG CHN	PHYS CHN	REQ PER SEC	AVG QUEUE LNGTH
1	1,7	71.1	0.56
4	2,8	42.6	1.01
6	3,9	40.2	0.22
10	5	8.4	0.19
11	5,9	5.7	0.17
12	6	0.3	0.00
17	B	5.9	0.13

(Other Logical Channels had insignificant usage.)

Table 17.1b — RMF Channel Activity Report

acquisition for each group and period: SRB, which roughly corresponds to directly attributable CPU overhead activity.) The third part indicates, again for each group and each period, the resource consumption rate of an average transaction (AVG ABS (absorption) RATE, measured in service units per second), the number of swaps, the average number of ready requests (AVG TRANSACTS, which includes swapped-out requests so is not a good estimator of multiprogramming level), the completion count (ENDED TRANSACTS), and the response time (AVG TRANS TIME).

17.3. Overview of RMF Reports

Direct Access Device Activity

DEV ADR	LOG CHN	DEVICE ACTIVITY COUNT	% DEV BUSY	AVG Q LNGTH
100	1	32796	39.1	.01
101	1	6072	9.1	.00
108	1	43912	35.5	.21
109	1	33008	42.1	.01
117	1	40136	28.6	.13
11B	1	40424	28.6	.15
1D2	1	35876	30.6	.02
298	4	6144	7.2	.00
2C1	4	18644	17.9	.13
2C8	4	20476	18.8	.24
2C9	4	16900	13.9	.07
2CB	4	18068	15.5	.08
2D3	4	10372	6.8	.02
2D8	4	10288	5.7	.02
2DB	4	5376	5.4	.01
2DC	4	23632	11.7	.03
330	6	12492	8.4	.00
332	6	8760	5.6	.06
335	6	23392	15.5	.03
339	6	34736	41.7	.01
33A	6	20288	12.7	.01
33B	6	24900	20.5	.07
33D	6	18128	14.6	.04

(Another 133 disk volumes had device busy percentages of less than 5%. The sum of their % DEV BUSY was 50.2, indicating a total of 1807 seconds of busy time.)

Table 17.1c − RMF Direct Access Device Activity Report

The *Paging Activity Report* provides a complete breakdown of paging activity from a system perspective. (Unfortunately, paging activity is *not* broken down by performance group.) A second part of this report summarizes swapping activity. It gives both logical and physical swap counts broken down by type. It also gives the total number of swaps, the swapping rates, and the average number of pages involved in each page in and page out. The *Page / Swap Dataset Activity Report* indicates the devices

Workload Activity

SERV DEF COEF:

    IOC = 5
    CPU = 10
    MSO = 3

		<--	INTERVAL	SERVICE	-->	
GRP NUM	GRP PER	IOC	CPU	MSO	TOT	PER SEC
000	1	62	147	42	251	70
001	1	681	966	1130	2777	771
001	2	25	109	142	276	77
001	ALL	706	1075	1272	3053	848
002	1	654	1749	416	2819	783
002	2	359	622	196	1177	327
002	3	644	1178	395	2217	616
002	ALL	1657	3549	1007	6213	1726

(IOC, CPU, MSO, and TOT service units are expressed in thousands.)

GRP NUM	GRP PER	AVG ABS RATE	NUM OF SWAPS	AVG TRANS-ACTS	ENDED TRANS-ACTS	AVG TRANS TIME (SECS)
000	1	34	4	2.00	4	.21
001	1	183	142	4.32	52	15.94
001	2	116	56	.65	12	482.18
001	ALL	174	248	4.97	64	103.36
002	1	210	24160	3.84	24292	0.46
002	2	242	836	1.34	660	8.72
002	3	353	496	1.76	356	27.54
002	ALL	253	25492	6.96	25308	1.05

Table 17.1d — RMF Workload Activity Report

used for various types of paging and swapping. Because our example model is not intended to be used for situations in which paging activity is expected to change substantially, our parameterization will not involve information from these last two reports, and they are not shown.

17.4. Customer Description

In Chapter 12 we indicated that the identification of customer classes in a model is based primarily on the workload components to be distinguished with respect to their performance.

In a queueing network model of an MVS system, a customer class sometimes represents a single performance group, sometimes a single performance period of a performance group, and sometimes an aggregation of several performance groups. For example, a TSO performance group might correspond directly to a class, or each TSO performance period might be represented as a separate class (for example, in order to be able to report response times for trivial TSO transactions). Performance groups corresponding to production batch and test batch might be aggregated into a single class.

Certain performance groups correspond to various started tasks. Some started tasks are significant workload components (e.g., CICS, IMS Control Region, IMS Message Processing Regions), and should be represented as customer classes in the model. Others can be treated as system overhead (e.g., JES, RMF, and TCAM); the resource usage of these "overhead" performance groups must be distributed carefully across the customer classes of the model.

When a customer class corresponds to two or more performance groups, the statistics in the Workload Activity Report must be aggregated. For most quantities, aggregation involves addition over the relevant performance groups. However, for those quantities that refer to a single transaction rather than to an entire performance group (AVG ABS RATE and AVG TRANS TIME are the two examples in Table 17.1d), an average weighted by throughput must be calculated. For example:

$$AVG\_TRANS\_TIME_{AGG} = \frac{\sum_{g \in G} \left[ENDED\_TRANSACTS_g \times AVG\_TRANS\_TIME_g \right]}{\sum_{g \in G} ENDED\_TRANSACTS_g}$$

where G is the set of performance groups that correspond to the customer class. In the rest of this chapter we will assume that such aggregation of Workload Activity Report data items has been carried out whenever necessary.

Once classes are identified and associated with performance groups, the next task is to specify the type (transaction, batch, or terminal) and workload intensity of each class. General guidelines for choosing the type of a class were given in Chapter 12. RMF treats TSO specially (in providing the average number of active terminals, for example). Consequently,

TSO can be represented as a terminal class. Without information from other sources (e.g., specialized subsystem monitors), all other classes must be treated as either batch (if the number of active tasks is known) or transaction (possibly with a memory constraint to limit the number of customers active simultaneously).

Values for workload intensity parameters can be calculated from RMF data items as suggested in Table 17.2, using data items from the CPU Activity and Workload Activity Reports. For transaction classes, the calculation is based on the assumption that throughput is equal to the arrival rate. The formula shown for batch type classes estimates the average number of ready requests residing in main memory. (As noted in Section 17.3, the RMF data item AVG TRANSACTS includes non-resident ready requests.) This formula is known to be less reliable than the other formulae presented in this chapter, often yielding a result that significantly over-estimates the actual number of "threads of control" that are concurrently active. This is an instance where "calibration" may be appropriate. One approach used frequently in practice is to represent the workload initially as a transaction class, calculating its arrival rate as ENDED_TRANSACTS / INTERVAL. After the model has been evaluated once, the workload can be converted to a batch class whose population is determined from the outputs of this initial evaluation.

Transaction:

$$\lambda = \text{ENDED\_TRANSACTS} / \text{INTERVAL}$$

Batch:

$$N = \frac{\text{PER\_SEC\_INTERVAL\_SERVICE}}{\text{AVG\_ABS\_RATE}}$$

$$= \frac{\text{total rate of service delivery}}{\text{rate of service delivery per active job}}$$

Terminal (TSO only):

$$N = \text{TSO\_AVG\_ASIDS} = \text{number of active terminals}$$

$$Z = \frac{\text{TSO\_AVG\_ASIDS}}{\text{ENDED\_TRANSACTS} / \text{INTERVAL}} - \text{AVG\_TRANS\_TIME}$$

$$= \frac{\text{number of active terminals}}{\text{throughput}} - \text{average response time}$$

Table 17.2 — Workload Intensity Parameter Value Calculation

17.5. Center Description

> **EXAMPLE:** Workload Component Identification
>
> (from the IPS and the Workload Activity Report of Table 17.1)
>
> We choose to treat performance group 0 as overhead. It will not be represented as a class; its resource consumption will be apportioned among the user classes.
>
> | Performance Group | Workload Component | Customer Class | Class Type |
> |---|---|---|---|
> | 000 | overhead | (none) | – |
> | 001 | batch | BATCH | batch |
> | 002 | interactive | TSO | terminal |

> **EXAMPLE:** Workload Intensity Calculation
>
> (from the IPS and Tables 17.1 and 17.2)
>
> $N_{BATCH} = 848 / 174 = 4.9$
>
> $N_{TSO} = 78.5$
>
> $Z_{TSO} = \dfrac{78.5}{25308 / 3600} - 1.05 = 10.1$ secs.
>
> TSO is assigned to a domain with a capacity of 8.

17.5. Center Description

The structure of the model is determined primarily from knowledge of the configuration.

The Device and Channel Activity Reports may reveal some system components that are so lightly utilized that they need not be included in the model. For example, of the hundreds of disk drives in a large installation, it typically is the case that less than 25% of them will have utilizations of 5% or more in any observation interval. (In our example system, 23 disks had utilizations of 5% or more, while 133 disks had utilizations of less than 5%.) Obviously, any disk with a utilization of zero can be omitted from the model. In addition, though, a single delay center can be used to represent the aggregate effect of all disks with utilizations of less than 5%. The service demand of a class at this delay center is calculated as the sum of the busy times attributed to the class at all these disks, divided by the total number of request completions for the class. Such an aggregation reduces the amount of work involved in constructing a model "by hand". Because queueing delays are insignificant at resources with utilizations of less than 5%, little error is introduced. When a program is used to obtain parameter values from measurement data, as described in Section 16.2.4, it is easiest to represent all devices in the model, no matter how light their use.

As recommended in Chapter 10, disk channels are represented implicitly in our example by "inflating" disk service demands to reflect path contention. Other channels are represented explicitly, while the devices to which they connect are not. The reason that other devices (e.g., tapes) need not be treated in as much detail as disks is that they do not have the same capability of concurrent activity independent of the channel. Unit record devices and their channels often are omitted from models since spooling allows their activity to be overlapped fully with other processing. Some other lightly used channels may be either omitted or represented as part of a single delay center for similar reasons.

17.6. Service Demands

Along with the workload intensity parameters, the most critical values that must be derived from measurement data are the service demands of the customer classes at each center. The most difficult step in doing this is allocating CPU and I/O busy times to customer classes. This can be done only roughly using RMF data alone. To apportion busy times more accurately, supplementary information from other sources, e.g., the System Management Facility (SMF) or the Generalized Trace Facility (GTF), is needed. For this discussion, however, we assume that only RMF data is available.

In Chapter 12 we presented several methods for allocating unattributed CPU activity (Section 12.5.1) and I/O activity (Section 12.5.2) to customer classes. The basic quantities required by these methods are available from RMF, as shown in Tables 17.3 (CPU) and 17.4 (I/O). Note that if RMF is reporting SRB INTERVAL SERVICE (see Section 17.3), this should be added to CPU INTERVAL SERVICE in calculating attributed CPU activity. Note also that a breakdown of most physical I/Os by device and by class can be obtained by the analysis of certain types of SMF records, although we restrict ourselves to RMF here.

After attributing CPU and I/O activity to customer classes, the service demands for each class at each device are calculated by dividing the busy time attributed to a class at a center by the number of completions observed for the class (ENDED TRANSACTS).

17.6. Service Demands

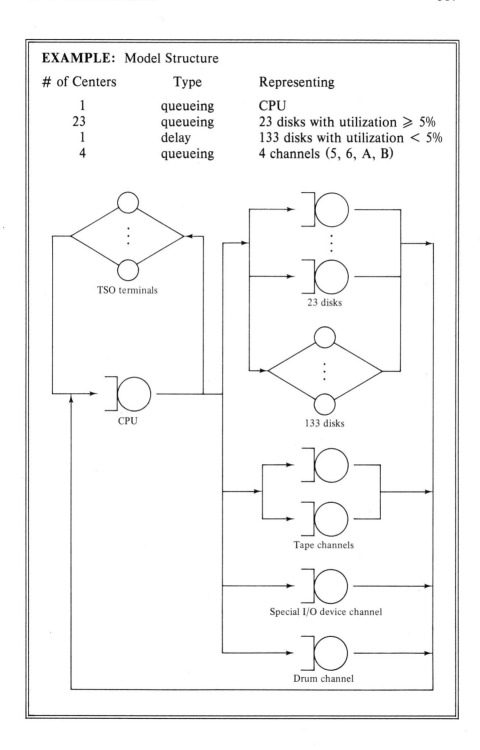

EXAMPLE: Model Structure

# of Centers	Type	Representing
1	queueing	CPU
23	queueing	23 disks with utilization $\geq 5\%$
1	delay	133 disks with utilization $< 5\%$
4	queueing	4 channels (5, 6, A, B)

TSO terminals

CPU

23 disks

133 disks

Tape channels

Special I/O device channel

Drum channel

Appendices: Constructing a Model from RMF Data

Measured CPU Busy Time:

$$B_{CPU} = \text{INTERVAL} \times (1 - (\text{WAIT\_TIME\_PERCENTAGE} / 100))$$

Accounted CPU Busy Time by Class:

$$A_{c,CPU} = \frac{\text{CPU\_INTERVAL\_SERVICE}_c}{\text{CPU\_SERV\_DEF\_COEF} \times \text{CPU\_speed\_factor}}$$

where CPU_speed_factor is determined by the model.

Swapping Overhead Factor by Class:

$$SW_c = \text{NUM\_OF\_SWAPS}_c$$

Table 17.3 — RMF Items for CPU Activity Allocation

Disk Device Busy Time:

$$B_k = \text{INTERVAL} \times (\%\_\text{DEVICE\_BUSY}_k / 100)$$

Physical I/Os by Device:

$$P_k = \text{DEVICE\_ACTIVITY\_COUNT}_k$$

Logical I/O's by Class:

$$L_c = \frac{\text{IOC\_INTERVAL\_SERVICE}_c}{\text{IOC\_SERV\_DEF\_COEF}}$$

Table 17.4 — RMF Items for I/O Activity Allocation

The two approaches to CPU activity allocation used below to treat the example system represent two extremes. While the TSO overhead factor certainly is higher than that for BATCH, it certainly is not as high as is indicated by the ratio of the reported swapping activity for the two classes.

17.7. Performance Measures

In order to validate a baseline model we need to determine from measurement data not only the input parameter values, but also the performance measure values. Table 17.5 indicates how various performance measure values can be obtained from RMF reports.

17.7. Performance Measures

EXAMPLE: CPU Activity Allocation to Classes

First approach of Section 12.5.1; assumes that overhead is proportional to accounted usage.

$B_{CPU} = 3600 \times (1 - .32) = 2448$ secs.

$A_{BATCH,CPU} = 1075000 \times factor \quad A_{TSO,CPU} = 3549000 \times factor$

where *factor* need not be calculated since this approach uses only relative, not absolute, accounted CPU time

$C_{BATCH} = 64 \quad\quad C_{TSO} = 25308$

$D_{BATCH,CPU} = 2448 \times \dfrac{1075000}{1075000+3549000} \times \dfrac{1}{64} = 8.89$ secs.

$D_{TSO,CPU} = 2448 \times \dfrac{3549000}{1075000+3549000} \times \dfrac{1}{25308} = 0.074$

EXAMPLE: CPU Activity Allocation to Classes

Second approach of Section 12.5.1; assumes that swapping is the primary source of overhead.

CPU_speed_factor of Amdahl 470 V/8 = 420

CPU_SERV_DEF_COEF = 10

$A_{BATCH,CPU} = \dfrac{1075000}{420 \times 10} = 256$ secs.

$A_{TSO,CPU} = \dfrac{3549000}{420 \times 10} = 824$ secs.

$SW_{BATCH} = 248 \quad\quad SW_{TSO} = 25492$

$f_{BATCH} = 1 + \dfrac{\dfrac{248}{248+25492} \times \left[2448 - (256+824)\right]}{256} = 1.05$

$f_{TSO} = 1 + \dfrac{\dfrac{25492}{248+25492} \times \left[2448 - (256+824)\right]}{824} = 2.64$

$D_{BATCH,CPU} = \dfrac{1.05 \times 256}{64} = 4.2$ secs.

$D_{TSO,CPU} = \dfrac{2.64 \times 824}{25308} = 0.086$ secs.

> **EXAMPLE:** Disk Activity Allocation to Classes
>
> In the absence of information to the contrary, assume that each class uses each disk in proportion to its overall I/O activity.
>
> Note that the I/O activity of performance group 0 is allocated implicitly to the two classes in proportion to their accounted usage.
>
> $$\text{BATCH\_share\_of\_IOC\_INT\_SERV} = \frac{706}{706 + 1657} = .298$$
>
> $$\text{TSO\_share\_of\_IOC\_INT\_SERV} = \frac{1657}{706 + 1657} = .702$$
>
> Calculations for disks 100 and 101 (the other 21 individually represented disks are treated similarly):
>
> $$D_{BATCH, D100} = \frac{.298 \times .391 \times 3600}{64} = 6.576$$
>
> $$D_{TSO, D100} = \frac{.702 \times .391 \times 3600}{25308} = 0.039$$
>
> $$D_{BATCH, D101} = \frac{.298 \times .091 \times 3600}{64} = 1.531$$
>
> $$D_{TSO, D101} = \frac{.702 \times .091 \times 3600}{25308} = 0.009$$
>
> Calculations for the aggregate disk center (the total busy time of the 133 other disks is 1807 seconds):
>
> $$D_{BATCH, DAGG} = \frac{.298 \times 1807}{64} = 8.442$$
>
> $$D_{TSO, DAGG} = \frac{.702 \times 1807}{25308} = 0.050$$

17.8. Summary

In this chapter we have illustrated the application of the general techniques presented in Chapter 12 to a specific case. Our example treated data obtained from RMF reports concerning an MVS system. We saw that many queueing network model inputs and outputs are provided directly by RMF, while others must be calculated indirectly, with varying degrees of reliability. Similar techniques are applicable and similar difficulties are encountered in dealing with other computer systems.

17.8. Summary

EXAMPLE: Non-disk Channel Service Demands

Tape Channels (assume BATCH is responsible for all tape usage):

$$D_{BATCH,CH5} = \frac{.1258 \times 3600}{64} = 7.076 \qquad D_{TSO,CH5} = 0$$

$$D_{BATCH,CHA} = \frac{.0525 \times 3600}{64} = 2.953 \qquad D_{TSO,CHA} = 0$$

Electronic Drum (assume that it is used for swapping and hence should be attributed to the TSO class):

$$D_{BATCH,CHB} = 0 \qquad D_{TSO,CHB} = \frac{.0183 \times 3600}{25308} = .0026$$

Special I/O Device (assume that it is used only by BATCH):

$$D_{BATCH,CH6} = \frac{.1227 \times 3600}{64} = 6.902 \qquad D_{TSO,CH6} = 0$$

Throughput by Class:

$$X_c = \frac{\text{ENDED\_TRANSACTS}_c}{\text{INTERVAL}}$$

Response Time by Class:

$$R_c = \text{AVG\_TRANS\_TIME}_c$$

CPU Utilization:

$$U_{CPU} = 1 - (\text{WAIT\_TIME\_PERCENTAGE} / 100)$$

Device Utilization:

$$U_k = \text{\%\_DEV\_BUSY}_k / 100$$

Device Queue Length:

$$Q_k = \text{AVG\_Q\_LNGTH}_k + (\text{\%\_DEV\_BUSY}_k / 100)$$

Table 17.5 − RMF Items Giving Performance Measures

> **EXAMPLE:** Performance Measures
>
> $X_{BATCH} = 64/3600 = .018$ $\quad X_{TSO} = 25308/3600 = 7.03$
> $R_{BATCH} = 103.4$ $\quad\quad\quad\quad\quad R_{TSO} = 1.05$
> $U_{CPU} = 68\%$
> $U_{D100} = 39.1\%$ $\quad\quad\quad\quad\quad Q_{D100} = .40$
> $U_{D101} = 9.1\%$ $\quad\quad\quad\quad\quad\; Q_{D101} = .09$

17.9. References

The first published discussion of determining queueing network model parameters from measurement data was given by Rose [1978]. Since then, Levy [1980] and Irwin [1983] have presented more detailed treatments of RMF specifically. Lindsay carried out some experiments to investigate the accuracy of some of the data items obtained from RMF [Lindsay 1980].

A special issue of the the *IBM Systems Journal* concerning capacity planning contains several papers relating to the use of measurement data from MVS systems. Bronner surveys capacity planning techniques and describes the relationship among various sources of measurement data for MVS systems [Bronner 1980]. Schardt describes some techniques for using measurement data to tune MVS systems [Schardt 1980]. Schiller describes the SCAPE model, an alternative to queueing network models which is tailored to deal with many special aspects of MVS environments [Schiller 1980].

Information on MVS and its System Resources Manager (SRM) can be found in papers by Beretvas [1978] and by Lynch and Page [1974], respectively. Chiu and Chow [1978] describe a hybrid model of the internals of MVS (see Section 15.6), while Buzen describes a queueing network model based on the MVS external interface [Buzen 1978].

[Beretvas 1978]
Thomas Beretvas. Performance Tuning in OS/VS2 MVS. *IBM Systems Journal 17*,3 (1978), 290-313.

[Bronner 1980]
LeeRoy Bronner. Overview of the Capacity Planning Process for Production Data Processing. *IBM Systems Journal 19*,1 (1980), 4-27.

[Buzen 1978]
Jeffrey P. Buzen. A Queueing Network Model of MVS. *Computing Surveys 10*,3 (September 1978), 319-331.

[Chiu and Chow 1978]
Willy W. Chiu and We-Min Chow. A Performance Model of MVS. *IBM Systems Journal 17*,4 (1978), 444-462.

[Irwin 1983]
Robert T. Irwin. RMF Equations: Obtaining Job Class Results from RMF. *Journal of Capacity Management 1*,3 (1983), 230-261.

[Levy 1980]
Allan I. Levy. Introduction to Practical Operational Analysis: An MVS Perspective. *Proc. CMG XI International Conference* (1980), 208-214.

[Lindsay 1980]
David S. Lindsay. RMF I/O Time Validation. *Proc. CMG XI International Conference* (1980), 112-119.

[Lynch and Page 1974]
H.W. Lynch and J.B. Page. The OS/VS2 Release 2 System Resources Manager. *IBM Systems Journal 13*,4 (1974), 274-291.

[Rose 1978]
Clifford A. Rose. A Measurement Procedure for Queueing Network Models of Computer Systems. *Computing Surveys 10*,3 (September 1978), 263-280.

[Schardt 1980]
R.M. Schardt. An MVS Tuning Approach. *IBM Systems Journal 19*,1 (1980), 102-119.

[Schiller 1980]
D.C. Schiller. System Capacity and Performance Evaluation. *IBM Systems Journal 19*,1 (1980), 46-67.

17.10. Exercises

1. Use the multiple class mean value analysis implementation in Chapter 19 to evaluate the model for which parameters were derived in this chapter. (This will require extending the Chapter 19 implementation to handle terminal classes and delay centers. Note that you will not be able to represent the non-integer customer populations or the TSO domain capacity using this implementation.) Compare the results to the RMF performance measure values shown in Section 17.7.

394 Appendices: Constructing a Model from RMF Data

2. "Calibrate" the value of N_{BATCH} as suggested in Section 17.4. (This will require extending the Chapter 19 implementation to handle transaction classes.) How are the various performance measure values affected?

3. Use Algorithm 9.2 to represent the TSO domain capacity. (This will require extending the single class mean value analysis implementation in Chapter 18 to handle load dependent service centers, as described in Chapter 20.) Note that the iteration is simplified by the fact that the batch population is fixed. How are the various performance measure values affected?

4. Based on information contained in Section 16.3, describe how you would specify this model using the queueing network modelling software package MAP.

Chapter 18

An Implementation of Single Class, Exact MVA

18.1. Introduction

In this appendix we provide a Fortran implementation of the most basic queueing network evaluation technique: the use of mean value analysis to obtain the exact solution of a separable queueing network model consisting entirely of queueing centers and containing a single class of batch type.

The algorithm on which this program is based is described in Chapter 6. The interested reader will find it educational to extend the program to accommodate two other characteristics described in that chapter: delay centers, and choice of batch, terminal, or transaction class types. The extension to multiple classes is given in Chapter 19. The extension to load dependent service centers is discussed in Chapter 20.

As noted in the overview of Part VI, our intention is that this program be used for educational experimentation with simple models. Its value as a capacity planning tool in no way approaches that of commercial queueing network modelling software. For a better idea of the interactions possible with that type of software, consult Chapter 16.

18.2. The Program

The program appears on the next two pages. Two statement labels (2001 and 2003) are included for reference in Chapter 20 and are not used in the program.

Note that some Fortran implementations impose restrictions on formatted I/O. It is best to include an explicit decimal point in real-valued input (but not integer-valued input) when using the program.

Appendices: An Implementation of Single Class, Exact MVA

```
              program single
c
c A maximum of 25 centers are allowed.
c
              integer Ncusts,Ncents,n,center
              real demand(25)
              real qlen(25)
              real rtime(25)
              real tput,sysr
c
              write (6,5)
5             format (27h Input number of customers:)
              read (5,10) Ncusts
10            format (i4)
              write (6,15)
15            format (25h Input number of centers:)
              read (5,10) Ncents
              write (6,20)
20            format (25h Input service demand for)
              do 25 center=1,Ncents
                  write (6,30) center
30                format (10h  Center ,i2,1h:)
                  read (5,35) demand(center)
35                format (f8.4)
25            continue
c
c Now that the network is described, we perform the evaluation.
c Begin by initializing to the trivial solution for zero customers.
c
              do 40 center=1,Ncents
                  qlen(center) = 0.0
40            continue
c
c The algorithm solves successively for each population.
c
              do 45 n=1,Ncusts
c
c First, compute the residence time at each center.
c
                  sysr = 0.0
                  do 50 center=1,Ncents
2001                  rtime(center) = demand(center)*(1.0+qlen(center))
                      sysr = sysr + rtime(center)
50                continue
```

```
      c
      c Next, use Little's law to compute system throughput.
      c
                  tput = n / sysr
      c
      c Finally, use Little's law to compute center queue lengths.
      c
                  do 55 center=1,Ncents
2003                  qlen(center) = rtime(center) * tput
55                continue
      c
45            continue
      c
      c Print results.
      c
                  write (6,60) tput
60                format (20h System throughput: ,f8.4)
                  write (6,65) Ncusts/tput
65                format (23h System response time: ,f8.4)
      c
                  write (6,70)
70                format (22h Device utilizations: )
                  do 75 center=1,Ncents
                      write (6,80) center,tput*demand(center)
80                format (i5,2h: ,f5.3)
75                continue
      c
                  write (6,85)
85                format (23h Device queue lengths: )
                  do 90 center=1,Ncents
                      write (6,95) center,qlen(center)
95                format (i5,2h: ,f8.4)
90                continue
      c
                  end
```

Chapter 19

An Implementation of Multiple Class, Exact MVA

19.1. Introduction

In this appendix we provide a Fortran implementation of the exact mean value analysis algorithm for separable queueing network models consisting entirely of queueing centers and containing three classes of batch type.

The algorithm on which this program is based is described in Chapter 7. The interested reader will find it educational to extend the program in three ways: to allow more than three classes, to allow delay centers, and to allow a choice of batch, terminal, or transaction types independently for each class. The extension to load dependent service centers is discussed in Chapter 20.

As with the program in Chapter 18, our intention is that this program be used for educational experimentation with simple models. Its value as a capacity planning tool in no way approaches that of commercial queueing network modelling software. For a better idea of the interactions possible with that type of software, consult Chapter 16.

19.2. The Program

The program appears on the next four pages. Two statement labels (2001 and 2003) are included for reference in Chapter 20 and are not used in the program.

Note that some Fortran implementations impose restrictions on formatted I/O. It is best to include an explicit decimal point in real-valued input (but not integer-valued input) when using the program.

19.2. *The Program* 399

```
              program multpl
c
c A maximum of 3 classes and 25 centers are allowed.
c Classes 2 and 3 are limited to a maximum of 10 customers each.
c Class 1 has no limit on its population.
c
              integer Ncents,center,class
              integer n1,n2,n3
c
c Ncusts(c) is the population of class c.
c N is a temporary population vector required by MVA.
c
              integer Ncusts(3),N(3)
c
c demand(c,k) is the service demand of class c at center k.
c
              real demand(3,25)
c
c qlen(class 2 pop.,class 3 pop.,k) is a 3-dimensional array
c containing the queue length at each center k for each possible
c combination of class 2 and class 3 network populations. The
c population of class 1, the outermost class in the iteration,
c need not appear as an index. Population indices run from 1 to
c 11 to represent populations from 0 to 10 because some Fortran
c implementations restrict the base of array dimensions to be 1.
c
              real qlen(11,11,25)
c
c rtime(c,k) is the residence time of class c at center k.
c
              real rtime(3,25)
c
c tput(c) and sysr(c) are the throughput and response time of class c.
c
              real tput(3),sysr(3)
c
              write (6,5)
5             format (30h Input number of customers for)
              do 10 class = 1,3
                write (6,15) class
15              format (9h   Class ,i1,1h:)
                read (5,20) Ncusts(class)
20              format (i4)
10            continue
```

```
              write (6,25)
25            format (25h Input number of centers:)
              read (5,20) Ncents
              write (6,30)
30            format (25h Input service demand for)
              do 35 center=1,Ncents
                  write (6,40) center
40                format (10h   Center ,i2,1h:)
                  do 45 class=1,3
                      if (Ncusts(class) .eq. 0) goto 45
                      write (6,50) class
50                    format (11h    Class ,i1,1h:)
                      read (5,55) demand(class,center)
55                    format (f8.4)
45                continue
35            continue
c
c Now that the network is described, we perform the evaluation.
c The algorithm iterates through all possible population vectors.
c
              do 60 n1=0,Ncusts(1)
                  do 65 n2=0,Ncusts(2)
                      do 70 n3=0,Ncusts(3)
                          if (n1+n2+n3 .eq. 0) goto 70
                          N(1) = n1
                          N(2) = n2
                          N(3) = n3
c
c First, compute the residence time at each center.
c
                          do 80 class = 1,3
                              sysr(class) = 0.0
                              if (N(class) .eq. 0) goto 80
c
                              N(class) = N(class) - 1
                              do 85 center=1,Ncents
2001                              rtime(class,center) =
     x                                demand(class,center) *
     x                                (1.0+qlen(N(2)+1,N(3)+1,center))
                                  sysr(class) = sysr(class) +
     x                                rtime(class,center)
85                            continue
                              N(class) = N(class) + 1
```

```
c
c Next, use Little's law to compute system throughput.
c
                         tput(class) = N(class) / sysr(class)
80                       continue
c
c Finally, use Little's law to compute center queue lengths.
c
                  do 90 center=1,Ncents
                      qlen(n2+1,n3+1,center) = 0.0
                      do 95 class=1,3
                          if (N(class) .eq. 0) goto 95
2003                      qlen(n2+1,n3+1,center) =
     x                        qlen(n2+1,n3+1,center) +
     x                        rtime(class,center) * tput(class)
95                        continue
90                    continue
c
70                continue
65            continue
60        continue
c
c Print results.
c
          do 100 class=1,3
              if (Ncusts(class) .eq. 0) goto 100
c
              write (6,105) class
105           format (7h Class ,i1,1h:)
              write (6,110) tput(class)
110           format (22h   System throughput: ,f8.4)
              write (6,115) Ncusts(class) / tput(class)
115           format (25h   System response time: ,f8.4)
c
              write (6,120)
120           format (24h   Device utilizations: )
              do 125 center=1,Ncents
                  write (6,130) center,tput(class)*demand(class,center)
130               format (i7,2h: ,f5.3)
125           continue
```

Appendices: An Implementation of Multiple Class, Exact MVA

```
        c
                        write (6,140)
    140                 format (25h  Device queue lengths: )
                        do 145 center=1,Ncents
                            write (6,150) center, tput(class)*rtime(class,center)
    150                 format (i7,2h: ,f8.4)
    145                 continue
        c
    100                 continue
        c
                    end
```

Chapter 20

Load Dependent Service Centers

20.1. Introduction

The mean value analysis (MVA) algorithms developed in Chapters 6 and 7 allow service centers of only the queueing and delay types. As noted in Chapter 8, though, it is possible to extend these algorithms to evaluate models containing *load dependent* service centers — centers at which the service rate (the reciprocal of the service time) varies with the number of customers present. These extensions are the subject of the present appendix.

On occasion, individual components of computer systems are represented most naturally using load dependent centers. An example is a disk device where accesses are served in an order that attempts to minimize head movement. The greater the number of requests queued at such a device, the smaller the time required to satisfy each, on average, since the effectiveness of the scheduling policy increases with queue length.

The most important use of load dependent centers, though, is to implement *flow equivalent service centers* (FESCs). The construction and use of FESCs was detailed in Chapter 8, and numerous applications were noted in Chapters 9 and 11.

In discussing the modifications to MVA necessary to accommodate load dependent centers, we restrict our attention to *closed* queueing networks (batch or terminal workload types) and to the *exact* MVA algorithms (Algorithm 6.2 for the single class case and Algorithm 7.2 for the multiple class case). We begin by recalling the three principal steps of mean value analysis:

1. Compute the residence time at each center for each class, based on the service demand of the class and the average number of customers seen upon arrival to the center by a customer of that class.

2. Compute the throughput of each class as the number of customers of that class divided by the sum of its residence times at all centers (plus the think time, if the class is of terminal type).
3. Compute the queue length of each class at each center as the product of its throughput and its residence time at that center.

The exact MVA algorithms involve the iterative application of these steps at increasing populations, with the results of Step 3 at one iteration used to compute the queue lengths needed in Step 1 of the next iteration.

The load dependent versions of the algorithms involve revisions to Steps 1 and 3 — modified equations that are applied to load dependent centers:

- Consider Step 1, the estimation of the service center residence times. For load independent centers, this quantity is calculated using the (load independent) service demand and the average number of customers seen upon arrival to the center. For load dependent centers, service rates vary with queue length, so the residence time equation used in Step 1 must be augmented by terms reflecting the varying queue lengths and corresponding service rates.

- Consider Step 3, the estimation of the service center queue lengths. For load independent centers, only the average queue length is required by Step 1, so only this quantity is calculated in Step 3. For load dependent centers, the *queue length distribution* — the proportion of time that each possible customer population exists at a center — is required, so must be calculated in Step 3.

Load dependent service rates indicate the rate of customer completions at a center as a function of its current customer population. Because these rates inherently are *per visit*, while the result of the residence time equation is the total time spent at a center (i.e., the time per visit multiplied by the number of visits), service center visit counts appear as multiplicative factors in the load dependent version of the residence time equation. Thus, it appears that load dependent centers are more complicated to parameterize than load independent centers not only because of the need to give many service rates instead of a single service demand, but also because of the need to provide service center visit counts. Fortunately, this latter complication can be avoided: it is possible to rewrite the residence time equation in a way that obviates explicit visit count information. This transformation is shown in the last section of this appendix, where implementation considerations are addressed. We have chosen to include the visit count factors in the initial presentation because intuition is sacrificed in the transformation.

As in Part II of the book, our presentation is organized as a discussion of the single class case, followed by a discussion of the multiple class case. Implementation issues are discussed in a final section.

20.2. Single Class Models

We consider models with K service centers and a single customer class of batch or terminal type. Let $\mu_k(j)$ be the service rate of center k when there are j customers there. Let $p_k(j \mid n)$ be the proportion of time that center k has j customers present when the number of customers in the entire model is n. The following expressions are substituted for Steps 1 and 3 of the load independent MVA algorithm for each load dependent center k. (The load independent equations still are used for all load independent centers in the model.)

1'. Compute the residence time at load dependent center k:

$$R_k(n) = V_k \sum_{j=1}^{n} \frac{j}{\mu_k(j)} p_k(j-1 \mid n-1)$$

where V_k is the number of visits each customer makes to center k. (As noted earlier, this term is required since R_k represents the total time spent at a center, while the μ_k are service rates per visit.)

3'. Compute the queue length distribution for load dependent center k:

$$p_k(j \mid n) = \begin{cases} \dfrac{X(n)}{\mu_k(j)} p_k(j-1 \mid n-1) & j = 1, \ldots, n \\ 1 - \sum_{i=1}^{n} p_k(i \mid n) & j = 0 \end{cases}$$

20.3. Multiple Class Models

We consider closed, multiple class models with K service centers and C customer classes. There are two ways in which service centers in multiple class models can exhibit load dependent behavior:

- The simpler is for the service rates of all classes to vary in an identical manner as functions of the total number of customers at the center. For instance, suppose that the service rate of class A at a particular center with four customers (of any class) present is 1.5 times the service rate of class A at that center with two customers present. Then this simpler form of load dependence would require that the service rate of class B at that center with four customers present be 1.5 times its rate with two customers present.

- The more complex form of load dependence, required for the implementation of FESCs, allows the service rates of the classes to vary independently of one another, and to be functions not of the total number of customers at the center, but of the actual mix of customers there. (In this case the service center is scheduled using the fictitious *composite queueing* discipline, discussed in Chapter 8.)

We begin with the first form of load dependence. The service rate $\mu_{c,k}(j)$ indicates the rate at which class c customers would complete at center k if they were in service alone (i.e., any customers of other classes were queued but not in service) and there were a total of j customers at the center. (Again, these are rates *per visit*.) For this form of load dependence, the modifications to load independent MVA are straightforward extensions of those used in the single class case. Let the population of the model be $\vec{n} \equiv (n_1, n_2, \ldots, n_C)$, so that $n \equiv \sum_{c=1}^{C} n_c$ is the total number of customers in the model. Then for load dependent centers, Steps 1 and 3 are replaced by:

1′. Compute the residence time of class c at load dependent center k:

$$R_{c,k}(\vec{n}) = V_{c,k} \sum_{j=1}^{n} \frac{j}{\mu_{c,k}(j)} p_k(j-1 \mid \overrightarrow{n-1_c})$$

where $V_{c,k}$ is the number of visits made by each class c customer to center k.

3′. Compute the queue length distribution for load dependent center k:

$$p_k(j \mid \vec{n}) = \begin{cases} \sum_{c=1}^{C} \frac{X_c(\vec{n})}{\mu_{c,k}(j)} p_k(j-1 \mid \overrightarrow{n-1_c}) & j = 1, \ldots, n \\ 1 - \sum_{j=1}^{n} p_k(j \mid \vec{n}) & j = 0 \end{cases}$$

Now we consider the second form of load dependence, in which the service rates of each class depend on the number of customers of each class present at the center. (As explained in Section 8.4, only certain such sets of rates are valid. Further details can be found in that section.)

Let $\vec{n}$ be the customer population of the model, and let $\vec{n_k} \equiv (n_{1,k}, n_{2,k}, \ldots, n_{C,k})$ be the customer population at center k, where $n_{c,k}$ is the number of class c customers at center k. The load dependent service rates of class c at center k are denoted $\mu_{c,k}(\vec{n_k})$. As with the simpler form of load dependence, the MVA algorithm for centers of this type involves the substitution of new expressions for Steps 1 and 3 of the load independent algorithm:

1'. Compute the residence time of class c at load dependent center k:

$$R_{c,k}(\vec{n}) = V_{c,k} \sum_{\text{all } \vec{n_k}} \frac{n_{c,k}}{\mu_{c,k}(\vec{n_k})} p_k(\overrightarrow{n_k - 1_{c,k}} \mid \overrightarrow{n - 1_c})$$

3'. For each class c compute its queue length distribution at load dependent center k:

$$p_{c,k}(\vec{n_k} \mid \vec{n}) = \begin{cases} \dfrac{X_c(\vec{n})}{\mu_{c,k}(\vec{n_k})} p_k(\overrightarrow{n_k - 1_{c,k}} \mid \overrightarrow{n - 1_c}) & \vec{n_k} > \vec{0} \\ 1 - \sum_{\text{all } \vec{n_k} > \vec{0}} p_k(\vec{n_k} \mid \vec{n}) & \vec{n_k} = \vec{0} \end{cases}$$

20.4. Program Implementation

Fortran implementations of mean value analysis for closed models with load independent queueing centers are given in Chapters 18 and 19 for the single and multiple class cases, respectively. These programs can be modified to accommodate load dependent centers as follows:

- Alter the model definition section to allow load dependent centers to be identified and to allow load dependent rates to be provided for these centers.
- Alter the model definition section to allow service center visit counts to be provided.

As noted earlier, it is possible to rewrite the residence time equations in a way that obviates explicit visit count information, thus reducing the number of input parameters required. If this were done, the two steps outlined above would be modified. This will be discussed shortly.

- Initialize the queue length distributions at all load dependent centers for the zero population case. The distribution values should be set to one for the empty queue and to zero for all other queue populations.
- Substitute the appropriate Step 1' for the calculation of *rtime* (statement 2001 in the Fortran programs) for each load dependent center.
- Substitute the appropriate Step 3' for the calculation of *qlen* (statement 2003 in the Fortran programs) for each load dependent center.
- The output sections of the programs print queue lengths for each center assuming that *qlen* has been set by statement 2003. This will not be the case for load dependent centers; their average queue lengths will need to be calculated at the conclusion of the iteration, and these values assigned to *qlen*.

For many applications of FESCs, it is most convenient to avoid the specification of visit count information. Two examples of this follow:

- If the visit counts are determined by the structure of the model, they can be written into the residence time equations as constants, rather than being input by the user. For example, the techniques suggested in Section 9.3 for evaluating memory constrained queueing networks replace the central subsystem with a single FESC. It is clear that customers make one visit to this FESC per interaction, and so the visit count must be one. (See the example in Section 9.3.1.)

- Sometimes it is most convenient to define the FESC by specifying the rate at which a single customer would complete all of its service, plus a set of *service rate multipliers* that indicate the speed of the service center with a certain customer population relative to its speed with a single customer. For instance, in modelling a tightly-coupled dual processor (see Section 11.2) it is most natural to describe the processor by giving the service demand of a single customer (say, 10 seconds) and the relative rate at which instructions are executed as a function of the number of customers present (say, 1.0 with one customer and 1.8 with two or more customers). This information can be used by applying the following transformation to the residence time equations 1' (here we show the single class case for ease of notation):

Let the service rate multiplier for center k with j customers in its queue be denoted $\alpha_k(j)$, which is defined by:

$$\alpha_k(j) \equiv \frac{\mu_k(j)}{\mu_k(1)}$$

Then we can rewrite the single class residence time equation 1' as:

$$R_k(n) = \frac{V_k}{\mu_k(1)} \sum_{j=1}^{n} \frac{j}{\alpha_k(j)} p_k(j-1 \mid n-1)$$

Since the reciprocal of the service rate with one customer in the queue is simply the service time per visit (S_k), this leads to:

$$R_k(n) = D_k \sum_{j=1}^{n} \frac{j}{\alpha_k(j)} p_k(j-1 \mid n-1)$$

The required inputs now are the nominal service demand and the set of service rate multipliers.

Index

A, see Arrivals
Abstraction in modelling, 20-22, 217
Accumulated time in system (W), 42
Accuracy of models, 14, 16, see also Separable models, robustness
Active customer (memory constrained system), 185
ADEPT (tool for modelling proposed systems), 329-32
Aggregate (hierarchical modelling), 152-58, see also Flow equivalent service center
Aggregate performance measures for multiple class models, 62-63
Algorithms
  asymptotic bounds, 79
  balanced system bounds, 93
  disk I/O
    non-RPS, 228
    RPS, 232
    multipathing, 239
  hierarchical decomposition, 161
  mean value analysis
    single class exact, 115
    single class approximate, 118
    multiple class exact, 141
    multiple class approximate, 143
    multiple class mixed exact, 146
  memory constraints
    single class, 188
    multiple class, independent, 193-94
    multiple class, shared, 195-96
  open models
    single class, 111
    multiple class, 137
  priority scheduling, 259
  swapping
    dedicated device, 199
    shared device, 200
    variable overhead, 307
Amdahl (case studies) see IBM
  IMS capacity, 106-9
  using modelling software, 360-69
  using RMF, 376-94
Appropriateness of queueing networks, 14

Approximate solution techniques
  for separable models, see Mean value analysis, approximate
  for non-separable models, see Disk I/O; Memory; Processors
Approximation transformations (modelling software), 356
Arrival instant queue length in separable models, 112, 114, 139-40
Arrival instant theorem for separable models, 114, 140
  and balanced system bounds, 126
Arrival rate (λ), 41
Arrivals (A), 40
Assumptions, testing of, 33-36, see also Validation
Assumptions of separable models, see Separable models, assumptions
Asymptotic bounds, 72-87, see also Bounds on performance
  Algorithm 5.1, 79

B, see Busy time
Balanced system bounds, 86-93, see also Bounds on performance
  Algorithm 5.2, 93
Balanced systems, utilizations in, 90
Baseline model, see Existing systems
Basic observable quantities, 40-42
Batch workload, 58, see also Closed models
Benchmarking, 3, 30-31
Block size (I/O), 290-91
Books on performance, 18-19
Bottleneck, 72-73
  effect of removing (example using asymptotic bounds), 82-84
  secondary, 82-83
Bounds on performance (Chapter 5), 70-97
  advantages, 70-71, 94
  asymptotic bounds, 72-87
    Algorithm 5.1, 79
    examples and case studies, 77-87

409

Bounds on performance (*cont'd.*)
  balanced system bounds, 86-93
    Algorithm 5.2, 93
    and arrival instant theorem, 126
  optimistic and pessimistic bounds, 71
Busy time (B), 41

C, *see* Completions; Customer classes
Cached I/O devices, 244-45
CAD/CAM system (performance projection of proposed system), 329-32
Capacity planning, *see* Evolving systems
Capture ratio (measurement), 285
Carrier sense multiple access with collision detection (CSMA-CD), 340
Case studies and examples
  Amdahl
    IMS capacity, 106-9
    using modelling software, 360-69
    using RMF, 376-92
  asymptotic bounds, 78-87
  bottleneck removal (asymptotic bounds), 82-84
  CAD/CAM system (proposed), 329-32
  capacity planning
    Amdahl 470, IMS, 106-9
    PDP-10, 132-33
  communication network (proposed), 324-27
  concurrency control (database), 343-47
  CPU replacement, 27-29, 33-36
  CTSS, 102-3
  Cyber 173, 133-34
  database concurrency control, 343-47
  disk I/O
    non-RPS, 227-29
    RPS, 232-33
  disk load balancing
    asymptotic bounds, 85-87
    Univac 1100, 312-13
  Ethernet, 339-41
  evolving systems, 309-15
  FCFS scheduling
    class-dependent service times, 263-64
    highly variable service times, 266-67
  FESC for memory constrained system, 189-90
  forced flow law, 48-50
  global balance, 163-69
  hierarchical modelling, 163-66
  hierarchical workload characterization (instructional computing acquisition, Prime, VAX), 30-34, 51-52

Case studies and examples (*cont'd.*)
  IBM
    early virtual memory system, 206-9
    processing complex (workload balancing), 24-27
    360, 104-6
    3790, 8130, 8140 (insurance company), 35-37, 78-82
  IMS (Amdahl 470, capacity planning), 106-9
  instructional computing acquisition (Prime, VAX), 30-34, 51-52
  insurance company (IBM 3790, 8130, 8140), 35-37, 78-82
  Little's law
    and forced flow law, 48-50
    and memory constrained systems, 49-50, 55-56
    at various levels, 44-46
  loosely-coupled multiprocessor (Cyber 173), 133-34
  memory, 206-17
  memory constrained system
    FESC, 189-90
    Little's law, 49-50
  modelling software, use of (Amdahl 470), 360-69
  modification analysis, 309-15
    asymptotic bounds, 85-87
    IBM processing complex, 24-27
    IBM 360, 104-6
  multiple class models, 129-34
  multiple class solution techniques, 137-38, 142, 144-45, 147
  multiprogramming level variability, 182-84
  MVS
    Amdahl 470 (using RMF), 376-92
    IBM processing complex, 24-27
    SRM, 347-50
  non-RPS disks, 227-29
  paging, 202-5
  PDP-10, 132-33
  Prime (instructional computing acquisition), 30-34
  priority scheduling, 259-61
  proposed systems, 324-27, 329-32
  RMF usage (Amdahl 470), 376-92
  RPS disks, 232-33
  single class models, 102-9
    of heterogeneous workloads, 130-31
  single class solution techniques, 112-13, 116-17, 119
  single service center, 5-8

Index

Case studies and examples (cont'd.)
  SNA flow control, 336-39
  software resources, 342-43
  swapping
    moving from drum to disk (Univac 1100), 314-15
    to a shared device, 200-201
    tightly-coupled multiprocessor (Univac 1100), 310-11
  Univac 1100, 309-15
  upgrade from single to dual processor (Univac 1100), 310-11
  variation in multiprogramming level, 182-84
  VAX
    instructional computing acquisition, 30-34, 51-52
    memory modelling, 209-17
    virtual memory, 202-5
    workload balancing (IBM processing complex), 24-27
Center description, 59
  for models of existing systems, 283
  from RMF, 385-86
Central subsystem, 45, 58, 185-86
Channel Activity Report (RMF), 379-80
Channel contention (I/O)
  non-RPS, 225-30
  RPS, 230-33
Channel holding time (I/O), 227
Classes, see Customer classes; Customer description
Closed classes in mixed models, 144-48
Closed models, 58, 62, 112-19, 139-45
Communication networks, see Computer communication networks
Complementary network (hierarchical modelling), 152-55
Completions (C), 40, 47
Components of modelling software, 354-60
Composite queueing (FESCs), 157
Computational algorithms, see Algorithms
Computer communication networks (case studies)
  Ethernet, 339-41
  projecting performance of proposed system, 324-27
  SNA, 336-39
Concurrency control (database) (example), 343-47
Conducting a modelling study (Chapter 2), 20-39
  list of specific points, 38-39
Conferences on performance, 17-18

Contention time (I/O), 224
Controller (I/O), 234-35
Core computational routine for separable models, see Mean value analysis
  evolution, 369-70
  in modelling software, 354
CPU, see Processors
CPU Activity Report (RMF), 378-79
CPU replacement, parameterization to reflect, 301
  case study, 27-29, 33-36
CRYSTAL (tool for modelling proposed systems), 327-28
CSMA-CD, 340
CTSS (case study), 102-3
Customer classes (C), 62
Customer description, 57-58
  for models of existing systems, 279-82
  from RMF, 383-85
Cyber 173 (case study), 133-34

D, see Service demand
DASD, see Disk I/O
Database modelling
  approaches, 343-47
  high-level front ends, 359-60
Decomposition, see Flow equivalent service center; Hierarchical modelling
Definition of models, 9-12
Delay service center, 59
Derived observable quantities, 40-42
Device homogeneity, 120, 147
Direct Access Device Activity Report (RMF), 379, 381
Disk I/O (Chapter 10), 222-52, see also Algorithms, disk I/O
  alternate modelling approaches, 248-49
  inferring parameter values from measurement data, 245-47, 249
  load balancing
    asymptotic bounds example, 85-87
    Univac 1100 case study, 312-13
  non-RPS example, 227-29
  representation using modelling software, 247, 367-68
  RPS example, 232-33
  service time, components of, 224
Domains, 191-96
  in MVS, 378
  representation using modelling software, 363
Dynamic reconnection (I/O), 237

Effective disk service time, 224
End effects in measurement, 43-44, 277
Ethernet, 339-41
Evaluation of models, 13-14
　closed and mixed, *see* Mean value analysis
　open, *see* Open models
Event monitor, 276
Evolution of core computational algorithms for separable models, 369-70
Evolving systems (Chapter 13), 296-319, *see also* Parameterization
　case studies and examples
　　simple example, 24-27
　　using asymptotic bounds, 85-87
　　IBM 360, 104-6
　　Univac 1100, 309-15
　　using modelling software, 365-69
　primary and secondary effects, 296-97
Examples, *see* Case studies and examples
Execution graphs (for workload characterization), 329-30
Existing systems (Chapter 12), 274-95, *see also* Parameterization
　RMF example, 376-94
Experimentation, 3
External arrival homogeneity, 120, 149

FCFS, *see* First-come-first-served
FESC, *see* Flow equivalent service center
File placement
　techniques for modelling, 303-4
　Univac 1100 case study, 312-13
First-come-first-served scheduling, 128
　class-dependent service times, 262-64
　highly variable service times, 263-66
Flow balance assumption, 51-52
Flow control (SNA case study), 336-39
Flow equivalent service center (FESC), 152-75, *see also* Load dependent service center
　and separability, 159-60, 191-92
　and transaction workloads, 157
　applications
　　Ethernet, 339-41
　　memory constrained system, 185-96
　　SNA flow control, 336-39
　　tightly-coupled multiprocessor, 254-56
　cost of evaluating a model containing, 157, 159, 191-92
　desired characteristics, 155-58
　evaluating high-level models, 159-60
　obtaining parameters, 158-59

Flow equivalence and hierarchical modelling (Chapter 8), 152-175, *see also* Flow equivalent service center; Hierarchical modelling; Load dependent service center
Forced flow law, 47-50
Front ends for modelling software, 358-60
Fundamental laws (Chapter 3), 40-56
　table of, 53

Global balance, 162-69
　cost, 162-63
　details, 166-69
　to evaluate model of SNA, 336-39
　to model priority scheduling, 163-66
Goal-oriented scheduling, 262

Hardware modifications, *see* Evolving systems
　parameterization to reflect, 300-303
　representation using modelling software, 366-68
Heads of string (I/O), 235-36
Hierarchical modelling, 152-75, *see also* Flow equivalent service center; Hybrid modelling
　Algorithm 8.1, 161
　applications
　　Ethernet, 339-41
　　memory constrained systems, 185-96
　　SNA flow control, 336-39
　detailed example, 163-66
Hierarchical workload characterization, 30-34
Homogeneity assumptions, 120, 147-49
　and memory modelling, 179-80, 191-92
Hybrid modelling, 16, 170-73
　of MVS SRM, 347-49

IBM (case studies), *see* Amdahl
　early virtual memory system, 206-9
　processing complex (workload balancing), 24-27
　360, 104-6
　3790, 8130, 8140 (insurance company), 35-37, 78-82
IMS (Amdahl 470 capacity planning case study), 106-9
Inflation of service demands, *see* Load concealment

Index 413

Inputs and outputs of models (Chapter 4), 4-14, 49, 57-68, 99-101, *see also* Parameterization
Insight, sources of, 35-37
Installation Performance Specification (IPS) (MVS), 377-78
Instructional computing acquisition (case study, Prime, VAX), 30-34
Insurance company case study (IBM 3790, 8130, 8140), 35-37, 78-82
I/O subsystem, *see* Disk I/O
I/O subsystem modifications, parameterization to reflect, 302-3
IPS (IBM MVS Installation Performance Specification), 377-78
Iterative solution techniques
 for separable models, *see* Mean value analysis, approximate
 for non-separable models, *see* Disk I/O; Memory; Processors

Journals on performance, 17-18

K, *see* Number of centers

Last-come-first-served (LCFS) scheduling, 129
Latency time (I/O), 224
LCFS, 129
Little's law, 40, 42-46
 and flow balance assumption, 51-52
 and forced flow law, 48-50
 and mean value analysis, 112, 139
 and memory constrained systems, 49-50, 55-56
 application at various levels, 44-46
 used to derive asymptotic bounds on performance, 76-77
Load concealment
 in evaluating mixed models, 145
 in modelling shared disks, 243
Load dependent service center (Chapter 20), 120-21, 149, 156-58, 403-8 *see also* Flow equivalent service center
 evaluating models containing, 159-60, 403-8
 MVA algorithms for, 403-8
 use in modelling tightly-coupled multiprocessor, 254-56, 310-11
Load independent service center, 120-21, 149, *see also* Queueing service center

Local area networks, 339-41
Logical I/O operation, 287-91
Loosely-coupled multiprocessor, 237, 242-43, 253
 Cyber 173 case study, 133-34

M, *see* Memory constraint
MAP (queueing network modelling software package), 360-69
Mean value analysis (MVA), 108-19, 134-48, *see also* Algorithms, mean value analysis; Evaluation of models; Open models
 approximate, 117-19, 142-45
 calculating outputs from, 116, 142
 exact, 112-17, 139-42
 implementations, 395-402
 key equations, 112, 139
 load dependent centers, 403-8
 mixed models, 144-48
 population precedence, 115-16, 140-41
 residence time equation, 112, 139
Measurement, 21
 choice of interval, 43-44
 event recording, 276
 sampling, 276
 to parameterize FESCs, 158
 useful data items, 275-79
 using RMF, 376-94
Measurement data, inadequacy of, 34-36
Measurement interval duration (T), 40
Memory (Chapter 9), 179-221, *see also* Memory constraint; Multiprogramming level; Paging
 admission policies, hybrid modelling of, 172-73
 assumptions made in separable models, 179-80
 case studies
  early IBM virtual memory system, 206-9
  VAX/VMS, 209-17
 robustness of separable models, 180
Memory constraint (M), 184-96, *see also* Algorithms, memory constraints
 Little's law used to evaluate, 49-50, 55-56
 representation using modelling software, 363
Memory expansion, parameterization to reflect, 209-17, 301-2
Method of stages (global balance), 268-69

Mixed models, 62, 144-48
  Algorithm 7.4, 146
Modelling, alternatives to, 4
Modelling cycle, 22-27
Modelling methodology, *see* Conducting a modelling study
Modification analysis, 12-13, *see also* Evolving systems
Multipathing I/O, 237-41
  Algorithm 10.3, 239
Multiple class models (Chapter 7), 127-51, *see also* Algorithms
  advantages, 127
  case studies illustrating use, 129-34
  contrast with single class models, 130-31
  disadvantages, 127-28
  examples illustrating evaluation, 137-38, 142, 144-45, 147
  inputs and outputs, 62-64
  scheduling disciplines in, 128-29
Multiprocessor, *see* Loosely-coupled multiprocessor; Tightly-coupled multiprocessor
Multiprogramming level
  non-integer, in mean value analysis, 144
  variability (case study), 181-84
MVS (case studies), *see* SRM
  Amdahl 470 (using modelling software), 360-69
  Amdahl 470 (using RMF), 376-92
  IBM processing complex (workload balancing), 24-27

N, *see* Number in system
Non-integer multiprogramming level in mean value analysis, 144
Non-RPS disks, 225-30, *see also* Disk I/O
  Algorithm 10.1, 228
Non-separable models, *see* Disk I/O; Memory; Processors
  approaches to evaluating, 162, 177-78
  need for, 65-67
Notation, table of, 53
Number in system (N as input, Q as output), 60-61, *see also* Little's law
Number of centers (K), 58, 63

Objectives of modelling study, importance of understanding, 27-30
Observation interval, 43-44
One step behavior, 120, 147
Open classes in mixed models, 144-48

Open models, 58, 62, *see also* Algorithms, open models
  evaluation, 109-13, 134-38
Operating systems
  modelling detailed algorithms, 347-49
  modelling upgrades, 305-6
Operational analysis, xii-xiii, 123, 150
Optimistic bounds on performance, 71
Outputs of models, *see* Inputs and outputs of models
Overhead
  Algorithm 13.1, 307
  attribution of, 285-87
  using RMF, 386, 388-91
  modelling changes in, 306-8
  of paging activity, 202-5

Packages, queueing network modelling, *see* Queueing network modelling software
Paging, 201-5
Parameterization, 12-13, *see also* Evolving Systems; Existing systems; Inputs and outputs of models; Proposed systems
  RMF example, 376-94
  software assistance, 358-60
  types and sources of information, 275-79
Path (I/O), 222-23, 231
Path busy, estimating probability of (I/O), 240-41
Path elements, estimating utilizations of (I/O), 238, 240
Path selection (I/O), 237
PDP-10 (case study), 132-33
Performance database, 279
Performance group (MVS), 377
Performance measures, 22, *see also* Inputs and outputs of models
  from RMF, 388, 391-92
Performance-oriented design, *see* Proposed systems
Performance period (MVS), 377
Performance projection, *see* Evolving systems; Proposed systems
  simple example, 24-27
Pessimistic bounds on performance, 71
Physical I/O operations, 287-91
Population, *see* Number in system; Little's law
Population precedence (MVA), 115-16, 140-41

Index

Primary and secondary effects of modifications, 12-13, 20-21, 248, 296-97
Prime (case study, instructional computing acquisition), 30-34
Priority scheduling, 256-62
 Algorithm 11.1, 259
 representation using modelling software, 362-63
 an approach using global balance, 163-66
Processing capacity, see Bounds on performance
 of open models, 109, 134-35
Processor sharing scheduling, 129
Processors (Chapter 11), 253-71, see also CPU replacement; Multiprocessor; Scheduling discipline
Program listings
 single class exact MVA, 395-97
 multiple class exact MVA, 398-402
Projecting performance, see Evolving systems; Proposed systems
Projection phase (modelling cycle), 22-27
Proposed systems (Chapter 14), 320-34
 case studies, 324-27, 329-32

Q, see Number in system; Queue length
Queue length (Q), 60-61, see also Little's law
 at channel (I/O), 227
 in open models, 111, 136
Queue length distribution, 62
 for load dependent centers, 405-7
Queueing network modelling software (Chapter 16), 354-73
 approximation transformations, 356
 core computational routine, 354
 example use, 360-69
 high-level front ends, 358-60
 user interface, 356-58
Queueing service center, 59
Queueing theory, 16

R, see Residence time; Response time
Ready customer (memory constrained system), 185
Reconnect (I/O), 230
 algorithms, 237
 failure, estimating probability of, 241
Relationships among observed quantities, table of, 54
Remote terminal emulation, 27, see also Benchmarking

Residence time (R), 42, 61, see also Mean value analysis; Open models; Response time
Residence time equation (MVA), 112, 139, see also Mean value analysis
 for load dependent centers, 405-7
 modifications
  for biased processor sharing, 261-62
  for FCFS with class-dependent service times, 262-63
  for FCFS with highly variable service times, 265
  for priority scheduling, 258
Residual service time, 265
Resource Measurement Facility (RMF) (IBM), use of, 376-94
Response time (R), 42, 61, see also Mean value analysis; Open models; Residence time
Response time law, 46
 to estimate think times, 282
 to evaluate memory constrained system, 207
Retry (I/O), 230
RMF (IBM Resource Measurement Facility), use of, 376-94
Rotation time (I/O), 230
Rotational position sensing (RPS), 230-33, see also Disk I/O
 Algorithm 10.2, 232
Routing homogeneity, 120, 147
RPS, see Rotational position sensing

S, see Service time per visit
Sampling monitor, 276
Saturated system, 72, 109, 134-35
Scheduling discipline, see First-come-first-served; Last-come-first-served; Priority; Processors; Separable models, scheduling disciplines in
Secondary bottleneck, 82-83
Secondary effects of modifications, 12-13, 296-97, 306-9
Seek time (I/O), 224
Sensitivity analysis, 33-36
Separable models, 15
 advantages, 121-22
 arrival instant theorem, 114
 assumptions, 20-22, 62, 147, 149
 inputs and outputs, 57-68
 limitations, 64-66, 121
 robustness, 121, 180, 222-25, 248, 261, 264

Separable models (*cont'd.*)
  scheduling disciplines in, 59, 62, 128-29, 253-54
  theoretical foundations, 119-21, 147, 149
Service center, 4-8, *see also* Center description
Service center flow balance, 120, 147
Service center types, 59
Service demand (D), 48-49, 59-60
  estimation of
    for models of existing systems, 283-91
    CPU, 285-87
    I/O, 287-91
    difficulties, 284
    using RMF, 386, 388-91
  inflation, *see* Load concealment
Service objective (MVS), 377
Service rate (μ) (FESC), 156-58, *see also* Flow equivalent service center; Load dependent service center
Service rate of tightly-coupled multiprocessor, 255-56
Service time per visit (S), 41
Service time homogeneity, 120, 149
Shadow CPU technique (priority scheduling), 258-61
Shared disks, 242-43
Simulation, 15-16
  in hybrid modelling, 170-73
  to parameterize FESCs, 158
Simultaneous resource possession, 185
Single class models (Chapter 6), 98-126, *see also* Algorithms
  advantages, 98, 122, 127-28
  case studies illustrating use, 102-9
  examples illustrating evaluation, 112-13, 116-17, 119
  inaccuracy when workload is heterogeneous, 130-31
  inputs and outputs, 57-62
  limitations, 98-99
Single server queue, 4-8
SNA (IBM System Network Architecture), 336-39
Software modifications, parameterization to reflect, 303-6
Software resources, 342-43
Software specifications, refinement of (modelling proposed systems), 324-36
Solution of models, *see* Evaluation of models

SRM (IBM MVS System Resources Manager)
  description, 377-78
  hybrid model of, 346-49
State, *see* Global balance, details of
States of a customer (memory constrained system), 185
Static reconnection (I/O), 237
Stochastic analysis, xii-xiii, 123, 150
Swapping, 196-201
  moving from drum to disk (Univac 1100 case study), 314-15
  to a dedicated device, 197-99
    Algorithm 9.4, 199
  to a shared device, 198-201
    Algorithm 9.5, 200
System Network Architecture (SNA) (IBM), 336-39
System Resources Manager (IBM MVS) *see* SRM

T, *see* Measurement interval duration
Tape I/O, 290-91, 386, 387, 391
Terminal workload, 58
Think time (Z), 46
  estimating, 282, 384
Thrashing, 202-5
Throughput (X), 41, 61
  bounds on, 72-73, 76-77
  in memory constrained systems, 187-88
  in open models, 109, 134-35
  versus multiprogramming level, 181-84
Tightly-coupled multiprocessor, 253, 254-56
  upgrading from single to dual processor (Univac 1100 case study), 310-11
Tolerances, in validation, 292
Transaction (database), 344-46
Transaction workload, 58, *see also* Open models
  calculating utilization, 51-52
Transfer time (I/O), 224
TSO (IBM Time Sharing Option) (case studies), 24-27, 360-69, 376-94

U, *see* Utilization
Unattributed busy time, apportioning, 285-91
Univac 1100 case studies, 309-15
User interface for modelling software, 356-58

Index

Utilization (U), 41, 61
Utilization law, 41-42, 45
  and flow balance assumption, 51-52

V, see Visit count
Validation, 22-27, 291-92
Validation phase (modelling cycle), 22-27
Variability in multiprogramming level, 181-84
Variance in service time, 265-66
VAX (case studies)
  instructional computing acquisition, 30-34, 51-52
  memory modelling, 209-17
Verification phase (modelling cycle), 22-27
Virtual memory, 201-5
Visit count (V), 47-50

W, see Accumulated time in system
Waiting customer (memory constrained system), 185
Workload Activity Report (RMF), 379-80, 382
Workload balancing (IBM processing complex case study), 24-27
Workload characterization
  hierarchical, 30-33
  in modelling proposed systems, 323-34
  using RMF, 376-94

Workload components
  identification, 279-81
  modifications, parameterization to reflect, 399-400
Workload intensity, 57-58, 62, 63, 100-101
  determining for existing systems, 281-82
  from RMF, 383-85
Workload measures, 22
Workload modifications
  parameterization to reflect, 297-300
  PDP-10 case study, 132-33
  representation using modelling software, 365
Workload representation in single class models, 99-101
Workload types, 57-58
  selecting for models of existing systems, 279-81
  from RMF, 383-85

X, see Throughput

Z, see Think time

λ, see Arrival rate

μ, see Service rate